Metaphor, Sustainability, Transformation

This book offers an eclectic range of transdisciplinary insights into the role of metaphor, myth, and fable in shaping our understanding of the world and how we interact with it and with each other.

Drawing on innovative perspectives from widely different fields, this book explores how metaphor might facilitate and underpin transformative change towards environmental, ecological, and societal sustainability. It illustrates the ways in which contemporary metaphors lock us into patterns of thinking, modes of behaviour, and styles of living that reproduce and accentuate our current socio-environmental problems. It sets itself the task of finding new metaphors and myths that might help move us towards sustainability as societal flourishing. By examining the use of metaphor in diverse fields such as energy use, the food system, health care, arts, and the humanities, it invites the reader to reflect on the deep-seated influence of language in general, and metaphor in particular, in shaping how we understand and act upon the world.

Re-imagining the use of language in framing both the problems we face and the solutions we devise, this novel contribution is a vital source of ideas for those aiming to change how we think and act in pursuit of more sustainable futures.

Ian Hughes is Senior Research Fellow at the MaREI SFI Research Centre for Energy Climate and Marine. His research interests are in deep institutional innovation, sustainability, and human development. He is author of *Disordered Minds: How Dangerous Personalities Are Destroying Democracy*, and contributing author to *The Dangerous Case of Donald Trump*.

Edmond Byrne is Chair Professor of Process and Chemical Engineering at University College Cork. His research interests include transdisciplinary approaches around sustainability. He chairs the 10th Engineering Education for Sustainable Development conference (EESD2021) and co-edited *Transdisciplinary Perspectives on Transitions to Sustainability* published by Routledge in 2017.

Gerard Mullally lectures in Sociology at University College Cork, specializing in environment, community, climate, energy, and sustainable development. He leads 'Imagining 2050' at UCC's Environmental Research Institute, a transdisciplinary research consortium which engages with civic society using innovative approaches to explore and co-develop future visions of, and pathways to, a low carbon and climate resilient future. He co-edited *Transdisciplinary Perspectives on Transitions to Sustainability*.

Colin Sage is an independent scholar based in Portugal who works on the interconnections of food systems, environment, and prospects for greater civic engagement around food. He is the author of *Environment and Food*, 2012; and co-editor of four books, including *Food System Transformations: Social Movements, Local Economies, Collaborative Networks*, 2021; and *Transdisciplinary Perspectives on Transitions to Sustainability*, 2017.

Metaphor, Sustainability, Transformation

Transdisciplinary Perspectives

Edited by Ian Hughes, Edmond Byrne, Gerard Mullally, and Colin Sage

Routledge
Taylor & Francis Group
LONDON AND NEW YORK

First published 2022
by Routledge
2 Park Square, Milton Park, Abingdon, Oxon OX14 4RN

and by Routledge
605 Third Avenue, New York, NY 10158

Routledge is an imprint of the Taylor & Francis Group, an informa business

British Library Cataloguing-in-Publication Data
A catalogue record for this book is available from the British Library

Library of Congress Cataloging-in-Publication Data
A catalog record has been requested for this book

ISBN: 978-0-367-69855-3 (hbk)
ISBN: 978-0-367-69856-0 (pbk)
ISBN: 978-1-003-14356-7 (ebk)

DOI: 10.4324/9781003143567

Typeset in Times New Roman
by SPi Technologies India Pvt Ltd (Straive)

MIX
Paper from responsible sources
FSC™ C013985

Printed in the United Kingdom
by Henry Ling Limited

We dedicate this book to the memory of our colleague Brendan Richardson

Contents

Figures

Tables

Contributors

Evan James Boyle is a PhD researcher in the Energy Policy and Modelling Group, MaREI/ERI and the Sociology Department, UCC. Current research interests include the diffusion of sustainability, participatory research methods, and collaborative governance. Evan is also involved in lecturing/tutorial roles with a current focus on Social Theory and Sociology of the Environment.

Edmond Byrne is Chair Professor of Process & Chemical Engineering at University College Cork. His research interests include transdisciplinary approaches around sustainability and (engineering) education for sustainability. Chair of the 10th Engineering Education for Sustainable Development conference (EESD2021) in Cork, he co-edited Transdisciplinary Perspectives on Transitions to Sustainability (Routledge, 2017).

Shane V. Crowley is a lecturer in the School of Food and Nutritional Sciences at University College Cork. He has degrees in both science and philosophy. His research interests include the science of food (especially the relationship between processing and structure) and the philosophy of food (broadly the ontological and axiological aspects of food design).

Jools Gilson is an artist, writer, and educator. She works across multiple disciplines including choreography, performance, public and installation art, creative writing, and radio broadcasting. She directed half/angel from 1995 to 2006 and makes creative radio for RTÉ and the BBC. Her work has been performed, exhibited, and broadcast internationally. She is Professor of Creative Practice at University College Cork.

James Glynn is a senior research scholar at the Center on Global Energy Policy in Columbia University SIPA. He is an energy systems engineer and works at the interface between energy technologies and data-driven policy, building integrated energy systems models to map future pathways to mitigate climate change and enable sustainable development.

Lidia Guzy is social anthropologist and scientist of religions, specialising in the anthropology of religions and global indigenous studies. She is a Director of the Marginalised and Endangered Worldviews Study Centre (MEWSC) at

University College Cork (UCC) and works since many years on indigenous concepts of sustainability.

Ian Hughes is a Senior Research Fellow at the SFI MaREI centre for energy climate and marine and has worked as a Senior Policy Advisor in innovation policy for the Irish government. He has trained in physics and psychoanalytic psychotherapy and is author of Disordered Minds: How Dangerous Personalities are Destroying Democracy.

Kieran Keohane is Senior Lecturer in Sociology and a member of the Moral Foundations of Economy & Society research centre, in the School of Society, Politics and Ethics, University College Cork.

Connor McGookin is a PhD Student in MaREI's Energy Policy and Modelling Group. His primary research focus is on the use of participatory methods in energy system modelling; in particular exploring how a transdisciplinary approach may facilitate regional energy transitions.

Gerard Mullally lectures in Sociology at University College Cork, specializing in environment, community, climate, energy, and sustainable development. He co-edited *Transdisciplinary Perspectives on Transitions to Sustainability* (2017). He leads 'Imagining 2050' at UCC's Environmental Research Institute, a transdisciplinary research consortium which engages with civic society using innovative approaches, to explore and co-develop future visions of, and pathways to, a low carbon and climate resilient future.

Maureen O'Connor lectures in English in University College Cork. Author of *The Female and the Species: The Animal in Irish Women's Writing* (2010), she has co-edited a special issue of the *Canadian Journal of Irish Studies*, 'Irish Studies and the Environmental Humanities' (2018), as well as several volumes of essays. *Edna O'Brien and the Art of Fiction*, is forthcoming from Bucknell in 2021.

Brian Ó Gallachóir is Professor of Energy Engineering at University College Cork and Director of Ireland's MaREI Centre for Energy, Climate and Marine Research. Brian has published extensively on energy transition pathways to a low carbon future and has directly informed energy and climate policy decisions. He has a B.Sc. from TCD and a PhD from UCC.

Fionn Rogan is a Senior Research Fellow at the Environmental Research Institute and MaREI Centre at University College Cork in Ireland. He is interested in energy and climate policy, energy systems modelling, societal transitions, philosophy of technology, and innovation systems.

Colin Sage is an independent research scholar who works on the interconnections of food systems, environment, and on the prospects for greater civic engagement around food. He was the founding Chair of the Cork Food Policy Council. He is the author of *Environment and Food* (2012); and co-editor of *Food System Transformations: Social movements, local economies, collaborative networks* (2021); *Food Transgressions: Making sense of contemporary* *food politics* (2016); and *Transdisciplinary Perspectives on Transitions to Sustainability* (2017). He is now based in Portugal.

Cormac Sheehan is a member of the Department of General Practice, UCC. His research focusses on varied health care including dementia, epilepsy, eating disorders, and diabetes. He works closely with the HSE and acts as an advisor on patient-focused research.

1 Metaphor, transformation, and transdisciplinarity

Colin Sage, Ian Hughes, Edmond Byrne, and Gerard Mullally

> Yes, metaphor. That's how this whole fabric of mental interconnections holds together. Metaphor is right at the bottom of being alive.
>
> Capra (1988: 79; quotation attributed to Gregory Bateson)

The language of transformation

This book is concerned with the ways in which language, metaphor in particular, but also myth, fable, parable, allegory, and other literary devices, can shape how we think about, and respond to, the environmental challenges that humanity is currently facing. It recognises that language makes sense of our world and shapes how we think and act: individually and collectively (Fløttum, 2014). Narratives – constructed stories – frame problems and issues in ways that are meaningful, creating an "architecture" for understanding the state of the world and what might be done to improve it (Jepson, 2018). Metaphor, myth, and fable influence how we frame problems and set agendas (Lakoff, 2004), affect whether or not we are motivated to act, and can play a role in bringing about needed transformations in deeply held beliefs, social norms, and institutions (Moser, 2006). These feed into narratives and discourses in ways which recursively and interdependently influence social, political, and economic institutions across a range of fields – psychological, philosophical, cultural, historical – thereby shaping the ways in which we engage with the natural world (Harré et al., 1999). This is why, for Bateson, metaphor was the "language of nature", the "logic upon which the entire living world is built" and the basis for establishing the "pattern which connects" (Capra, 1988: 84; Olds, 1992).

This is a book that is consequently interested in language but that has been written, largely, by people who do not regard themselves as linguists, rhetoricians, or cognitive scientists. On the contrary, we present ourselves – editors and contributors alike – as *transdisciplinarians* motivated by a shared interest in sustainability and concerned by the threats to the stability of Earth system processes. As environmental, social, and natural scientists, engineers, eco-humanity scholars, and with other academic influences, we have been engaged in a deliberative process of dialogue and discussion that has enabled us to reflect upon the role of language in shaping our understanding of current problems and possible

DOI: 10.4324/9781003143567-1

solutions. We have come to appreciate – both individually and collectively – just how powerfully stories of, and about, the present lock us into styles of living, modes of behaviour, and patterns of thinking that can constrain our imaginations for the future. Consequently, we embarked upon a process of interrogating the role of metaphor, but also myth and fable, in shaping the ways in which we as individual academics think about our particular research "problem", but also how such devices might help or hinder in devising ways for transformative change.

Some chapters in this volume face this challenge head on, while in other instances, the approach is more circumspect, or even oblique. However, all share the same appreciation for the value and efficacy of the aforementioned language tools, not just for communication but in effecting transformational change. While many explore the historical underpinnings of language, again there is a shared recognition that humankind needs to mindfully employ these tools at the current point in our history to effect positive change, now more than ever. This is as a result of the significant challenges humanity faces, principally of our own doing. These challenges may be collectively termed as challenges of (un)sustainability; for while we appear to be flourishing on many levels, such as with respect to technological prowess or global population or connectivity, the flip side reveals systems at, or heading towards collapse. This is supported by current evidence of the interconnected crises in the areas of biodiversity loss, depletion of freshwater stocks, climate breakdown, and across the fields of food, energy, health, and well-being. Of these challenges, climate breakdown has generally stood out as the most pressing, though the COVID-19 pandemic has highlighted the precarious nature of many of our globalised human systems (while the likelihood of such pandemics appears now to have been significantly enhanced by increased human encroachment on the natural world). This book consequently includes contributions from authors representing a variety of disciplinary backgrounds engaged in research across a range of domains where sustainability has come sharply into view. These stretch from the global food and energy systems, through reflections on what we mean by healthy human development to our very existence in the world and the meanings we ascribe to our experience. While the book's focus is primarily on metaphor, it also contains contributions on myth and fable, as constructs that similarly condition our thoughts and actions, and indeed how these may be directed into narrative and discourse, in precipitating required transformative change.

This volume is based on the conviction that in such a fundamental rethinking, we need to pay attention to the language we use in framing both the problems we face and the solutions we devise. It is based on a belief that the metaphors we currently use can lead us to act inappropriately and that an active reimagining of our language is needed. Which metaphors will we need to address the deep era of transformation we are currently navigating? What myths might we need to overcome? What fables could act as a signpost to the direction we need to be travelling? This book, in its transdisciplinary exploration of metaphor, myth, and fable, and its inclusion of authors from across a diverse range of disciplines, aims both to highlight the critical role of language in bringing about the transformations required for sustainability, and to act as a source of ideas for others aiming to change how we think and act in pursuit of those transformations.

Given the importance of metaphor in helping us to make sense of the natural world, and recognising that some readers of this volume will be less familiar with the roles performed by language and literary devices in framing our understanding, we have sought to provide a simple explanation in the following section. Those already familiar with the work of George Lakoff and especially of Brendon Larson (Larson, 2011) may choose to pass quickly over this and head for the following section 'Metaphor and structure of the volume', which explores the significance of transformation. This is a word – much like sustainability – prone to misuse where claims for its operability disguise or conceal interests seeking to maintain business as usual or incremental improvements in efficiency. For us, transformative change cannot rest on supply-side solutions without also interrogating the nature of demand: how it is constituted, by whom and at what cost – socially and ecologically? This is followed by an overview of the rest of the book which is structured into four parts, with the rationale for this explained, together with a short summary of each chapter.

The importance of metaphor

Aristotle's Poetics provides a classic definition of metaphor as the thing that gives a name to something else (Kuusi et al., 2016). Metaphor enables us to understand one thing in terms of another, and to think of an abstraction in terms of something more concrete. So, far from being superfluous, metaphors have profound purpose. We rely on metaphors to understand the world around us, to think, and to communicate. Recent scholarship in the field of cognitive linguistics shows that not only are metaphors essential to human thought and communication, but that they have intense influences on how we conceptualise and act with respect to important societal issues. Indeed, as Byrne describes in Chapter 6, some propose that metaphors actually frame the very basis of human understanding of the world around us (through the right hemisphere of the brain) which in turn is translated (in the left hemisphere) into words and language (McGilchrist, 2009).

According to Ison, Allan, and Collins,

> Humans understand and relate to the world around them with the help of frameworks that mediate what is observed, what it means and what is considered as wise action… they are developed and maintained via discourse, and in particular, spoken and written language … While all language is important, metaphor has been shown to be disproportionately influential in developing and reinforcing frames…
>
> (Ison, Allan and Collins, 2015, pp. 1699–1700)

Metaphors can define a problem, delineate the scope of analysis, and suggest hypotheses for testing theoretical propositions (Marks, 2011). In academic enquiry, metaphors have been described as "one of the deepest and most persistent phenomena of theory building and thinking" (Paprotte and Dirven, 1985). Exposure to different metaphors has also been shown to induce substantial differences in opinion about how to solve social problems, while demonstrating that the power of framing by metaphor is covert (Thibodeu and Boroditsky, 2011).

By way of example, they show how metaphorical framings of urban crime as respectively an attack by a virus or by a beast elicits contrasting responses along a spectrum from social reforms to catching and caging.

Discourse metaphors are a particular type of metaphor that draw upon and reflect the cultural and social preoccupations of their time. According to Zinken et al. (2008), discourse metaphors employ cultural knowledge and function as key framing devices within a particular discourse. They function both to express a particular understanding of an issue, as well as evoking an emotive response. Once established, they can frame public discourse and policy responses. In this way, discourse metaphors extend beyond individual cognition and influence both public policy and society.

Atanasova and Koteyko (2017) provide the examples of war and religion as dominant sources of discourse metaphors in the ongoing debate around climate change. In their analysis of opinion pieces and op-eds, published between 2006 and 2013, they contrast how "Guardian Online" predominantly used war metaphors to advance pro-climate change arguments, while "Mail Online" primarily used religion metaphors to advance climate-sceptic arguments.

War metaphors were present in around 44% of all Guardian Online editorials and op-eds that contained metaphors. These metaphors spoke of "fight", "retreat", and "the battle" to prevent climate change. These war metaphors were used to evoke images of collective effort, to instil a sense of unity and patriotism, and to appeal to shared goals and collective action.

Religion metaphors appeared in over 78% of all "Mail Online" editorials and op-eds that contained metaphors over the period to advance anti-climate change arguments. Climate change activists were depicted as "zealous fanatics", "medieval preachers", proclaiming that the end of the word is nigh, while climate change was described as "a creed", "a faith", a subject that went beyond the rational. Climate sceptics were depicted as being treated like "heretics", and the science behind climate change was likened to a religious text that cannot be questioned, and thus not "real science".

As discourse metaphors, these metaphors of war and religion draw upon a reservoir of cultural myths and social representations readily available in social memory, namely memories of past wars and religious conflicts. In doing so, they implicitly evoke past memories as a potential guide for current action. For example, military thinking, according to Annas (1995), "concentrates on the physical, sees control as central, and encourages the expenditure of massive resources to achieve dominance", and so suggests particular kinds of responses. However, as Mullally reminds us, context matters as his analysis of climate narratives in Irish print media revealed (Mullally, 2017). Here, war metaphors were deployed to a very limited extent while religious metaphors were invoked by discourse coalitions for climate action as well as by much less prominent climate sceptics.

Given the ubiquity of metaphors and the profound role they play in problem solving, provoking new understandings (Brown, 1976), and providing tools for effective communication, this book serves as an entry point to explore the potential uses of metaphor, and their different cultural forms, e.g. myth, fable, parable, etc. in furthering research into, understanding of, and communication on the transformative changes necessary for the transition to a more sustainable society.

Properties of metaphors

Metaphors draw on our physicality

As Lakoff and Johnson (1980) pointed out in *Metaphors We Live By*, we interact and interpret the world as embodied human beings and we use this basic fact of our embodiment in the world to make sense of new phenomena and to communicate with one another. To provide just one example, metaphors that draw on our physicality include those of balance, which we apply to balance of nature, well-balanced personality (as opposed to being imbalanced), balancing the books, and tipping the balance.

Metaphors are embedded in shared cultural contexts

Metaphors are not merely shorthand for facts. As Larson (2011) points out, scientific metaphors embed facts within webs of social, moral, political discourses, and webs of meaning. The selfish gene metaphor (Dawkins, 1976) provides one example of how scientific metaphors can communicate particular values, as can that of survival of the fittest (Spencer, 1864). Scientists, and other academics, consciously and subconsciously draw upon culture for metaphors with which to describe and communicate their work.

Metaphors influence our social reality and our perception of the natural world

Metaphors are part of the frames or cognitive structures that organise ideas. In *Metaphors We Live By*, we are told that much of cultural change results from the introduction of new metaphorical concepts and the loss of old ones (Lakoff and Johnson, 1980). In the preface to *Don't Think Like an Elephant*, Lakoff comments that: "our frames shape our social policies and the institutions we form to carry on our policies. To change our frames is to change all of this. Reframing is social change" (Lakoff, 2004: xv).

Of particular importance are so called feedback metaphors. According to Larson (2011), feedback metaphors are metaphors that harbour social values and circulate back into society to bolster those very values. They have been widely adopted and structure thought and action along one particular line rather than another. They resonate with widely held cultural values and form part of the cultural meta-narrative. A feedback metaphor has become naturalised so that we forget that it is a metaphor and we live according to it.

Two examples of feedback metaphors which Larson examines in detail are the metaphors of progress and competition. Progress and competition are powerful ideological metaphors that justify how we act in relation to each other and to the natural world. These metaphors reinforce status quo values of progressive idealism and competitive capitalism, rather than alternative values of sufficiency, co-operation, and interconnection. Feedback metaphors can become naturalised, eventually becoming entrenched as natural and true.

Metaphors can suggest specific modes of action

Larson (2011) again points to two prominent metaphors that environmentalists have adopted to illustrate how metaphors can suggest particular courses of action. Gaia, derived from the name of the Greek Earth goddess, is used to envision the earth as a living organism, with self-regulatory capacity and stability over enormous periods of time. The other, Spaceship Earth, conjures up a quite different image of our planet as a finite system hurtling through endless space, encouraging us to better manage its resources. Gaia suggests we need do nothing (or at least minimise unnecessary intervention) as mother earth can look after herself. Spaceship Earth, on the other hand, suggests that it might be better to leave environmental decisions to expert technocrats.

Metaphors can enhance or inhibit effective communication

Metaphors can play a number of roles in communication, namely in transferring knowledge, enhancing open dialogue, and prompting action.

(i) The role of metaphor in transferring knowledge: Traditionally, scientists have understood the purpose of their communication to be remedying the public's knowledge deficit. Metaphoric resonance provides a revised view of science communication. Metaphors communicate not only facts, but a web of facts and values, and resonate within a particular framing. If a metaphor does not resonate within the frame of the listener, communication will be ineffective, no matter how compelling the factual case may be.
(ii) The role of metaphor in enhancing open dialogue: Metaphors can prompt dialogue between people with different perspectives. No single metaphor can capture a phenomenon in its entirety because every metaphor highlights certain elements while backgrounding others. Using diverse metaphors to introduce differing perspectives can allow opposing views and different ways of viewing a problem to remain in dialogue. Multiple opposing metaphors may also be necessary to grasp the complexity of reality. The psychologist William James, for example, promoted the benefits of holding diverse metaphors in mind, an ensemble of metaphors to characterise human psychology (see Kress, 2000).
(iii) Role of metaphor in prompting action: One of the key aims of communication in sustainability is to prompt governments and citizens to take action to avoid further environmental and social damage arising from unsustainable practices. A wide range of metaphors are being used to evoke emotional responses in the hope of motivating action. One issue of particular importance in regard to climate change is the use of metaphors of impending catastrophe. It is a matter of some debate as to whether such metaphors are effective in promoting change, or whether non-threatening and positive imagery (e.g. Ehrenfeld's sustainability-as-flourishing (Ehrenfeld, 2008)) may be more effective in motivating genuine change.

As Schön (1979, p. 255) has written, "The essential difficulties in social policy have more to do with problem setting than with problem solving, more to do with ways in which we frame the purposes to be achieved than with the selection of optimal means for achieving them". It is important that metaphors provide the new frames as well as contribute solutions within existing frames. If our current crises are partly crises of thinking, and thus of language and metaphor, one of the solutions is also to be found in new metaphors and with the stories they might help to construct.

Minding our language

Larson's (2011) injunction to *mind* our metaphors operates on a number of different levels. On a cognitive level, metaphors act as a framing device or a conduit for communicating meaning. On a normative level to *mind* connotes a responsibility for care, or *care of* something. Taken together Larson refers to these colloquially as *metaphoric resonances* and the "connections that this creates among different cultural realms as a *metaphoric web*" (Larson, 2011: 12 [emphasis in original]). Metaphors frame communication by selecting specific associations between language and experience (Castells, 2009). As such, the conative or emotional level (Norgaard, 2011), as in to *mind* as to love or *care for* or care about as articulated in *Laudito Si, Caring for Our Common Home* also needs to be foregrounded (see Chapters by Hughes, Byrne, Sheehan this volume). The value of a metaphor also therefore resides "... in its placing in a web of new complex relations, through which it is brought into a new light, receives peculiar emotional values and is comprehended more vividly and completely than before" (Rickards, 2015: 281).

Consequently, we require new memes to facilitate cultural evolution and language can play a critical role as a form of social disturbance. A language that is environmentally adequate should enable us to talk about environmental matters in an informed manner and promote the well-being of humans and the environment. Yet the language we use is shaped by the society in which we live, while at the same time that that society is shaped by the language that is used: a dilemma likened to fish reflecting on water in which they have lived all their lives. Yet the purpose of seeking novel language constructs is neither to reject entirely the ones we have nor to provide a single alternative. Instead, new forms of language reveal limitations with the dominant view and thereby point to elements of more creative thinking: they enrich our perception and can break the grip of entrenched thinking.

Metaphor, then, reminds us of the fundamental interconnection between things. This extends across dualities of fact-value, science-society, literal-figurative. With metaphor we see one thing in terms of another and the key question is whether they enhance our sense of interconnection, and in what way. We want language to connect us to one another, and to connect us to the world. As this book will explore, it is perhaps on metaphors, myths, and fables that embody subjective and empathic relations, rather than mechanistic values, that a new and enduring ethic can be built.

This book aims to encourage the development and adoption of language that will be more conducive to the conceptual shift we require for sustainability. We will require more novel, poetic images that help us see the limits of ordinary language; ones that express the depths of our fears and sorrows, and the loftiness and possibility of our hopes and dreams. We require images that are hopeful as well as catastrophic, that bring us closer to the world rather than separating us from it. We require language that reforms human relations and that harnesses hope rather than fear. It is in this spirit that the book is compiled; the chapters do not aim to provide the definitive word or coherent structure ("a tamed beast") on metaphor for transformation; rather they seek to facilitate and embrace a range of approaches and perspectives, emanating from a diverse range of disciplinary fields to provide some playful explorations and enticing glimpses of, at the very least, a less unsustainable future.

In the spirit of Brendon Larson (2011), then, this book poses the question: "If our metaphors do not encourage us to maintain a world in which we can live, what is the point of understanding?"

Metaphor and structure of the volume

We have organised this volume into four parts: Metaphors of Reason; Myths and Metaphors of Unreason; Metaphor, Myth and Mind; and Metaphors of Creativity and Practice. Further on, we summarise the key arguments of each chapter but here provide a brief explanation of the metaphors (and myths and fables) that are drawn upon.

The first section of the book, then, focusses on metaphors of science, technology, and reason and examines the metaphors that are used to describe technological change. Do such metaphors really reflect the complex technological and social processes involved? Do they minimise the potentially destructive unintended consequences that new technologies often bring? Climate change is the most obvious of such unintended consequences, but this section also looks at both positive and negative consequences of technology in the global agriculture and food systems. The metaphors of planetary boundaries and metabolic rift are explored as images that can aid our understanding of the remarkable advances and the daunting problems that agricultural science and technology have created. Metaphors of food also influence how we view ourselves as individuals and as societies. This section also charts how food metaphors are evolving in step with technology, often in a direction that points to the replacement of "bothersome" food with supplements and nutrients that will enable us to transcend the limitations of our energy hungry bodies. Or alternatively, how can metaphors develop a narrative capable of portraying food as being more than mechanistic building blocks required to fuel the body and support physical growth and, in transcending this purely functional role, convey its enormous potential as the basis for social cohesion at a number of levels (family, community, regional culture, etc.). How are metaphors enabling or resisting this transition to a cyborg future? Finally, the section returns to the challenge of climate change and explores how the simple metaphor of carbon budgeting is helping individuals and governments grapple with the fundamental problems of equity and responsibility in the context of our continuing carbon prolificacy.

The second section in the book, Myths and Metaphors of Unreason, aims to widen the scope of debate on sustainability by challenging the primacy of purely scientific, technological, and reason-based approaches. It opens by exploring the famous metaphor of Yin and Yang, or complementary duality, as an invitation, not to reject scientific reasoning, but rather to complement rational thought with holistic, symbolic, "right brain" thinking. This section also argues that sense making in terms of myth and fable has been marginalised, with Goethe's Faust (who sold his soul in return for knowledge) being perhaps the sole remaining myth of modernity. It explores the work of neuropsychiatrist Ian McGilchrist who asserts that while science and reason are primarily left-brain functions, the right hemisphere of the brain thinks more integratively in terms of metaphor, myth, and fable. This section too argues for the urgent necessity for the restoration of myth as a tool for sense making in order to provide direction amid the radical uncertainty of unsustainability. It concludes by exploring the seemingly unreasonable proposition that society should proceed slowly (though wisely) in charting the future course towards sustainability. This proposition, seemingly unreasonable because of the urgency of climate change and environmental and species destruction, is explored in a sequence of variations of the fable of the Hare and the Tortoise. When it comes to the challenges of sustainability, could "slow and steady", paradoxically, really win the race?

The next section, Metaphor, Myth and Mind, explores metaphors and myths from psychology, psychoanalysis, illness, mental health, and dreams. It asks how metaphors of mind can help to reframe the challenges we face in overcoming unsustainability and the mindsets that underpin it. It looks at the myths of Narcissus and Oedipus and suggests that, as a collective, humanity has still to overcome some of the most fundamental challenges of early human development, namely the development of capacities for altruism and concern. The section also broadens our attention, again, away from environmental concerns to explore metaphors of illness and the challenge of dementia. Metaphors of decline and decay, of war, flood and epidemics are common in discourses about dementia. The section argues that such metaphors increase suffering and calls for new and more imaginative metaphors as being an integral part of our human response to the illness. Finally, the section returns to an earlier theme of the book, that a fundamental reorientation of our current dominant view of reality is integral to the transition to sustainability. It explores the Shamanic dream, a cultural practice that spans from Siberia to the Amazon, as a means of gaining insight into the radically different worldviews of some indigenous cultures. It argues that the shamanic dream provides metaphoric insight into humanity's deep interconnection with and responsibilities towards our ancestors, our environment, and our future.

The final section in the book, Metaphors of Creativity and Practice, turns to literature and creative practice and the metaphors that suffuse these domains. The section examines the thoughts and writings of Irish writers James Joyce and Mike McCormack. It explores how the metaphors of "portal" and "diffraction" were used by these authors as devices in their work to change how we, as readers, experience the world. Echoing the earlier metaphor of agonistic dualism, these writers use techniques of ambivalence, over-determination, free association, and antitheses left unresolved, as means to disturb narrative and thought, and, in doing so, open up space for new imaginative possibilities. The metaphor of the portal, as a space which transforms us as we pass through, was one which deeply infused Joyce's work and reflected Joyce's intention that Ulysses and Finnegan's Wake be transformative works. McCormack's writing similarly seeks to bring about a fundamental change in worldview, through literature that does not simply reflect reality by mirroring experience, but rather by diffracting and breaking up reality, making it multiple and discontinuous, thereby enabling a renewal of experience. The section closes by further illustrating how creative writing, dance, theatre, music, film, and visual art practices are knowledges which challenge the hegemony of critical analytical modes of meaning-production. It closes with the metaphor of the "Rain Box", a box within which it is always raining, as an image of a place within which stories are dissolved, stories suffused with meanings and metaphors – meanings and metaphors that hold the potential to lead us towards a brighter and more sustainable future.

Transformation: The role of language

This section will discuss the meaning of transformation and explain why it represents such a vital issue for our era. For if we are to fully comprehend the scale, complexity, and deeply interconnected nature of the global environmental crisis that we face then, inevitably, it also reveals the profound shortcomings of prevailing operating procedures through which this crisis is currently being addressed. Such procedures most often originate in the realm of governance where an array of economic instruments, including taxes and incentives, and regulatory measures covering environmental standards, are accompanied by a range of modest social policy initiatives seeking to "nudge" the behaviour of citizens. Yet, wherever we look, we see governments largely incapable of engaging with the scale of the problems faced and all appear to be beholden to the pursuit of economic growth, seemingly irrespective of its social and environmental consequences. We return to the matter of growth below. Policy must, of course, be informed by scientific evidence; but the nature of the environmental problems with which we now grapple have presented a profound challenge to the scientific community and its traditions. While a rear guard defence of positivist methods girded by a corresponding "techno-optimistic reductionist scientism" (Barry, 2017) has informed a broader scientific scepticism, others have worked tirelessly to develop a new ontology capable of embracing notions of complexity, uncertainty, and non-linearity, as well as to appreciate that facts and values are less clearly demarcated. Labels such as Mode 2, post-normal, and triple-helix approaches alert us to new ways of conducting research involving complex system analysis that recognises the dynamic nature of coupled social-ecological systems where values and judgements shape human action with real-world consequences (Funtowicz and Ravetz, 1993; Gibbons et al., 1994; Etzkowitz and Leydesdorff, 2000; Voß and Bornemann, 2011; Carayannis et al., 2012). Moreover, by the very nature of conducting research in a field labelled sustainability is to also understand the challenge of generating "actionable knowledge", that is, findings informed by non-academic stakeholders with a view to ameliorating, improving, or resolving current socio-environmental problems and crises.

It is in this context that transformation emerges as an antidote to the production-consumption treadmill of the global economic system and speaks of structural, qualitative change. In relation to, say, the effort required in the context of climate change mitigation involving the wholesale replacement of a fossil-fuel-driven energy regime by a low-carbon alternative system then, at one level, we can see that this involves a deep process of socio-technical transition. However, we might ask whether this in itself constitutes transformative change if end user practices remain as before with little engagement with the governance of such new arrangements. Switching from fossil fuels to nuclear for energy provision or supplementing fossil fuels with renewables hardly encompasses the necessary qualitative transformational change necessary in the context of a growth-fuelled, consumption-driven society (of material, energy, and data). The development of new data centres in the Republic of Ireland, for example, has the potential to add over 40% to energy demand from 2028 (Eirgrid/SONI, 2018), from a base whereby they drew on less than 6% of energy demand just over a decade earlier in 2015 (Coyne and Denny, 2018). On the other hand, transformation might arguably have occurred where energy supply was now largely provided by distributed small-scale, community-owned generation schemes and where demand was in line with the availability of local resources.

Our understanding of transformation consequently reflects a preoccupation with the need to secure effective, equitable (i.e. socially just) and durable solutions to our global environmental predicament, one that represents qualitative change in the human experience. This means that technical innovations, though vital, are but just one part of a complex set of inter-connected changes and where the biggest challenge of all is likely to involve a re-calibration – a re-boot if you will – of the popular imagination: human hearts and minds. This is where metaphor, myth, and fable have such a vital role to play.

For example, as Tim Jackson reminds us, the metaphor of Adam Smith's "invisible hand" has proven extraordinarily powerful and has been central to modern economics. As Jackson argues, "this one single metaphor has motivated a ferocious defence of the virtues of an unbridled 'free market' in which self-interest is given full rein". (Jackson, 2016: 132). Economics subsequently served to conflate self-interest with human nature but as Jackson argues, there is plenty of evidence of altruism and this was fundamental to our evolution as social beings. While there is evidence for individualism and novelty-seeking in human adaptation, so there is also for altruism and conservation. Indeed, contra Dawkins' metaphor of the selfish gene, Ulanowicz, informed by his work on ecosystem networks would forcefully contend that in the natural world symbiosis and mutuality comes before and below competition (not the other way around, as is evidenced by autocatalysis, the centripetal action that underpins ecosystems; indeed, competition only emerges when there is a scarcity of resources) (Ulanowicz, 2009: 73–76). Larson too (2011: 83–88) reflects on this traditional bias in science and ecology towards competitive instincts in nature over mutualism, while citing the possibility of individual and cultural contexts in such scenarios, including the dominance of male researchers (Keddy, 1989). However, the prevailing social paradigm of consumerism privileges and encourages selfish individualism and its success in embedding personal hedonism has served to jeopardise conditions for a shared prosperity.

Challenging the primacy of such economic metaphors as the "invisible hand" of the market as well as a whole array of mainstream economic "sacred cows" has long been a struggle for those who fall under the umbrella of heterodox economics. Sharing a rejection of the neoclassical "homo economicus" model of individual behaviour (rooted in self-interest) heterodox approaches, while highly diverse in philosophy and method, share an appreciation of institutions, social structures, and evolutionary change. While this is not the place to embark upon a detailed disaggregation of this highly pluralist turn in economics, it is necessary to recognise that the metaphors of the economic incumbency are being challenged by new concepts, theories, and metaphors from those who recognise the need for transformation. While this might once have referred only to Marxist economics, there is now established fields of innovative thinking taking place across the fields of feminist, environmental, and ecological economics with many other currents besides. What this has helped to do is to establish critical mass and momentum in cross-disciplinary dialogues that are challenging the dominant metaphors, such that notions of "degrowth", surely regarded as blasphemy in mainstream economics, has become a "hot topic" in sustainability circles (D'Alisa et al., 2015; Kallis and Vansintjan, 2018).

One brief example here to illustrate the degree to which transformation will require a fundamental rethink of human agency in the world can be garnered from a consideration of the emergence of contemplative social science. This may seem not only a world away from conducting an "objective analysis" of our current predicament, but an abnegation of responsibility by promoting an inward – rather than outward – perspective. However, if we are serious about transformation then we must break once and for all the "homo economicus" model that drives hedonistic consumption and find a way of (re-)connecting ourselves to the natural world. Never has there been such an important moment to undertake such a change of direction – and not solely due to our environmental crisis. As two of the chapters in this book explore, our current global food system is both deeply unsustainable and creating environmental and human catastrophe. Many western societies appear to be beset by a rising epidemic of mental health problems, the causes of which are obviously complex and deep-seated, but which are starkly manifest in numbers of suicides. The crisis of dementia, also explored in a later chapter, illustrates the unsustainabililty of our dominant model of caring (or un-caring). Is there evidence to suggest that consumer culture plays its part? The market appears unrelenting in its capacity to offer tantalising novelties for our enjoyment: if only we had the means to pay for them. More worryingly, young people are under enormous pressure to conform to contemporary standards of appearance and beauty that are pushed through social media by corporate interests fronted by glamorous representatives and brand ambassadors such as the Kardashians. An always-on mode of ever-increasing consumption of material, energy, and data (e.g. through social media and via smart society) increasingly pervades, feeding into a throwaway consumptive culture. Unsurprisingly, health professionals are urgently exploring solutions to this predicament and one measure that has come into popular public view is that of mindfulness (Kabat-Zinn, 2004). We wonder if this, or what is more widely termed as contemplation, an ethic which recognises a need to go beyond consumptive materialism, is perhaps foundational to achieving authentic transformation. A challenge, then, is how this can be brought into the realm of policy making, replacing negative economic incentives and their associated mental states with positive inducements that stimulate sustainable economic mindsets and behaviour.

The legacy of enlightenment philosophy rooted in the separation of humans from nature, such that we could better objectify and thereby measure the Earth and its life forms, remains a persistent influence even within contemporary sustainability science. This gives rise to the formulation of technical solutions to our environmental crisis without ever truly grappling with the underlying mindsets (beliefs, values, attitudes), and their systemic structures and behavioural patterns which they underpin. A relational epistemology, in contrast, begins from recognising subjectivity across all forms of life – human and non-human – which are increasingly visibly entangled. If we are to take steps to reduce our impact upon the Earth and all its species, then perhaps we need to find ways to move beyond narrowly scientific understandings of cause and effect and draw upon a moral code of practice rooted in compassion and empathy. Arguably it is only by being open to the vulnerability and suffering of the poorest, least responsible victims of climate change that will enable us to appreciate the embodied responsibility of hyper-consumerism in the West and the urgent need to address levels of consumption. In this regard, the need for a new language through which to mobilise a collective enlightenment is truly pressing (Wamsler, 2018; Walsh, 2018a, 2018b).

Overview of chapters

Metaphors of reason

Technologies associated with energy have had some of the most significant environmental impacts with the most alarming consequence of carbon-based energy technologies being climate breakdown. In his chapter, Fionn Rogan is concerned about the way in which metaphors used to describe technological change can often divert attention from the detrimental consequences that such change can bring about. His chapter examines four key metaphors of technological change – technological fix, technological determinism, technological dialogue, and technological momentum – and assesses how these metaphors can bring attention to, or shift attention away from, the unintended consequences of new technologies. The metaphor of technological fix, for example, gives primacy to technology as the decisive factor in the solution of society's problems. It implies that technologies are distinct from, rather than embedded within, broader society. The metaphor of technological fix, Rogan argues, is likely to fuel techno-optimism and divert attention away from the wider societal consequences of technology. Other metaphors of technological change, by contrast, such as the metaphor of technological momentum, better reflect the social embeddedness of technological change. The impact of a new technology can begin slowly and acquire momentum as complex social and technological dynamics lead to the widespread adoption of the technology, often accompanied by profound social changes. The metaphor of technological momentum therefore situates technologies not as single discrete artefacts, but as parts of interconnected complex systems. According to Rogan, this metaphor better reflects the complexity of socio-technological change and allows for a deeper engagement with the potential unintended consequences that new technologies invariably bring. The metaphor of technological momentum also opens up the debate on technology to wider non-specialist audiences who may be profoundly impacted by technological change.

Colin Sage draws attention to the fact that climate breakdown is just one of the many unintended consequences that technologies and their accompanying social practices are having on our environment. The metaphor of planetary boundaries, as set out by Rockström et al. (2009a, 2009b) and by Steffen et al. (2015), identifies nine key Earth system processes, each of which are being affected by human activity. These nine planetary boundaries delimit the space which must not be breached if we are to maintain Earth as a safe habitat for humanity. For three of the nine processes however – climate change, biodiversity loss, and nitrogen and phosphorous flows – we are already breaching these boundaries.

In his chapter on the Global Food System, Planetary Boundaries and the Metabolic Rift, Sage concentrates on one of these breached boundaries, namely the global biogeochemical flow of nitrogen. Although not as widely acknowledged as the challenge of climate change, the disruption of the global nitrogen cycle by human activity is a serious environmental challenge facing humanity. While climate breakdown can be understood as an unintended consequence of energy technologies, the disruption of the global nitrogen cycle is largely an unintended consequence of agricultural technologies, in particular the development and widespread use of nitrogen-based fertilisers. As Sage explains, the development of the Haber-Bosch process early in the twentieth century to synthesise atmospheric nitrogen (N_2) into the form of fertiliser has had a transformative effect on both population growth and on humanity's ability to feed itself. The astonishing impact of nitrogen-based fertiliser can be seen from the fact that nitrogen fertilisers are responsible for feeding almost half of the world's population today (Erisman et al., 2008).

The large-scale use of nitrogen fertilisers, however, is also having a transformative effect on the global nitrogen cycle (Galloway et al., 2002). The detrimental consequences of this are becoming ever more apparent in the form of nitrate leaching into groundwater, a loss of terrestrial and marine biodiversity, agricultural emissions of ammonia that create toxic air quality, and contributions to climate change through N_2O, a potent greenhouse gas. Sage explores these multiple destructive impacts using the metaphor of the metabolic rift, a metaphor that highlights the disturbed metabolic interaction between human society and the environment under the current global agri-food system.

In the chapter "Defamiliarisation of Food", Shane Crowley stays with the global agri-food system by exploring how contemporary metaphors mediate our relationship with food. That everyday language is rich with food metaphors – food for thought, an unsavoury character, half-baked plans, a bitter taste in our mouth – is not surprising, given the centrality of food in our lives. Such examples indicate that food as a signifier is involved in the transmission of meanings far beyond its immediate function of ensuring human survival (Stajcic, 2013). Cooking, for example, is a means by which raw food (nature) can be transformed, following Levi-Strauss, into culture, and traditional dishes form the basis of cuisine as part of a cultural identity which can be shared.

Metaphors concerning food influence how we view ourselves as individuals and as societies. Metaphors of food are also closely linked to metaphors of the body. The metaphor of food as fuel, for example, is connected to the metaphor of the body as a machine. As Crowley outlines, today's metaphors increasingly compare the human body to a computer, with food as a bug in the system. Food metaphors are moving towards a narrative in which food functions solely as a vehicle for energy and nutrients, a reductionist philosophy that views both food preparation and the body to be problematic aspects of life (Miles and Smith, 2016). Crowley takes this development further using the vision of the cyborg – future humans as hybrids of organism and machine – and sees contemporary developments such as smart foods and meal replacement nutritional drinks and pills as steps that will eventually enable the transcending of bodily limitations altogether. However, he points out that many of the techno-utopian metaphors that suffuse contemporary discourse on food and diet are in direct tension with a lived experience of food that spans cultural and social domains. Can traditional food, the ill-disciplined pursuit of which is linked to disease and waste, survive future developments in food technology? Ultimately, Crowley concludes, food scientists and engineers are not only manipulating organic matter but also metaphor, giving themselves considerable power to shape future conceptions of food, the body, and the social world.

In the final chapter in this section on metaphors of science and reason, James Glynn returns to the challenge of climate breakdown by discussing the metaphor of carbon budgets. Carbon budgets give a meaningful and simple method to understand and communicate the scale of action required if we are to limit the dangers of climate change. Global temperature increase is driven by the cumulative greenhouse gas emissions from the combustion of fossil fuels, emissions from industry, and land use change for agricultural purposes. The remaining cumulative carbon dioxide emissions that would result in a 1.5°C or 2°C temperature increase with a given probability is referred to as a Carbon Budget.

If we are to limit future warming to 2°C with a high probability, we can only emit around 1 trillion tonnes more of CO_2 into the atmosphere; that is, we have a remaining carbon budget of 1 trillion tonnes of CO_2. To limit warming to well below 2°C and towards 1.5°C, as is the goal of the Paris Agreement, we can only emit between 200 Gt CO_2 and 700 Gt CO_2, so our remaining carbon budget is much lower. For comparison, global CO_2 emissions in 2017 were estimated to be 41 Gt CO_2 and are rising.

As Glynn explores, the metaphor of a carbon budget succinctly communicates a number of vital messages about climate change. First, that long-term temperature rise does not depend on CO_2 emissions at a specific time, but on the cumulative emissions that have been made, largely since the beginning of the Industrial Revolution. Second, that our current CO_2 emissions are critically important as they are exhausting our remaining carbon budget. Third, that CO_2 emissions will need to be phased out to net zero eventually in order to achieve global temperature stabilisation.

The chapter also explores how the metaphor of carbon budgets is being used as an instrument of equitable climate action. For example, because they industrialised earlier, the wealthier countries of the world, western Europe, Canada, Japan, and the USA, have produced considerably more CO_2 than their fair share carbon budget. It can be reasonably argued, therefore, that these countries should enable developing countries to decarbonise by providing financial aid, access to financial capital, knowledge transfer, and technology transfer (CSO Equity Review, 2018). The utility of the powerful metaphor of carbon budgets to issues of equity and responsibility also extends to sub-national and individual levels, as Glynn also explores.

Myths and metaphors of unreason

In this first chapter of the second section, Edmond Byrne poses the fundamental proposition that the challenges of unsustainability will require much more than technological change. They require a reorientation in our current conception of reality, and for such a reorientation to occur, changes in the metaphors that reflect our view of reality are also urgently needed.

Our current dominant paradigm, which has been incredibly successful over the past four centuries, inspired by neo-Cartesian rationality, is one characterised by antagonistic dualism seeking reduction, separation, control, and certainty, and which now transcends all of our globalised societies. The challenges of sustainability, however, Byrne argues, require a radical shift from such monopolistic thinking towards a complementary view of reality characterised by the metaphor of agonistic dualism, whereby polar opposite tendencies are seen as mutually obligatory. Such a metaphor has been used across many human cultures throughout history, most prominently in Eastern traditions such as Taoism and Zen Buddhism which espouse the complementary opposites of Yin and Yang.

Byrne's chapter reflects on parallels between the metaphor of agonistic dualism and the physical make-up of the brain and its two hemispheres, as elaborated in the work of neuropsychiatrist Ian McGilchrist (2009). McGilchrist describes the two asymmetrically different though complementary cerebral components: with the left hemisphere taking a literalist, rationalist, explicit, decontextualised, and "either/or" approach to the world around it; while the right hemisphere takes an integrative (facilitating "both/and") approach, while seeking both context and (inter)connection.

The chapter argues that metaphor – together with narrative, story, and myth, as well as other right hemisphere constructs such as art and music – are really the only means of sufficiently moving individuals, communities, and societies to embark upon the type of transformational change that is required to achieve authentic sustainability. Byrne explores, in particular, how metaphors of "sustainability as flourishing" and "nature as sacred" might radically reorient our collective worldview as a precondition to bringing about the fundamental changes in our lifestyles and behaviours that sustainability challenges demand.

In the chapter "Myth beyond Metaphor: Myths in Transition", Evan James Boyle builds on Byrne's thesis by arguing that myth has largely been extinguished from modernity as a means of sense making. While myths previously established social customs and moral lessons, with modernity they came to be a representation of irrationality, unable to stand up to scientific reasoning. Quoting Nietzsche in The Birth of Tragedy, Boyle argues that "Man today, stripped of myth, stands famished among all his pasts and must dig frantically for roots". The directionality provided by myth, however, is necessary if we are to navigate the challenges of sustainability.

Modernity, Boyle argues, does have one dominant surviving negative myth in the form of Goethe's Faust. In Goethe's tale, the protagonist signs a blood oath with Mephistopheles, to gain all knowledge and power during his life in return for his soul in death. The contemporary citizen, hypnotised by technological acceleration and the commodification of social life, like Faust, has sold her/his soul in return for material progress. The contemporary absence of positive myth, however, mitigates against what complexity biologist Stuart Kauffman terms an emerging "global ethic" (Kauffman, 2010). According to Kauffman, we lack a transnational mythic value structure that can expand our consciousness and sustain our emerging global civilization.

But rebuilding a new mythology to sustain and accelerate our transition towards such an ethical globalised future is not a simple task. Hyman (1955) goes as far as to suggest that no one can invent myths. Instead mythology is a pantheon, a cumulative creation, borne out of many generations (Schorer, 1960). The creation of a new pantheon of myths, Boyle argues, requires revisiting the myths of old to seek guidance. In suggesting a beginning to such a Herculean task, Boyle suggests the myth of Oedipus as one candidate for mythic revival. In consulting the Oracle at Delphi, Oedipus was informed that he would kill his father and marry his mother. Taking the maternal to represent the earth, or "mother nature", we are faced with a quandary not dissimilar to the one faced by Oedipus in having to choose what aspects of the father (masculinity, dominance, control) we must leave behind if we are to accept that our fate is married to that of mother Earth.

Fables have been used for centuries to relay important messages in a playful manner. In the fable of the hare and the tortoise, the tortoise declares that he will beat the boastful hare in a race. As the fable recounts, the complacent hare wakes from a nap just in time to see the tortoise crossing the finish line. "Slow and steady wins the race" is the moral of the story. Yet in terms of pressing contemporary issues around sustainability, "slow and steady wins the race" appears paradoxical in the context of the often-panicked sense of urgency being expressed by those intimately involved in the area. In their chapter, McGookin, Ó Gallachóir, and Byrne explore a number of narratives around issues of sustainability, in particular concerning climate change, by casting the hare and tortoise in a variety of roles. With the hare in the role of radical environmentalists, experts in climate science who have raced ahead and seen the coming catastrophe, and entrepreneurs marketing new technologies they are convinced will solve the problem, the chapter presents a series of narratives that argue that there is value to the tortoise's "slow and steady" approach. While innovative individuals, be they activists, scientists, or inventors, are necessary to disrupt societal structures and begin the proverbial race to sustainability, only through a widespread change in beliefs and values in wider society (represented by the tortoise) can the race be won. And while there is urgent need for transformative and dramatic change to the course of our civilisation, we must be careful to ensure that the changes we are making do not have unintended consequences highlighted in earlier chapters.

In their final narrative, the authors present the hare as neoliberal society racing relentlessly on in pursuit of economic growth and material accumulation though eventually to be overtaken by an as-yet emerging society based on more collective and sustainable values. The hare, which appears to be making most progress, has principal concern around personal attainment and material possessions, though accepts the high and rising costs of unsustainability. The more purposeful (and ultimately wiser) tortoise, by contrast, can see a greater value in restraining its personal desire for consumption and seeks instead to flourish through interpersonal connections. If the tortoise wins this race, we will have constructed a global society whose priorities lie in achieving harmony between itself and its environment.

Metaphor, myth, and mind

In the opening chapter of this section concerned with aspects of human consciousness, Ian Hughes explores how metaphors in psychoanalysis can contribute to our understanding of how we might achieve the necessary changes in relationships towards greater empathy and equity, required for transition to sustainability. The chapter explores the psychoanalytic concept of development and the myths and metaphors that are used therein.

Hughes explores two myths that are foundational in psychoanalysis, namely the myth of Narcissus and the myth of Oedipus, which capture the two primary psychic challenges every child must face during the first years of life. The child's first challenge is to overcome its infantile state of primary narcissism and accept the reality that they are not the sum total of existence, a realisation brought on by the gradual withdrawal of the intensity of care provided by the child's parents in the first few weeks and months of life. If negotiated successfully, the child establishes an internal capacity for the containment of emotion and an internalisation of the world as a benign and supportive place. If this crucial initial stage of psychic development is not negotiated successfully, the child, like Narcissus, is at risk of remaining in an infantile state characterised by concern for self-preservation, paranoia, and an ability to only love itself. The infant's second major psychic challenge follows on quickly, as the child realises that not only is it not omnipotent, but that it must also share the world with many other people who have wishes of their own. For Freud, the myth of Oedipus encapsulated the child's primitive wishes to deny the dawning reality that he is not the sole focus of his mother's love, and the powerful feelings of envy and rage that accompany the infant's recognition of this painful fact.

The myths of Oedipus and Narcissus, Hughes argues, and humanity's collective failure to deal successfully with the challenges of early psychic development, are reflected in the dominant zeitgeist of contemporary society. This zeitgeist is characterised by humanity's grandiose sense of self-importance and uniqueness; our exhibitionistic need for constant attention and admiration; our lack of empathy and disregard for rights of future generations and other species; and relationships with one another (and with nature) marked by a sense of entitlement and exploitation.

In his chapter, Cormac Sheehan shifts our attention towards another growing challenge for societies around the world, namely increases in the prevalence of dementia. In 2015, it was estimated that dementia affected 45 million people. By 2050, this is set to rise to 131 million, with the largest increases predicted in low- to middle income countries, due both to ageing populations and improvement in diagnosis and awareness. Sheehan explores metaphors of dementia and how such metaphors can either increase or alleviate suffering. His point of departure is Sontag's seminal work "Illness as Metaphor and AIDS and Its Metaphors" (Sontag, 2009). Sontag wrote (metaphorically) that when we are born, we "hold dual citizenship in the kingdom of the well and in the kingdom of the sick" and that, on becoming sick, we take up citizenship in that "other place". Sontag criticised the, often malicious, metaphors of illness and argued that we must understand illness metaphors in order to be liberated from them.

In this chapter, Sheehan describes how metaphors of decline and decay, of war, flood and epidemics are common in discourses about dementia. Terms like "dissolution" and "unbecoming of the self" are commonly found, along with descriptions of dementia as an "extinction of personhood" and a death-in-life or living death. The metaphor of "losing one's mind", which is associated with culturally constructed notions of personhood, even raises questions about a person's eligibility for moral membership of the human social environment (Johnstone, 2011: 382). In response, Sheehan counters that while the dominant metaphors may deem a person to be within a "living death", those close to the affected person actively challenge such an imposition. Carers do not allow the physical appearance of a person with dementia – their hair, clothes, and cleanliness – to become unkempt. This care for the body is the antithesis of social death. A shift in social attitudes in recent years is discernible as a host of films, plays, and novels explore the human side of dementia. However, if we are to be further liberated from the current pantheon of illness metaphors, Sheehan argues, new more imaginative and more humane metaphors are urgently needed.

Dreams, visions, revelations, oracles, and prophecies have shaped societal, cultural, and religious changes throughout the history of humankind. Dreams are imaginative, relying on other worldly visualisations and images of the world. Dreams question, represent, or sustain the world and they have the power to transform. Consequently if, as earlier chapters have argued, a fundamental reorientation of our current dominant view of reality is integral to the transition to sustainability, then there may be much to learn from understanding other eco-cosmological worldviews. In "The Dream as Metaphor of Transformative Change", Lidia Guzy explores the Shamanic dream to provide an insight into the radically different worldviews of some indigenous cultures. Throughout history, indigenous peoples have been marked by marginalisation, colonisation, and by the general devaluation of their knowledge systems. As a result, indigenous explanations appear largely "meaningless for the modern world" (Brabec de Mori, 2016: 80–81).

As Guzy outlines, eco-cosmologies are worldviews relating the human with the non-human, the cosmos and the other-than-human sphere such as trees, animals, rivers, mountains, and spirits. An important element is the absence of the dualistic separation between the human and the surrounding geography and landscape, such as a mountain, a river, or a tree. "Identity" is not ego or body centred but is spread and integrated with the surrounding ecology, geography, and territory. In a cultural practice that spans from Siberia to the Amazon, the shaman in this eco-cosmological setting is the local intellectual and spiritual leader with the capacity to transcend different dimensions of existence through the shamanic dream and its ritual communication. The shamanic dream marks a different form of knowledge system that does not anthropomorphise everything by explaining all existence in terms that are relevant to humans. Indigenous peoples are also deeply aware of the fact, only recently acknowledged by science, that every aspect of their forested territory has been transformed by their ancestors – that their primal home is not "wild" but has been human-influenced according to the ethno-agricultural and ethno-biological knowledge of these hunter gatherer societies. The shamanic dream, reflecting humanity's deep interconnection with ancestors and environment, past and future, can be read as a fundamental dismissal of global anthropocentrism, modernity, and materialism.

Metaphors of creativity and practice

Kieran Keohane begins his chapter, Joyce's Arches, with the assertion that the methods and mind-sets of Modernity are fully implicated in the malaises of our times: from existential threats of climate breakdown and species extinction to the insidious and pervasive dissolution of traditions, values, and ideals that are essential to well-being and human flourishing. In our time which demands renewal, Keohane turns to James Joyce for a rich source of metaphors that create spaces of ambiguities and ambivalences, paradoxes and dialectical antitheses, and which open up new possibilities for imagining future horizons. Keohane agrees with Byrne's assertion in his earlier chapter that our current crises of sustainability require metanoia – a change in the trend and action of the whole inner nature, intellectual, affectional, and moral; a transmutation of consciousness; a conversion.

The chapter describes how specific objects and artefacts for Joyce held the prospect of enabling transition, metanoia, transfiguration, and transformation. The ancient Roman arches that Joyce knew well when he lived in Pula, Trieste, and Rome were, for him, objects that in his imagination could connect disparate times and places. The Arch of Constantine, for example, was for Joyce over-determined and ambiguous, standing in a thoroughfare in ancient Rome while simultaneously having all of modern humanity flow through it, while reminding him of his good friend Constantine Curran and the city of Dublin. Joyce's arches had the effect, by enabling this mind shifting through multiple perspectives of reality, of transformation. As Keohane describes, Joyce intended Ulysses and Finnegan's Wake, with their over-determinations, portmanteaus, and metaphors, to have just such effects. Joyce was concerned with how modern people generally suffer from conventional morality, solipsism and inept maladaptation to the demands of changing cultural, political, psychological, and moral contexts, and that this willing conformity to the flow of prevailing discourse, an inability and unwillingness to turn around is one of the pathogenic social currents of our time. His writing was intended as a portal, a transformational object, which once encountered or passed through, would awaken the reader to a new perspective on reality and a renewed resilience to sustain civilization through a contemporary Dark Age, such as the late-modern antediluvian eve of the Anthropocene.

Maureen O'Connor's chapter draws attention to the role of the ecological humanities in seeking to destabilise conventional notions of subjectivity. Here she argues against the idea of independent entities that forms the basis of a separation between the human and the nonhuman and which has proven so disastrous for the environment and the planet's inhabitants. In her chapter, O'Connor challenges the metaphor of the mirror that science has claimed it holds up to reality, establishing and perpetuating the idea of objective, scientific "truth" regarding the natural world and natural phenomena. In its place, an alternative dynamic metaphor of diffraction, first proposed by Donna Haraway (1992), is explored for the relationship between the observing human consciousness and the finally unknowable world around. The metaphor of diffraction suggests a different approach to understanding physical phenomena, one that accepts the incomplete nature of human knowledge, that recognises and even embraces the unstable, the plural, and the partial.

The 2016 prize-winning novel, Solar Bones, by Irish writer Mike McCormack, which is the focus of the chapter, not only directly addresses the ecological destruction brought about by late capitalism in twenty-first century Ireland, but also, in the text's innovative form, enacts the crisis through a diffractive aesthetics of fragmentation and heterogeneity that reveals unsuspected but vital continuities and connections. The text's narrative form evokes a decentralising and disassembling view of reality and experience while simultaneously painting a picture of a universe in which everything is deeply interconnected. The "luminous bones" of the title refer, inter alia, to sections of a disassembled wind turbine, a symbol to the narrator, Marcus, of a failure of imagination, "the world forfeiting one of its better ideas", defeat in the struggle to imagine a new and better world.

The text, which is haunted by James Joyce, ranges over themes of ghosts, death, illness, and mechanical and societal breakdown. It laments an imagined better world but one not brought to fulfilment and, by radically disturbing our engagement with the world, holds out the possibility of a more ethical response, more accepting of the partial and fragmented, than the patriarchal model of omniscience and control.

At the beginning of her chapter, Jools Gilson asserts that metaphor and creative writing, dance, theatre, music, film, visual art practices (as well as multiple combinations of these) contribute blue sky/outside of the box thinking, and that if we want to retain our blue sky and not end up in boxes, we need the world-making intelligences of creative practices. The chapter proposes that creative practice disciplines have a powerful and critical role to play in collaborative research models for sustainability. These are knowledges, Gilson argues, which challenge the hegemony of critical analytical modes of meaning-production. They engage with the world differently than many academic disciplines, are fluent at embodiment, affect, visual literacy, imagination, and engaging with communities.

Using rain and water as the primary metaphor, Gilson demonstrates the power of creative practice disciplines by describing a series of artistic installations. Gilson draws upon the writings of writer and theologian John Hull which recounts his journey into blindness. Hull, after being deeply moved by the sound of rain outdoors, makes the wish, "If only there could be something equivalent to rain falling inside, then the whole of a room would take on shape and dimension". Film makers Peter Middleton and James Spinney (Middleton and Spinney, 2016) make John Hull's dream of an internal raining, literal: as he sits quietly at his kitchen table before a teacup, the panning camera and sound pick out distinctly the sound of rain as it falls on its delicate china surfaces. Like a bat's sonar, the sound of rain and its complexity of different tones, cadences, and rivulets allow Hull to see once again, but in a radically different manner.

In the chapter, Gilson also describes her own art works, in what she calls a drenching lineage, the most recent of which is The Rain Box (Adams and Gilson, 2017), a playful piece of radio that explores the science and poetry of rain through the tale of a child who finds a hidden box with rain falling inside. As Gilson describes it, The Rain Box is a whole folklore of rainy tales, of a people who sing the rain out of the sky, who weep when it rains, who mark in its spaces all species of yearning. In Gilson's work, Ireland itself, and the wider world too, is presented as a box of rain that holds stories suffused with meanings and metaphors that hold the potential to create a brighter future.

Conclusion: Language as catalyst for sustainable, transformative change

The respective chapters of this book possess a genesis which in itself indicates more than a disparate collection of myth-infused and metaphor-imbued offerings from across a range of disciplines. Each chapter is derived from the authors' commonly fermented understanding of the meaning, power, and value of language, in particular, how humans interact with each other through metaphor and myth in the process of societal development and change. Secondly, the book represents offerings from a collective which considers that such societal change and development should be, and indeed in our current phase of societal development, must be, strongly and rapidly directed towards "sustainability", and by that meaning something which is more akin to environmental and societal flourishing, rather than narrowly and technologically focussed on, for example, atmospheric carbon reduction. Thirdly, each of us would strongly profess that such an enormously ambitious aspiration can only be achieved when science and the humanities are driven by a transdisciplinary ethos. Indeed, recognition of the centrality of metaphor (and its offspring – myth, narrative, and story) necessarily posits the humanities and social sciences as front and centre in any quest for such a sustainable society and environment. Clearly, such positioning effectively removes any illusions of a "one-eyed" reductionist scientism, associated narrow "mode 1" conceptions of science or the holy grail of a prevailing techno-optimistic "science" which possesses the "solutions" which only need to be uncovered. Nor does it relegate the humanities and social sciences to a role whereby they are accorded a neat window-dressing role to speak for "society". For, as we hope this book makes clear, science is both built and communicated through metaphor, and the stories that it offers are as politically and ideologically present in its disciplines, including above all the metaphor (the myth?) of the "objective observer", which is central to the scientific method. The great advantage that humans possess over other species (we believe) is the self-aware consciousness that makes us beings of stories, narratives, and metaphor – and all categories of scientific and technological advance, as well as societal progress, depend on this fact.

It is within this context, and on this explicitly recognised more level playing field that the authors have engaged in this particular project and its associated (and ongoing) dialogue. A generous spirit of "disciplinary humility" (Tripp and Shortlidge, 2019) ensconced within enquiring scholarship therefore both brought us together and informed our work as we proceeded through the preparation and presentation of draft chapters and a willingness to revise in response to review and critique. This ongoing and constructive conversation, informed by a strong sense of trust in each other's motives, enabled and encouraged us to seek value within and across each other's disciplinary perspectives. Far from diminishing our individual disciplinary integrity, we have found added value from widening the lens and bringing new perspectives back into our "home" disciplines.

It is through this ongoing, inclusive and transdisciplinary spirit that we have forged this work. As with our earlier publication, "Transdisciplinary Perspectives on Transitions to Sustainability" (Byrne, Mullally and Sage, 2017), the current book arises from an initial gathering that sought to widen the conversation within the academy; this time drawing in the arts and humanities as well as medicine. While each of us – as editors and contributors – are deeply motivated by the question of sustainability, we are equally driven by the question of transdisciplinary practice and knowledge. On this occasion, however, the notion of metaphor served as an entry point – or boundary object – that allowed for communication across disciplinary silos. For metaphors and myths have a universal power capable of helping us collectively to achieve the scale of transformative change necessary to inflect us from the course of global ecological catastrophe and towards societal and environmental flourishing.

References

Adams, S. and Gilson, J., 2017. *The Rain Box. Lyric FM.* Limerick: Radio Feature. [online] Available at: https://soundcloud.com/the-lyric-feature/the-rain-box [Accessed 3 July 2020]

Annas, G., 1995. Reforming the debate on health care: Reform by replacing our metaphors. *The New England Journal of Medicine*, 332, pp. 744–747.

Atanasova, D., and Koteyko, N., 2017. Metaphors in Guardian online and mail online opinion-page content on climate change: War, religion, and politics. *Environmental Communication*, 11(4), pp. 452–469.

Barry, J., 2017. Bio-fuelling the Hummer? Transdisciplinary thoughts on techno-optimism and innovation in the transition from unsustainability. In E. P. Byrne, G. Mullally, and C. Sage, (eds) *Transdisciplinary Perspectives on Transitions to Sustainability*. Abingdon: Routledge, pp. 106–123.

Brabec de Mori, B., 2016. What makes natives unique? Overview of knowledge systems among the world's indigenous people. *Taiwan Journal of Indigenous Studies*, 8, pp. 43–61.

Brown, H.R., 1976. Social theory as metaphor: On the logic of discovery for the sciences of conduct, *Theory and Society*, 3(2), pp. 169–197.

Byrne, E. P., Mullally, G., and Sage, C., 2017. *Transdisciplinary Perspectives on Transitions to Sustainability*. Abingdon: Routledge.

Capra, F., 1988. *Uncommon Wisdom: Conversations with Remarkable People*. London: Century Hutchinson.

Carayannis, E.G., Barth, T.D. and Campbell, D.F.J., 2012. The Quintuple Helix innovation model: Global warming as a challenge and driver for innovation. *Journal of Innovation and Entrepreneurship*, 1(2), pp. 1–12.

Castells, M., 2009. *Communication Power*. Oxford and New York: Oxford University Press.

Coyne B. and Denny, E., 2018. An Economic Evaluation of Future Electricity Use in Irish Data Centres. TRiSS Working Paper Series No. TRiSS 02 – 2018 Version 1. Trinity College Dublin. [online] Available at: https://www.tcd.ie/triss/assets/PDFs/wps/triss-wps-02-2018.pdf [Accessed 2 September 2019].

CSO Equity Review, 2018. Equity and the Ambition Ratchet: Towards a Meaningful 2018 Facilitative Dialogue. *Figshare*. [online] Available at: https://doi.org/10.6084/m9.figshare.5917408 [Accessed 3 July 2020].

D'Alisa, G., Demaria, F., and Kallis, G., 2015. *Degrowth: A Vocabulary for a New Era*. Abingdon: Routledge.

Dawkins, R., 1976. *The Selfish Gene*. Oxford: Oxford University Press.

Ehrenfeld, J. R., 2008. *Sustainability by Design*. New Haven: Yale University Press.

Eirgrid/SONI, 2018. All-Island Generation Capacity Statement 2019–2028. Dublin: EirGrid Plc, Belfast: SONI. [online] Available at: http://www.eirgridgroup.com/site-files/library/EirGrid/EirGrid-Group-All-Island-Generation-Capacity-Statement-2019-2028.pdf [Accessed 3 July 2020].

Erisman, J., Sutton, M., Galloway, J., Klimont, Z. and Winiwarter, W., 2008. How a century of ammonia synthesis changed the world. *Nature Geoscience*, 1, pp. 636–639.

Etzkowitz, H. and Leydesdorff, L., 2000. The dynamics of innovation: From national systems and "Mode 2" to a triple helix of university–industry–government relations. *Research Policy*, 29, pp. 109–123.

Fløttum, K., 2014. Linguistic mediation of climate change discourse. *ASp: la revue du GERAS*, 65, pp. 7–20.

Funtowicz, S.O. and Ravetz, J.R., 1993. The emergence of post-normal science. In: R. von Schomberg, ed. *Science, Politics and Morality*. Dordrecht: Springer, pp. 85–123.

Galloway, J., Cowling, E., Seitzinger, S. and Socolow, R., 2002. Reactive nitrogen: Too much of a good thing?. *Ambio*, 31(2), pp. 60–63.

Gibbons, M., Camille, L., Nowotny, H., Schwartzman, S., Scott, P. and Trow, M., 1994. *The New Production of Knowledge: The Dynamics of Science and Research in Contemporary Societies*. London: Sage.

Haraway, D. J., 1992. *Promises of Monsters: A Regenerative Politics for Inappropriate/d Others*. Abingdon: Routledge.

Harré, R., Brockmeier, J. and Mühlhäusler, P., 1999. *Greenspeak: A study of environmental discourse*. London: Sage.

Hyman, S.E., 1955. The ritual view of myth and the mythic. *The Journal of American Folklore*, 68(270), pp. 462–472.

Ison, R., Allan, C. and Collins, K., 2015. Reframing water governance praxis: Does reflection on metaphors have a role? *Environment and Planning C: Government and Policy*, 33(6), pp. 1697–1713.

Jackson, T., 2016. *Prosperity Without Growth*, 2nd ed., Abingdon: Routledge.

Jepson, P. 2018 Recoverable Earth: A twenty-first century environmental narrative. *Ambio*, 48:123–130.

Johnstone, M.J., 2011. Metaphors, Stigma and the 'Alzheimerization' of the Euthanasia Debate. *Dementia*, 12(4), pp. 377–393.

Kabat-Zinn, J., 2004. *Wherever You Go, There You Are: Mindfulness Meditation for Everyday Life*. London: Piatkus.

Kallis, G. and Vansintjan, A., 2018. *In Defense of Degrowth: Opinions and Minifestos*. Uneven Earth Press. [online] Available at: https://indefenseofdegrowth.com/ [Accessed 3 July 2020]

Kauffman, S.A., 2010. *Reinventing the Sacred: A New View of Science, Reason, and Religion*. New York: Basic Books.

Keddy, P.A. 1989. *Competition*. New York: Chapman and Hall.

Kress, J.M., 2000. Contesting metaphors and the discourse of consciousness in William James. *Journal of the History of Ideas*, 61(2), pp. 263–283.

Kuusi, O., Lauhakangas, O. and Ruttas-Küttim, R., 2016. From metaphoric litany text to scenarios—How to use metaphors in futures studies'. *Futures*, 84, pp. 124–132.

Lakoff, G., 2004. *Don't Think of an Elephant! Know Your Values and Frame the Debate: The Essential Guide for Progressives*. White River Junction, Vermont, USA: Chelsea Green Publishing.

Lakoff, G. and Johnson, M., 1980. *Metaphors We Live By*. Chicago: University of Chicago Press.

Larson, B., 2011. *Metaphors for Environmental Sustainability: Redefining Our Relationship with Nature*, New Haven: Yale University Press.

Marks, M., 2011. *Metaphors in International Relations Theory*. Dordrecht: Springer.

McGilchrist, I., 2009. *The Master and His Emissary*. Yale: Yale University Press.

Middleton, P. and Spinney, J., 2016. *Notes on Blindness*. London: Artificial Eye/Arte. [film]

Miles, C., and Smith, N., 2016. What grows in silicon valley? The emerging ideology of food technology. In: H. L. Davis; K. Pilgrim and M. Sinha, eds. *The Ecopolitics of Consumption: The Food Trade*, Maryland, USA: Rowman & Littlefield.

Moser, S. C., 2006. *Communicating climate change-motivating civic action: Opportunity for democratic renewal. Climate Change Politics in North America, Wilson Center Occasional Papers, 2*, Washington D.C.

Mullally, G., 2017. Fear and loading in the anthropocene: Narratives of apocalypse and salvation in the Irish media. In *Byrne*, G. Mullally, C. Sage, (eds) *Transdisciplinary Perspectives on Transitions to Sustainability*. Abingdon: Routledge.

Norgaard, K., 2011. *Living in Denial: Climate Change, Emotions, and Everyday Life*. Cambridge, Mass: MIT Press.

Olds, L., 1992. *Metaphors of Interrelatedness: Toward a Systems Theory of Psychology*. Albany: State University of New York Press.

Paprotte, W. and Dirven, R., 1985. *The Ubiquity of Metaphor*. Amsterdam: John Benjamins.

Rickards, L. A., 2015. Metaphor and the anthropocene: Presenting humans as a geological force. *Geographical Research*, 53(3), pp. 280–287. doi:10.1111/1745-5871.12128.

Rockström, J., Steffen, W., Noone, K., Persson, A., Chapin, F., Lambin, E., Lenton, T., Scheffer, M., Folke, C., Schellnhuber, H., Nykvist, B., De Wit, C., Hughes, T., Van der Leeuw, S., Rodhe, H., Sörlin, S., Snyder, P., Costanza, R., Svedin, U., Falkenmark, M., Karlberg, L., Corell, R., Fabry, V., Hansen, J., Walker, B., Liverman, D., Richardson, K., Crutzen, P. and Foley, J. 2009a. Planetary boundaries: Exploring the safe operating space for humanity. *Ecology and Society*, 14(2), 32.

Rockström, J., Steffen, W., Noone, K., Persson, A., Chapin, F., Lambin, E., Lenton, T., Scheffer, M., Folke, C., Schellnhuber, H., Nykvist, B., De Wit, C., Hughes, T., Van der Leeuw, S., Rodhe, H., Sörlin, S., Snyder, P., Costanza, R., Svedin, U., Falkenmark, M., Karlberg, L., Corell, R., Fabry, V., Hansen, J., Walker, B., Liverman, D., Richardson, K., Crutzen, P. and Foley, J., 2009b. A safe operating space for humanity. *Nature*, 461(7263), pp. 472–475.

Schön, D.A., 1979. Generative metaphor: A perspective on problem-setting in social policy. In: *Metaphor and Thought*, A. Ortony, ed. Cambridge: Cambridge University Press, pp. 254–284.

Schorer, M., 1960. The necessity of Myth. In: H. Murray, ed. *Myth and Mythmaking*. New York: George Braziller, pp. 355–360.

Sontag, S., 2009. *Illness as Metaphor and AIDS and Its Metaphors*. London: Penguin Books

Spencer, H., 1864. *The Principles of Biology*, Volume 1. London: Williams and Norgate

Stajcic, N., 2013. Understanding culture: Food as a means of communication. *Hemispheres*, 28, pp. 5–14.

Steffen, W., Richardson, K., Rockström, J., Cornell, S.E., Fetzer, I., Bennett, E.M., Bigs, R., Carpenter, S.R., De Vries, W., De Wit, C.A., Folks, C., Gerten, D., Heinke, J., Mace, G.M., Ramanathan, V., Reyers, B. and Sörlin, S., 2015. Planetary boundaries: Guiding human development on a changing planet. *Science*, 347(6223). DOI: 10.1126/science.1259855.

Thibodeu, P.H. and Boroditsky, L., 2011. Metaphors we think with: The role of metaphor in reasoning. *PLoS ONE*, 6(2), e16782.

Tripp, B. and Shortlidge, E.E. 2019. A framework to guide undergraduate education in interdisciplinary science. *CBE—Life Sciences Education*, 18(2), 1–12.

Ulanowicz, R.E., 2009. *A Third Window; Natural Life Beyond Newton and Darwin*. West Conshohocken: Templeton Foundation Press.

Voß, J.P. and Bornemann, B., 2011. The politics of reflexive governance: Challenges for designing adaptive management and transition management. *Ecology and Society*, 16(2), p. 9.

Walsh, Z., 2018a. Navigating the great transition via post-capitalism and contemplative social sciences. In V. Giorgino, Z. Walsh eds. *Co-Designing Economies in Transition: Radical Approaches in Dialogue with Contemplative Social Sciences*. Cham, Switzerland: Palgrave Macmillan, pp. 43–61.

Walsh, Z., 2018b. Contemplating the more-than-human commons. *The Arrow: A Journal of Wakeful Society, Culture and Politics*, 5(1), pp. 4–17.

Wamsler, C., 2018. Mind the gap: The role of mindfulness in adapting to increasing risk and climate change, *Sustainability Science*. 13(4), pp. 1121–1135.

Zinken, J., Hellsten, I., & Nerlich, B. 2008. Discourse metaphors. *Body, Language and Mind*, 2, 363–385.

Part I

Metaphors of reason

2 Metaphors of technological change

Fionn Rogan

Introduction

That the modern world has been profoundly influenced, shaped, and transformed by technology can easily be expressed quantitatively: for example, a rise in the human population to over 7 billion (thanks to advances in medical care), a rise in domestic living standards (increased availability of household appliances and energy services), reduced share of the population doing physically dangerous or tiring work (due to mechanisation and automation), increased calorie intake and longevity (due to advances in food production), increased mobility (due to advances in transport technology with associated decrease in cost), increased global connectivity (due to advances in information and communication technology), and many more. The impact of technology on changing the broader environment is also apparent in metrics of climate change, sea-level rise, deforestation, ocean acidification, and global warming. But the impact of technology on change cannot be fully charted with quantitative metrics alone.

Technology has influenced, shaped, and changed human worldviews to an extent that is often not appreciated. Many ideas about technology and change that are today regarded as self-evident have altered their meaning considerably from their historical origins. For example, technology and innovation are today widely paired (as technological innovation) and almost always regarded as self-evidently positive; however, the current meaning of both words originated in the nineteenth century. Before then, the word *technology* did not exist (Williams, 2003) and *innovation* was regarded as "introducing change in the established order" and therefore "deviant" (Godin, 2015). Technology has been both an influencer of change and an influence on what we think of as change. Much of the language and metaphors that we use to characterise technology and change has, like the word technology itself, had very recent historical origins. By using metaphors of technology in our thinking, we have become more influenced by technology but perversely, this can mean we are less able to appreciate the full impact of technology, since technology's presence has become so normalised in our worlds. To understand the full impact of technology, it is therefore necessary to interrogate today's metaphors of technological change.

Some of the most significant impacts that technology has had on the modern world are associated with energy. For example, energy and the use of coal

DOI: 10.4324/9781003143567-3

was instrumental to the first industrial revolution in England in the nineteenth century (Wrigley, 2013); the widespread availability of oil enabled the new technology of the automobile to completely disrupt patterns of settlement in the USA in the twentieth century (Nye, 2001); the widespread availability of electricity, especially in homes, transformed the living standards of many rural lives, for example in Ireland through the Rural Electrification Scheme from 1946 to 1976 (Shiel, 2003); and today renewable energy technologies such as solar are providing energy access to parts of Africa where previously it was challenging to have electricity services (IEA, 2017). The processes of diffusion of these different energy sources and technologies have been described alternatively as *revolutions* (Perez, 2003) and as *transitions* (Smil, 2010) by two different scholars of energy history and technology. These two framings emphasise different aspects of the same history: the former (revolutions) emphasises the intensity and upheavals associated with energy technology while the latter (transitions) highlights the centuries over which the shift from biomass-based energy to fossil fuel-based energy (coal, oil, & natural gas) and electricity-based energy took place. The two different framings also show how different metaphors can highlight different aspects of the same thing.

The example of energy in the modern world shows how much technology can remake and restructure life, though sometimes with unintended consequences. The modern energy system is responsible for two-thirds of the world's GHG emissions which are responsible for climate change. Governments and societies around the world are being challenged to change their energy systems to be less carbon intensive, but this is not an easy task. There are many different views on how to change the energy system. This reflects a diversity of views on what the energy system is (is it technological? social? socio-technical?) and what change is (technological? social? socio-technical?). Different metaphors of technological change will reflect this diversity of views and will, for example, include or exclude different perspectives and possibilities. Some metaphors will posit more human agency and some will include the possibility of unintended consequences.

The energy system is not unique: technologies have always had unintended consequences. In complex sociotechnical systems, these unintended consequences are both more likely to happen and are potentially more catastrophic; however, the way we currently talk about technologies (including the metaphors we use) hides or ignores this. Therefore, it is important to look at the metaphors we use to describe technological change and assess whether or not they accurately convey the fact that unintended consequences (including dangerous ones) are part of technological change. This chapter will review how language and metaphor can reflect and capture different aspects of the technological world and of technological change. It will review four key metaphors of technological change and assess how these metaphors provide or don't provide insights to sustainability and unintended consequences. Finally, before concluding, it will examine these four metaphors in the context of changing the energy system.

Metaphors and the unintended consequences of technology

As an aid to understanding and communicating science, metaphors are a well-known and well-used tool. For example, the dynamics of the atmospheric system are usually described using the analogy of "the greenhouse effect" and electricity systems are often explained in terms of the metaphor of plumbing[1]. But metaphors don't just do specialist work. In their seminal text "Metaphors We Live By", authors George Lakoff and Mark Johnson contrast a widespread understanding of metaphor as "a device of poetic imagination" and "rhetorical flourish" with their contention that "metaphor is pervasive in everyday life" and "our ordinary conceptual system [...] is fundamentally metaphorical in nature" (Lakoff & Johnson, 2003). Throughout their book, the authors describe how the use of metaphor is one of the basic tools of thought and language.

Lakoff and Johnson (2003) are but one set of authors that describe how the nature of language and metaphor influences what we perceive of as reality. The adequacy of certain metaphors to comprehending particular issues is an important topic. In their study of environmental discourse, scholars Harré et al. (1999) apply their linguistic and philosophical expertise to what they describe as Greenspeak[2]. They contend that "the environmental case [...] is both more subtle and complex than much current ways of Greenspeak allow". One of their specific criticisms is the inability of certain environmental metaphors to articulate the interrelatedness of environmental processes. In their discussion on different metaphors of the environment, the authors note "a continual conflict between metaphors that set up human beings apart from the environment and others that display them as an integral part of the world system". I believe this also applies to the world of technology. So embedded has technology become in the modern world that a separation between nature and technology is inadequate: a better description is a "hybrid world... in which technology and nature are inextricably linked" (Williams, 2003). The concept of environmental interrelatedness overlaps heavily with the concept of unintended consequences of technology: in a technological process where many elements are tightly coupled, unexpected interactions can cascade rapidly into unanticipated states (Perrow, 1999).

While metaphor is the basis for how people understand the world, which metaphors are used to understand the world is very often influenced by the dominant technology of the day. Throughout history, the human brain has often been understood in terms of the dominant technology: in the past it was mechanical clocks, today it is the computer (Daugman, 2001). In a work of social history on how new forms and sources of energy changed America from the colonial period to the turn of the twenty-first century, historian David Nye describes how different energy technologies came to be used as metaphors in everyday speech. For instance when the dominant form of energy was horsepower, people spoke of "horsing around", they might advise "hold your horses", or could recount "he got the bit between his teeth"; once steam power grew ascendant, people might say "he got up a head of steam", "he blew a gasket", "let off steam" (Nye, 2001). These are just some ways that energy technologies can become metaphors[3] which in turn influences how energy and the broader energy system is perceived.

Professionals also make use of different metaphors and conceptions to understand different states and situations although sometimes these specialist perceptions can affect understanding of the broader system. The anthropologist Laura Nader describes interviewing "bankers, contractors, architects, building inspectors, lawyers and realtors" about a proposed set of building codes in California in the 1980s. She concluded that each group "had almost unique ways of looking at building codes" and that the only group able "to see the whole picture" were the general public (Nader, 2010). This underscores the need for an evaluation of metaphors of technological change and of the need for metaphors of technological change to get beyond specialist framings to capture the interrelatedness and potential for unintended consequences.

The literature on the unintended consequences of technology is abundant. In the energy literature, the rebound effect is the resulting increase in energy consumption that can occur after an energy-efficient investment makes the energy service cheaper (e.g. driving more due to increased fuel-economy in a more fuel-efficient car). In his book "Why Things Bite Back", author Edward Tenner cites six different categories of unintended consequence of technology: revenge effects (e.g. overconfidence in safety equipment leading to risky behaviour such as the Titanic); rearranging effects (e.g. summer air-conditioning that makes indoors more comfortable but outdoors hotter and more dangerous); repeating effects (e.g. labour-saving devices that lead to more work rather than more free-time); recomplicating effects (software and automation that mean manual jobs can be eliminated but an equivalent number of programmers need hiring to trouble-shoot and fix problems that arise); regenerating effects (anti-missile defence systems that lead to more damage from the shrapnel effects of intercepted exploded missiles); and recongesting effects (new road infrastructure built to ease congestion that leads to worse congestion due to the increased volume encouraged by the prospect of congestion-free roads) (Tenner, 1997).

There are several reasons why technology tends to have unintended consequences. Technologies are physical artefacts and tend to accrue over time. As a collection of contemporary texts show, the Industrial Revolution in England during the 18th and early nineteenth century was experienced as "The Machine Age" (Jennings, 1987). Following the growth of railway line infrastructure across the United States in the late eighteenth century, people first began thinking in terms of technological systems (Nye, 2001) and since the exponential diffusion of technology after the second world war, it has become standard (at least among scholars) to describes technology as a system rather than as individual machines or artefacts (Hughes, 2004). However, when technologies are connected in systems, they are much more likely to have knock-on effects, and as these systems grow in complexity, these knock-on effects are more likely to be unanticipated, even by the designers of these systems (Perrow, 1999). The cascading effect of multiple unanticipated events lies behind many well-known technology accidents and disasters, such as the Three Mile Island nuclear meltdown, the Challenger space shuttle explosion, and the Fukushima Daiichi nuclear disaster.

Metaphors of technological change

In this section, I will review four metaphors of technological change that are found across both popular and academic discourse on technology. In particular, I will explore how certain metaphors of technological change highlight or ignore the unintended consequences of technology. An overview of these metaphors is shown in Table 2.1. Each metaphor will be described and discussed in some detail in the following section.

Technological fix

As a metaphor of technological change, *technological fix* or *techno-fix* has two related and intertwined elements: (i) a framing of any given issue in terms of a problem-solution duality and (ii) primacy to technology as the decisive factor in the solution. The saying "*If all you have is a hammer, you treat everything as a nail*[4]" captures its essence. By giving primacy to technology as the solution, the context and setting of any given issue can be downplayed or minimised in a manner that can backfire.

An example described by Scott (1998) comes from eighteenth century Germany where the new practice of scientific forestry was being applied to reverse the declining timber yields from the natural temperate forests that had long pre-existed the local human population. The complex ecosystem of the forest (which was little understood at the time) was reduced to a problem-solution abstraction: timber yields are declining, how to increase them? The method applied was to clear the existing forest of all its biodiversity and plant rows of Scots pine and Norway spruce in its place. When the natural forest was replaced with a mono-crop forest, there was an initial higher economic yield of timber; however, an unanticipated disease struck and wiped out this new mono-crop leaving the managers of the forest with less timber than if they had done nothing to the original forests. The methods available to the foresters at the time were simply to cut and replant, but this practice was divorced from a full understanding of how a forest ecosystem functions. The solution turned out to have many unintended consequences that flowed from the application of a techno-fix approach.

Table 2.1 Metaphors of technology and technological change.

Metaphor	*Description*
Technological fix	Technology as a solution to problems
Technological determinism	Technology as a powerful and autonomous force that is a primary driver of cultural and historical change
Technological dialogue	Process by which technologies are changed as they spread from one environmental and cultural context to another
Technological momentum	Social and technical processes by which small sets of niche technologies become large incumbent technical systems over time

A technological fix approach is most likely to be found in settings with a focus on the internalist mechanism of a technology, as distinct from an approach that understands technology as embedded within broader society and subject to wider societal dynamics. For example, a mono-disciplinary team of only engineers would be more likely to adopt a techno-fix approach compared to a multi-disciplinary team comprising engineers, biologists, sociologists, historians, geographers, economists, etc. who could understand both the internalist mechanism, embedded context, and broader society surrounding a technology (Davies & Oreszczyn, 2012).

As we have already seen, the setting of the environment is where many techno-fix approaches have "come-a-cropper", due to the interaction of a simple problem-solution abstraction with a complex ecosystem setting. Though environmental activists often critique techno-fix approaches (e.g. Kingsnorth, 2017), there is also a conflicted relationship between environmentalist activists and science and technology, where environmental activists both reject and defer to the science that is both harming the earth and that has the potential to protect the earth (Harré et al., 1999). This has also been described as "the paradox of technology and the environment" since "technology is both a source and remedy of environmental change" (Grübler, 2003).

The techno-fix approach arises from a reductionist framing of technology. For example, the energy system is both a technological system (e.g. infrastructure, maintenance systems, standards, etc.) but is also connected to lifestyles, life choices, and livelihoods. In the words of Robert Fri, "the energy system is not simply a collection of autonomous pieces of plug-and-play technology. Rather, it is an integral part of our individual lives, influencing where we live and shop, shaping how we establish social networks, and molding countless other everyday habits [...] If the energy system itself changes, then all these individual and institutional links to it will have to change, too" (Fri, 2013). A techno-fix approach to technological change misses this broader context.

As a metaphor, "techno-fix" imposes an explicit problem-solutions framing on many complex issues. It also sometimes relies on oversimplifying metaphors for explaining away important context. For example, geoengineering solutions to climate change (such as solar mirrors in space to redirect the sun's rays) often employ metaphors such as "the planet is a machine" to justify their use (Nerlich & Jaspal, 2012) despite such a metaphor being such an over-simplification that it negates the understanding of the planet as developed through climate science as an predominantly chaotic (in a statistical sense) system (Hulme, 2009).

Because the problem-solution framing is very simple, it precludes the possibility of unintended consequences. Part of the reason a techno-fix approach can go awry has already been alluded to in terms of the unintended consequences of technology arising from tightly coupled systems (Perrow, 1999). Technologies today that have the potential to bring about results widely accepted as good (e.g. decarbonising the built environment) could have unintended consequences (e.g. worsening indoor air quality, energy poverty) if they were applied without an interdisciplinary approach (Davies & Oreszczyn, 2012). An approach to technological change would therefore be described as techno-fix if it framed issues in terms of problem solving, is reductionist, and gives little consideration to the possibility for unintended consequences (Huesemann & Huesemann, 2011).

Technological determinism

Technological determinism is a metaphor of technological change that posits technology as an autonomous and powerful force – outside society – that is the prime mover of all change (historical, cultural, etc.) within society. It also asserts that, in any given society, the characteristics of the available technologies will either heavily influence or completely determine the form and culture of that society.

A famous example is the stirrup theory of Lynn White which contends that the invention of the stirrup enabled mediaeval knights to become such an unbeatable military technology that the entire structure of feudal society (i.e. peasant vassals) was built around providing food and supporting the knights (Mokyr, 1992). Such an idea, that a society's structure can be explained on the basis of one technology, is now heavily critiqued. Among historians of technology, technological determinism is regarded as "self-evidently untrue" (Williams, 2003); however, technological determinism informs a lot of popular understanding of change and it therefore warrants investigation (Healy, 1998).

Technological determinists ideas can be found in popular writing about science, business, and technology. A prominent example is the author, Kevin Kelly, the title of whose books display an overt technological determinism: "What Technology Wants" and "The Inevitable". Another prominent example is Moore's Law, which states that the computing power of a transistor will double every 1.5 years. Careful empirical analysis has shown that "strictly speaking there is no such law" (Tuomi, 2002) and that the terms associated with Moore's Law have been alternatively cited as 1 year, 1.5 years, and 2 years (Morozov, 2013), rendering its status as a "law" questionable. Moore's Law is probably better described as a self-fulfiling prophecy that has given power and legitimacy to an industry. It has also served the shareholders of Intel – the company Gordon Moore himself founded – very well. The supposed truth of Moore's Law has seen it used as an analogy for numerous other industries, for example Al Gore has used Moore's law and the price of solar PV to extrapolate continuous reductions in the price of PV.

As a metaphor of change, technological determinism promotes a binary framework: one is either for or against "technology", whatever technology can be made to represent. In her book "Retooling" Rosalind Williams described the Reengineering Project at MIT whereby management consultants were brought in to reorganise the work of MIT staff to achieve efficiencies and reduce costs that would avert a university-wide budget shortfall. While none of the consultants talked about technological determinism or history, the process was framed in terms of change management and a dichotomy was established between culture (the established ways of working) and technology (the proposed ways of working, i.e. management software). The new binary logic became "In the language of change management, technology equals change, and culture equals resistance to change" (Williams, 2003). Technological determinist ideas thus contributed to minimising the agency of MIT staff and therefore co-opted them into co-operating with the Reengineering process, something which they initially resisted.

By characterising the spread of technology as inevitable, the reasons or factors that actually explain diffusion of technology such as infrastructure, peer effects, inertia, cultural practices and norms, etc. are concealed from view. Although technological determinism is sometimes presented as an explanation, it merely asserts that technology diffuses, it does not explain how or why. The example of Moore's Law shows that technological determinist ideas can have a hidden agenda, in this case commercial. Other authors have noted how well technological determinist ideas tend to align with a belief in free-markets and a laissez-faire attitude to economics (Nye, 2007).

Since technology determinism posits technology as an autonomous force, it has a disengaging effect. A future that is determined by technology closes off options, frames the future in narrow pathways, or simply precludes agency, priorities, and human goals. Individuals influenced by technological determinism are less likely to be interested in new technology, especially in adapting it. Unintended consequences tend to be ignored in accounts that rely on a technological determinist approach or when they aren't being minimised, they are asserted as inevitable side-effects, collateral damage, or externalities.

Technological dialogue

Technological dialogue is a metaphor of technological change that captures the processes of how technologies get modified, changed, and inspire fresh designs as they diffuse from the original site of their invention, where the original site could be a workshop, region, or country. The term was coined by historian of technology Arnold Pacey in his book "Technology in World Civilization" (Pacey, 1991). Pacey introduced the term *technological dialogue* as an alternative to the term *technology transfer*. According to Pacey, technology transfer "implies the recipients of a new technique passively adopt it without modification"; however, this contrasts with the historical record that shows the "transfer of technology nearly always involve modifications to suit new conditions, and often stimulate fresh innovations".

An example of technological dialogue is when the transfer of Italian silk to England stimulated local innovations, "the ingenuity expressed in machines from one culture evoked a response elsewhere, as in a dialogue". Another example from the historical record is when the technology of gun manufacturing in India improved in response to the initial arrival of European colonialists in the seventeenth century (Pacey, 1991).

Pacey uses the idea of technological dialogue as a metaphor for four different historical dynamics and processes, which include (1) the exchange of technological ideas between cultures and across time, (2) how the arrival of new technologies that represents an improvement in quality can challenge the recipients of the technology to explore new techniques of their own, (3) how the transfer of technological artefacts can lead to modification of the technology in a new environment, and (4) how the arrival of new technologies can stimulate new ideas that lead to entirely new technological artefacts. As an historian's metaphor, it has the advantage of having proven explanatory power. It is also more aligned with an interconnected understanding of environment, technology, and processes.

Technological dialogue is an overt metaphor. It takes an idea from one field (how humans communicate) and applies it in another field (technology studies), an approach that often leads to fruitful insights (Harré et al., 1999). Though it has been used to understand the past, the metaphor of technological dialogue is also useful to think about the future. The openness of the metaphor facilitates the possibility of unintended consequences and is a powerful aid to thinking about the full implications of technology.

Technological dialogue focusses attention not on the internal mechanism of the technology but at the interface between where a designer and user of a technology meet. It highlights the higher importance of how a technology is used in all its forms, rather than how it was designed. This broader perspective embraces how technology exists in a real-world context, not just in a lab. This is important, not just because the context or environment where a technology is used can differ from the lab where a technology is designed, but the original environment where a technology is first trialled can be different to the environment where the technology spreads. This is particularly the case for technology in agricultural or ecological environments, which often have very site-specific characteristics. It is also the case for cultural contexts. Some global technology companies such as Intel have even employed anthropologists to help understand how technologies are used in different cultural contexts.

Technological momentum

Technological momentum is a metaphor of technological change that describes how large technical systems change characteristics over time, from an early phase of being a small, incomplete, and niche system to a late phase of being a large, expansive, and incumbent system. The metaphor encompasses concepts and perspectives from both technological and sociological expert disciplines. In their early phase, young technical systems "tend to be more open to sociocultural influences" and can acquire momentum, which as they grow and develop carries them through to a late phase where they tend to be "more independent of outside influences and therefore more deterministic", i.e. they acquire inertia, which makes them difficult to change (Smith & Marx, 1994).

The development of the modern automobile transport system could be described in terms of technological momentum. In the early phase (1890–1910s), the automobile was a niche technology with very little supporting infrastructure (e.g. roads of sufficient quality, service stations) or people involved who weren't engineers or technicians. Despite a range of competing automobiles (internal combustion engine, steam engine, and electric car), all cars had to compete within the context of existing settlement patterns, e.g. internal combustion engines were more suitable for rural-urban and inter-urban journeys, whereas electric cars (due to their lower range) were more suitable for intra-urban journeys. In the early phase, the design of the automobile system was shaped by the environment (i.e. settlement patterns) and by social norms (i.e. city streets at the time were for human and not vehicular traffic and cars were initially considered an unwelcome and dangerous intrusion). During the early phase, there was considerable uncertainty about which would be the dominant car technology or even that there would a dominant car technology – at the time railways and horse carriage were the dominant transport mode. Much of the reason the internal combustion engine won out was the sheer contingency of it being the preferred choice of an engineer (Henry Ford) who would manufacture the internal combustion engine car at an unbeatable price thanks to the engineering and managerial innovation of the Ford assembly line (Nye, 2013). The reductions in price achieved by the Ford assembly line led to phenomenal demand for new cars such that it became the aspiration and the new normal to own a car. Over time, the system supporting the car gathered momentum and by the late phase cars were a shaper of the environment rather than being shaped by it. For example, the automobile gave rise to suburbia as cities became much more spread out and, as the city was thus redesigned, it locked in the car in a way that made it difficult to resist. Cars took over city streets after a concerted and successful campaign to reframe people on streets as jay-walking and therefore intruders to the car's domain (Norton, 2007). Engineers and scientists became less involved in decisive turning points of the car, it was now the domain of town planners, insurance experts, car sales' people, and government departments designing car tax schemes. Today, the automobile transport system is in an utterly dominant position and it is hard to imagine transport systems or even modern life without them.

Technological momentum was first coined and described by Thomas Hughes, an historian of technology. It was devised to accommodate insights from two different (and somewhat opposing) schools of thought: social constructivism[5] and technological determinism. It has been used by historians and other writers to explain dynamics of historical change (Pool, 1997; Nye, 2001). It has been less used as a concept to theorise about the future, but there is nothing to preclude that from happening.

As a metaphor, technological momentum accommodates some of the contradictory dynamics inherent in technological change (e.g. social forces and technological forces) and avoids some of the common pitfalls in thinking about the impact of technology, namely determinism (Nye, 2001). It is also an expansive metaphor that can apply to a range of settings: a young technical system can build up and gain momentum, e.g. as a niche development might in the multi-level perspective, or a mature technical system can acquire such momentum that rivals cannot compete, e.g. carbon lock-in of the energy system. The power of technological momentum as a metaphor is that it encompasses all these processes, yet isn't deterministic. Technological momentum leaves room for thinking about alternatives which makes it more amenable to creative imagining of future possibilities, including those favourable to sustainability. It is also a powerful framework for thinking about unintended consequences because it includes both social and technological dynamics, which can interact in unexpected ways.

Application of metaphors of technological change to sustainability transitions

At the start of chapter, I noted there are many different ways to change an energy system in part because there are many different views of what an energy system is and what technological change is. In this section, I will provide a sample of how these four metaphors of technological change have been applied or could apply to changing the energy system. The current energy system is heavily dependent on fossil fuels; as of 2015, over 80% of the world's energy supply comes from oil, coal, or natural gas. To mitigate climate change, this high-carbon energy system will need to become a low-carbon energy system and so a substantial transformation of the world's energy system will be required. How the energy system can be transformed is a question that can be addressed in many ways.

A *technology-fix* approach is commonly found in discourse that frames the energy system as a predominantly technological system. In this outlook, changing the energy system can be summarised as "we just need the right technologies"; in frustration at the perceived slow rate of progress, this is sometimes extrapolated to "we just need to persuade people to use the right technologies". It is impossible to sustain a view of the energy system as just a technological system for very long. The use of energy is embedded in the lifestyles of everyone and the diffusion of a new set of technologies will therefore need to include the agency of more than just a minority. A techno-fix approach that ignores the social context will also ignore unintended consequences such as rebound, backfire, or protests until these factors cannot be ignored. Separately (and briefly), techno-fix solutions are sometimes presented as part of the evidence base for decarbonising the energy system during international climate negotiations as part of the UNFCCC. The extent to which these techno-fix solutions create room for manoeuvre during negotiations, while forwarding the unintended consequences of these technologies to the future, is an underexplored area worthy of further research.

Technological determinist ideas are also often found in discourse about changing the energy system. An example of this is the exuberant trend extrapolations that are often applied to the cost reductions and diffusion rates of certain new energy technologies. In the past decade, the cost of electricity from solar PV has fallen across the world; the reasons behind this are numerous and the rate of reduction also varies by location. However, the reasons behind these reductions and the variety of the reductions are often ignored by assertions that posit a continued reduction in solar PV costs akin to a "Moore's Law for solar". We have already seen that Moore's Law is part flawed empirical observation and part commercial marketing strategy. The same caveats apply to a "Moore's Law for solar". A similar disinterest in specifics and details can be found in many projections of diffusion rates for electric vehicles that ignore the reasons behind the slower than expected diffusion rates in most countries and the faster than expected diffusion rates in one country (Norway). In both these examples of cost reductions and diffusion rates, understanding the complex reasons behind the trends is vital, but frequently there is a mere assertion of a trend as an autonomous force, which is technological determinism an alternative to understanding.

When the author Arnold Pacey explains the concept of *technological dialogue* and notes its usefulness in understanding history, he also asserts its usefulness for designers of technology and technology programmes, "to avoid such negative results is to introduce the new technology in a more flexible form to allow for a dialogue which may lead to modifications, possibly in equipment, but more especially in social arrangements affecting its use" (Pacey, 1991). Because technological dialogue is framed around the user of technology as much as the developer, unintended consequences are much less likely to occur since the initial response of users to a technology can be reincorporated into the design. The element of risk is acknowledged. In terms of changing the energy system, technological dialogue means including users of energy technologies (i.e. everyone) in the transition. This also means there is more uncertainty about the form of a decarbonised energy system (i.e. the final goal) because the emphasis is on the transition process rather than the transition destination.

One of the powerful aspects of *technological momentum* as a metaphor is that it can capture the fact that people change their minds about technology. For example, a technology that isn't widespread may be perceived as unpopular, but the same technology if it becomes widespread can then be perceived as popular. This can also apply to technologies such as nuclear which can go from being widely accepted to widely rejected[6]. This change can also apply to technology experts. When anthropologist Laura Nader interviewed engineers and technology experts about energy options in 1980 she encountered a scornful attitude to solar power and a consistent preference for nuclear power; one reason cited was "solar's not very intellectually challenging". Today, solar energy is widely seen as part of a suite of smart technologies and is often a visual shorthand for advanced high-tech (Nader, 2010). Technological momentum points to the importance of dialogue between developers of technology and the wider public.

Conclusions

It is a paradox that although technology is changing our world more and more, the results of this change are becoming more normalised and therefore the impact of technological change is becoming less visible. David Nye cites the term "horizoning of experience" to describe the way successive generations treat a radically new technology as completely normal since that is what they have grown up and have always known; this is in contrast to their parent's experience of upheaval (Nye, 2013). An example comes from the response of citizens to the energy crisis in the 1970s in the United States, who perceived the widespread availability of cheap energy as "natural" and the supply shortages as "artificial", despite the high-energy society being such a recent and technologically based phenomenon (Nye, 2001). This normalisation of technological change makes the process of dealing with sudden and enforced change more difficult. Analysis of the metaphors of technological change can help us re-see the processes of change and to reclaim agency over that change.

The analysis presented in this chapter argues that many widespread metaphors of technological change adopt a *technological-fix* or *technological determinist* outlook. These metaphors do not describe the true nature of technological change; rather, they conceal its true dynamics and either ignore or explain away the unintended consequences of technology. By contrast, *technological dialogue* and *technological momentum* more accurately capture the true dynamics of technological change (including unintended consequences) and also open up space for thinking about future technology management and adoption.

Technological momentum is the most valuable of the four metaphors because unlike the others it conveys the complexity of technological change, the stop-start nature of that change, the unintended consequences that can arise, and the need for human agency to make transition happen. Technological momentum gives rise to a number of powerful ideas that are useful for understanding technological change. For example, in the early phase of a technical system even though the dominant actors are often technical people, this is when the system design is most open to non-technical concerns; also, change is always possible, despite apparent system lock-in, all current mature technical systems started as early systems and displaced a dominant incumbent. Technological rigidities are not the same as technological determinism, which precludes human agency.

Healy (1998) discusses how different metaphors of change are embedded in different theories about the social mechanisms of change. In his discussion of technology as a source of social change, he notes a wide gap between the popular understanding of the genius inventor against the academic understanding which cites social context and forces, institutional setting, and sheer contingency. The analysis of metaphors of technological change as presented has tried to bridge this gap. Language changes slower than technology (Harré et al., 1999), but human culture is also adaptable, resilient, and capable of determining its fate.

Acknowledgements

Thanks to Ian Hughes and Lidia Guzy for helpful comments on an earlier version of this manuscript. The author would also like to acknowledge the support of Science Foundation Ireland and NTR Foundation under Grant No. 12/RC/2302.

Notes

1 https://www.rte.ie/eile/brainstorm/2018/0221/942392-how-renewable-energy-helps-to-power-your-kettle/ (accessed 26/02/2018)
2 The authors' define Greenspeak as a "catch-all term for all the ways in which issues of the environment are presented, be it in written, spoken or pictorial form" (Harré et al., 1999)
3 I personally participated in a conversation where the workplace competition between a limited number of EV charge points was described as "a turf war".
4 Though often attributed to Mark Twain, there is no evidence that he actually wrote or said this https://quoteinvestigator.com/2014/05/08/hammer-nail/ (accessed 04/06/2018)
5 Social constructivism is the idea that technological objects or artefacts are shaped, defined, designed, and given meaning by social groups. The theory is often explained with reference to how the modern "safety bicycle" emerged and acquired dominance from many the different competing and alternative designs.
6 I am grateful to Charlie Wilson for pointing this out.

References

Daugman, J. G. 2001. Brain metaphor and brain theory. In *Philosophy and the neurosciences: A reader* (Bechtel, W., Mandik, P., Mundale, J., Stufflebeam, R., eds.). London: Blackwell.
Davies, M., and Oreszczyn, T. 2012. The unintended consequences of decarbonising the built environment: A UK case study. *Energy and Buildings*, 46, 80–85.
Fri, R. W. 2013. The Alternative Energy Future: The Scope of the Transition. *Daedalus*, 142 (1), 5–7.
Godin, B. 2015. *Innovation contested: The idea of innovation over the centuries*. Abingdon, UK: Routledge.
Grübler, A. 2003. *Technology and global change*. Cambridge, MA: Cambridge University Press.
Harré, R., Brockmeier, J., & Mühlhäusler, P. 1999. *Greenspeak: A study of environmental discourse*. Thousand Oaks, CA: Sage.
Healy, K. 1998. Social change: Mechanisms and metaphors. Working Paper. Princeton, NJ: Department of Sociology, Princeton University.
Huesemann, M., & Huesemann, J. 2011. *Techno-fix: Why technology won't save us or the environment*. Gabriola Island, BC, Canada: New Society Publishers.
Hughes T.P. 2004. *Human-built world: How to think about technology and culture*. Chicago: University of Chicago Press.
Hulme, M. 2009. *Why we disagree about climate change: Understanding controversy, inaction and opportunity*. Cambridge: Cambridge University Press.
IEA 2017. *From poverty to prosperity: Energy access outlook 2017*. Paris: International Energy Agency.
Jennings, H. 1987. *Pandaemonium 1660–1886: The coming of the machine as seen by contemporary observers*. London: Picador Books.
Kingsnorth, P. 2017. *Confessions of a recovering environmentalist*. London: Faber & Faber.
Lakoff, G., & Johnson, M. 2003. *Metaphors we live by*. Chicago: University of Chicago Press.
Mokyr, J. 1992. *The lever of riches: Technological creativity and economic progress*. Oxford: Oxford University Press.
Morozov, E. 2013. *To save everything, click here: Technology, solutionism, and the urge to fix problems that don't exist*. Harmondsworth, UK: Penguin.
Nader, L. 2010. *The energy reader*. New Jersey: John Wiley & Sons.
Nerlich, B., & Jaspal, R. 2012. Metaphors we die by? Geoengineering, metaphors, and the argument from catastrophe. *Metaphor and Symbol*, 27(2): 131–147.
Norton, P. D. 2007. Street rivals: Jaywalking and the invention of the motor age street. *Technology and Culture*, 48(2): 331–359.
Nye, D. E. 2001. *Consuming power: A social history of American energies*. Cambridge, MA: MIT Press.
Nye, D. E. 2013. *America's assembly line*. Cambridge, MA: MIT Press.
Nye, D. E. 2007. *Technology matters: Questions to live with*. Cambridge, MA: MIT Press.
Pacey, A. 1991. *Technology in world civilization: A thousand-year history*. Cambridge, MA: MIT Press.
Perez, C. 2003. *Technological revolutions and financial capital*. Chelenham, UK: Edward Elgar Publishing.
Perrow, C. 1999. *Normal accidents*. Princeton, NJ: Princeton University Press.
Pool, R. 1997. *Beyond engineering: How society shapes technology*. Oxford: Oxford University Press.

Scott, J.C. 1998. *Seeing like a state: How certain schemes to improve the human condition have failed Yale*. New Haven, CT: Yale University Press.

Shiel, M. 2003. *The quiet revolution: The electrification of rural Ireland, 1946–1976*. Dublin: O'Brien Press.

Smil, V. 2010. *Energy transitions: History, requirements, prospects*. Santa Barbara, CA: ABC-CLIO press.

Smith, M. R., & Marx, L. 1994. *Does technology drive history? The dilemma of technological determinism*. Cambridge, MA: MIT Press.

Tenner, E. 1997. *Why things bite back: Technology and the revenge of unintended consequences*. London: Penguin Random House.

Tuomi, I. 2002. The lives and death of Moore's Law. *First Monday*, 7, 11. https://firstmonday.org/ojs/index.php/fm/article/view/1000/921

Williams, R. 2003. *Retooling: A historian confronts technological change*. Cambridge, MA: MIT Press.

Wrigley, E. A. 2013. Energy and the English industrial revolution. *Philosophical Transactions of the Royal Society A*, 371(1986), 20110568

3 Nitrogen, planetary boundaries, and the metabolic rift

Using metaphor for dietary transitions towards a safe operating space

Colin Sage

Human modification of the N cycle is profound.

Rockström et al. (2009a) p.13 of 33

Introduction

This chapter deploys two key metaphors as part of a critical examination of the global agri-food system, in particular, as a way of drawing attention to the wider environmental consequences arising from the deployment of technologies pre-occupied with increasing productivity. As Chapter 1 outlined, metaphors are a device, often used unconsciously, where we apply words or phrasing from quite different contexts in order to make sense of a situation. In relation to the food system, many of the prevailing metaphors used by business focus upon "growth", producing more in order to "feed the world" safely and cheaply. However, for those of us concerned with sustainability, the aspiration is to achieve a food "system" that produces nourishment for all, of course, but also one that operates within the parameters of the biosphere's capabilities. Paying attention to metaphors, then, reminds us to be alert to the different views of the world and the values that are implicit in the framing of problems and their proposed solutions. This chapter makes use of two metaphors that help us to do this, by critically evaluating a technology which has become an apparently indispensable element of the food system: synthetic nitrogen fertiliser.

The first metaphor is to propose that the global agri-food system be regarded as constituting a "planetary scale metabolic process". Metabolism is most commonly applied to the human body and refers to the process of converting food intake into energy. Developing this a little further metabolism encompasses the chemical and physical processes by which an organism – or a system - utilises materials to produce energy and other substances necessary for its maintenance and reproduction. Applying metabolism to agriculture allows us to consider the way in which a variety of chemical and biological resources – including soil nutrients and their supplements – are utilised (metabolised) in order to produce the primary foods (and fuels and fibres) on which the human population depends. As agriculture has developed, becoming ever more reliant on energy and mineral resources from the lithosphere and biosphere to sustain and expand production,

DOI: 10.4324/9781003143567-4

the metabolic process has grown in complexity and spatial extent such that flows of materials (inputs) and foodstuffs (outputs) cross all kinds of boundaries: ecological, national, and planetary.

Consequently, the second metaphor I wish to draw upon here is that of "planetary boundaries". The framework of planetary boundaries (PB), as set out by Rockström et al. (2009a, 2009b) and by Steffen et al. (2015a), draws our attention to nine key Earth system processes. The framework attempts to establish critical thresholds that represent limits to a "safe operating space" (another metaphor) for human societies. While we examine the scientific aspect of the PB model below, here we should note the metaphorical significance of the term. Clearly the notion of a boundary speaks to an outer limit or ceiling on what is permissible; transgressing this boundary then implies danger, or at least uncertainty about a possible outcome. There has been an important current of work within environmental science around the notion of boundaries and thresholds of which the Club of Rome report, "Limits to Growth" (Meadows et al., 1972), was the most prominent. Calculating precise boundaries on levels of resource extraction and "reserves" (consider "Peak Oil") or on volumes of waste streams entering the environment (e.g. greenhouse gas emissions) has been an increasingly important but scientifically vexed endeavour since that time (and as we will see below in relation to PB). However, it should be acknowledged that in contrast to a scientific view of "limits" stands the "cornucopian" metaphor (the "horn of plenty") representing a belief in human ingenuity and our innate capacity to develop new technologies that will overcome foreseeable constraints. This approach is especially prevalent among proponents of the existing agri-food system who point out that it has delivered more calories for a lower proportion of consumer spending than at any time in human history. Moreover, times of rising food prices (as during the financial crisis of 2008–12) present justification for new technologies – plant and animal genetic engineering, precision farming – in order to ensure food security. Here, the pursuit of productivity overshadows concerns for the environment as the problem is framed as a supply shortage rather than failings of the market (incapacity of the poor to register a demand for food). Each side, then, deploys their metaphors accordingly in order to gain ascendancy in the battle of ideas and moral suasion.

Today, the technological, economic, and political capabilities of the modern food system have enabled world population to grow to around 7.6 billion people, though at least two-fifths do not eat or absorb sufficient nutrients for a healthy life (IFPRI, 2016). Yet while more cultivated food is produced today than at any time hitherto, the global ecological costs of doing so are mounting quickly and these can be identified across a range of key biophysical processes.

For example, there are rising environmental concerns at the contribution of industrialised food production systems to land use change, habitat destruction, and loss of biodiversity (Foley et al., 2011 Campbell et al., 2017); as well as to climate change through emissions of greenhouse gases (Vermeulen et al., 2012; Tilman and Clark, 2014; Niles et al., 2018). A quite different anxiety is emerging from analysis that has followed the development of the notion of "virtual water" – i.e. the amount of water required to produce a certain volume of food and which might therefore be regarded as virtually embedded in it (Chapagain and Hoekstra, 2008; Allan, 2011). Here, the mapping of international flows has revealed the considerable risk posed by local water scarcity to downstream economies through globalised supply chains (Qu et al., 2018). Yet another, and less well-known, process has been the disruption to the global biogeochemical cycles of nitrogen and potassium, and it is this agri-food-driven development that forms the primary focus of this chapter.

The framework of planetary boundaries draws our attention to nine key Earth system processes each of which are being affected by human activity[1]; PB therefore makes evident the ways in which anthropogenic activity is pressing against biospheric limits. Of the nine processes, for three – climate change, biodiversity loss, and nitrogen and phosphorous flows – there is evidence of a clear breaching of their respective boundaries. However, there is reluctance on the part of the proponents of PB to suggest that we have passed a critical "tipping point" and, subsequently, there has been an effort to reframe the notion of "planetary limits" in relation to a "safe operating space" for humankind while recognising that these "planetary" processes have rather different regional characteristics. However, the metaphor of PB clearly serves to convey a sense that human activity is disrupting planetary scale processes.

While climate change has arguably become the single most important issue of our time and around which there is considerable scientific consensus about the process, this cannot be said to extend to the notion of a safe boundary limit. For while a 2 °C warming target may be perceived by the public as a universally accepted goal, no scientific assessment has clearly justified or defended this target as a safe level of warming (Knutti et al., 2016). This consequently raises profound questions regarding the atmospheric concentration of CO_2 and whether its planetary boundary should be set at 350 or 450 parts per million by volume (Steffen et al., 2015a). Other planetary scale processes are even less easily quantified, may be less directly connected to specific drivers, or their boundaries similarly under dispute (Jaramillo and Destouni, 2015). One process that <u>can</u> be attributed to anthropogenic activity, not least because it has come to underpin the survival of the species through food production, is that of the biogeochemical flow of nitrogen (N).

The development of the Haber-Bosch process early in the twentieth century to synthesise atmospheric nitrogen (N_2) into the form of fertiliser and thereby provide a source of reactive nitrogen (N_r) that could be taken up by crops has had a transformative effect on the global nitrogen cycle (Galloway et al., 2002). Together with the subsequent development of fertiliser-responsive high-yielding seeds, it has been argued that without the Haber-Bosch synthesis about 40 per cent of the world's current population could not be fed (Smil, 2002). Yet it has also been estimated that only 43 per cent of the reactive nitrogen applied to cropland worldwide is converted into harvested products and that more than half is lost to the environment (Billen et al., 2014). The global alteration of the N cycle leads to many different, cascading effects on human and ecosystem health (Galloway and Cowling, 2002). These include nitrate leaching into groundwater which is harmful for populations drawing domestic water supplies from such resources; eutrophication of aquatic ecosystems resulting in fish kills and the creation of coastal "dead zones" (Van Meter et al., 2018); loss of terrestrial and marine biodiversity; agricultural emissions of ammonia that combine with small particulate matter to create toxic air quality (Lelieveld et al., 2015); and contributions to climate change through N_2O, a potent greenhouse gas. Yet despite representing a large-scale biogeochemical experiment on planetary health, it is arguably one of the least recognised environmental issues among the general public.

These planetary scale processes, especially those that are approaching – or even transgressing – limits, bring us back to the metaphor of metabolism and a particular application that can be traced back to the nineteenth century: the notion of "metabolic rift". The term has its origins in the work of Karl Marx who used it to highlight the disturbed metabolic interaction between human society and the environment under capitalist production. Given Marx's original concern for the robbing of soil fertility by capitalist agriculture – discussed in more detail below - it seems appropriate to consider the way in which an apparently highly successful solution – the application of large volumes of nitrogen fertiliser capable of restoring levels of this key nutrient – has served to so dramatically disrupt a key global nutrient cycle. Thus, I seek to develop and extend the metaphor of metabolic rift by focussing upon the nitrogen cycle and the ways in which its global disequilibrium has been caused by an agri-inputs industry that has reshaped agriculture around the world. If we are to recover and restore planetary balance in this particular area, then it becomes clear that we must examine the potential for transition not only towards a different model of food production but also examine those dietary patterns that feature foods that are implicated in a high nitrogen loading.

Consequently, following a brief explanation of the notion of metabolic rift, the chapter proceeds to discuss the planetary boundaries framework, in particular examining the changing status of nitrogen within global biogeochemical cycles. This takes us, inevitably, into a discussion of the global food system and its embedded flows of nitrogen. Thus, this chapter is largely concerned with highlighting the fate of nitrogen by exploring the scientific basis of the PB and metabolic models and from here pays rather less attention to the metaphorical implications until we arrive at the conclusion.

The Metabolic Rift: Forms of ecological rupture

Influenced by the scientific writings of Justus von Liebig, regarded as the father of agricultural chemistry, Marx considered how the intensification of agriculture results in the depletion of nutrients from soils, where this failure to replenish nutrient loss – the "robbing of fertility" – was analogous to the exploitation of the worker (Foster & Magdoff, 2000). Von Liebig's analysis highlighted how nutrients drawn from the soil were carried away by harvested foods to be consumed far away from sites of production with those nutrients not directly utilised by the body then lost in waste streams and urban sewers. This view would suggest that human excrement, humanure, had provided an important – even primary – source of added fertility prior to the nineteenth century, a view challenged by McMichael and Schneider (2010) who highlight the role played by crop and field rotations in restoring nutrient balance. Whether the recovery and return to the land of "night-soil" would ever have kept pace with nineteenth century urbanisation is not a matter of concern here. But, in its absence, a rising demand for replacement nutrients led to the utilisation of Napoleonic battlefield bones, then to the lucrative trans-Atlantic trade in Peruvian guano – accumulated deposits of seabird droppings on islands off the Peruvian coast – and Chilean nitrates (Foster, 2013).

Von Liebig's experimental work resulted in his identification of nitrogen, potassium, and phosphorous as being essential plant nutrients and while his Law of the Minimum noted that yield was constrained by the scarcest mineral, he nonetheless highlighted the critical role performed by nitrogen. This element, N_2, is the most abundant of the gases in our atmosphere, yet in this form it remains chemically and biologically unusable. To create a reactive form of nitrogen, N_r, took the work of later German chemists, principally Fritz Haber and Carl Bosch. The development of the Haber-Bosch process of synthesising atmospheric nitrogen into ammonia and scaling this into a continuous industrial process has since provided for almost all of the inorganic nitrogen requirements of world agriculture, at around 100 Tg N per year (or 100 M tonnes)[2].

However, while the metaphor of a metabolic rift – as representing the theft of nutrients from the soil – may appear to be solved by the addition of chemical fertiliser, it is important to recognise that Marx considered the rift to extend beyond resource extraction and to serve as metaphor for a more profound alienation of society from nature under capitalism. For Marx, capitalist production concentrates the historical motive force of society while disturbing the metabolic interaction of humans and the earth (Foster, 2013). In Clark and York's (2005) interpretation of Marx's analysis, they argue that capitalism created a rupture in the metabolic interaction between humans and the Earth, one that was only intensified by large-scale agriculture, long-distance trade, and urban growth. They continue,

> Capitalism is unable to maintain the conditions necessary for the recycling of nutrients. In this capitalism creates a rift in our social metabolism with nature. In fact, the development of capitalism continues to intensify the rift in agriculture and creates rifts in other realms of the society-nature relationship, such as the introduction of artificial fertilizers.
>
> (Clark and York, 2005: 399)[3]

The broadening and deepening of global food production and supply under capitalism and the intensification of multiple rifts with nature extends beyond environmental disruptions. Indeed, particularly during the last half century, profound changes extend beyond the realm of agriculture into the very heart of consumption practices. In this respect, it is possible to develop and apply the metaphor of metabolic rift to represent a more thorough ecological, social, and cultural rupture around food systems. For example, Dixon et al. (2014) argue that the contemporary metabolic rift extends beyond the disrupted exchange between social and natural systems and is being propelled by four major ecological ruptures:

1. Depleted agro-ecologies – the environmental support systems to sustain food production – due to bio-physical system degradation including waste generation.
2. An erosion of food sovereignty at the nation state level, due to a reconfiguration of food system governance arrangements and foreign direct investments in food retailing systems.

3. The erosion of cuisines – consisting of knowledge, rules, cooking skills, and food provisioning strategies – due to the penetration of corporate interests.
4. Stressed human metabolic states, as a result of easy access to dietary energy together with a lack of physical activity, and the corporate restructuring of local food environments which, in turn, reduces options for obtaining "good nutritional" diversity. (Dixon et al., 2014: 135)

Although these wider interpretations of metabolic rift are not explored here[4], Dixon et al's typology serves to highlight the multiple disruptions generated by the industrial agri-food system and how these can serve as the basis for new metaphors. Point four, for example, speaks to the disruption of human metabolism, most visibly manifest in rising levels of body mass and obesity with direct consequences for well-being, arising from multiple changes in the global agri-food system. These stretch from huge increases in agricultural productivity through the development of cheap, processed foodstuffs to changes in retailing which makes energy-rich, nutrient-poor products so ubiquitous. Ultimately, these developments have been underpinned by changes in the realm of agricultural production and we explore these through the lens of planetary boundaries.

Planetary boundaries

As noted at the start of this chapter, the planetary boundaries model set out by Rockström et al. (2009a, 2009b) and further developed by Steffen et al. (2015a) highlighted nine key Earth system processes (listed in Endnote 1). Each of these processes are directly affected by human activity and for each the authors attempted to establish critical thresholds that represent limits to a safe operating space for human societies. The framework has emerged as a way of monitoring trends in Earth system processes with particular concern for deviation from the conditions that marked the past 12,000 year Holocene era which, Steffen et al. suggest, is the only state of the planet that has appeared capable of supporting contemporary human societies. It presents a precautionary, systems integrity approach in establishing boundary limits at a distance from possible tipping points, thus establishing parameters for stable and resilient global ecosystems to support human well-being (Häyhä et al., 2016). However, while planetary boundaries emphasise the urgency of global environmental problems – with three of the nine boundaries breached – the model has been subject to critique for using a language of universals that fail to speak to the differentiated interests, concerns, and capabilities of people (Hajer et al., 2015). Moreover, it deploys a narrow conception of planetary thresholds (Cornell, 2012) while different behaviours and practices can give rise to breaching some regional boundaries, which may not affect other regions (de Vries et al., 2013). Consequently, there is a need to translate global boundaries into specific national and regional contexts for effective policy making while the work of further defining boundary limits continues.

In the updated treatment of the model, Steffen et al. (2015a) further develop the PB framework, especially by introducing a two-tier approach for several of the Earth system processes to account for regional heterogeneity; and by updating boundary limits in light of new findings. Their paper also makes clear that the segmented architecture of the framework should not disguise the complex interactions between processes. They argue that two of the boundaries – climate change and biosphere integrity – are highly integrated, emergent system level phenomena that are connected to all of the other planetary processes that operate at the level of the whole Earth system and provide the planetary-level overarching systems within which the other boundary processes operate. Furthermore, large changes in the climate or in biosphere integrity would likely, on their own, push the Earth system out of the Holocene state and into a new era widely recognised as the Anthropocene (Steffen et al., 2011; Steffen et al., 2015b; Castree, 2017). A further refinement of the original framework is also made in relation to nitrogen and phosphorous where the paper proposes a more generic planetary boundary to encompass human influence on biogeochemical flows in general. Yet a lack of data and clarity regarding other elements and their respective independent and combined influence means that their focus remains on N and P. While, for the purpose of this chapter, our attention is focussed on the biogeochemical flows of nitrogen, it is recognised that phosphorous, too, is essential for life, is largely secured by mining of phosphate rock – which is finite – and represents a hazard as it accumulates in the aquatic environment (Wironen et al., 2018; Mew et al., 2018; Smil, 2000).

It is important to be clear, however, that for both N and P, the role of the agri-food system is central with the contribution from synthetic fertilisers responsible for the overwhelming anthropogenic perturbation of both cycles and which is most visible in the eutrophication of aquatic ecosystems. Yet the particular demand for nitrogen as the element required in the largest quantity for crop growth was responsible for driving an incessant search by nineteenth and early-twentieth century scientists for a way of creating a reactive form of N that would overcome the rift of nutrient extraction. Prior to the 1840s and the trans-Atlantic trade in Peruvian guano and then Chilean nitrates, soil nitrogen enhancement was only possible by planting leguminous cover crops that support symbiotic rhizobium bacteria that fix N in the soil but which reduce the area for staple food crops. With deposits of guano exhausted by the 1870s, the search for a chemically reactive form of nitrogen was a scientific priority.

In the most authoritative text on the topic, Vaclav Smil (2001) carefully traces the history of Fritz Haber's eventual success in synthesising atmospheric nitrogen at high pressure and temperature into the useable form of ammonia (NH_3). Unfortunately, Haber's misplaced patriotism drove the production of ammonia for the manufacture of explosives such that he had a significant influence on both World Wars and subsequent conflicts and perhaps can be held responsible for the deaths of up to 150 million people during the twentieth century (Erisman et al., 2008). However, the post Great War intervention of Carl Bosch brought further developments in high-pressure chemical engineering and enabled the large-scale continuous production of ammonia – the Haber-Bosch process – to be used for industrial fertiliser manufacturing. By 1930, according to Smil, worldwide ammonia synthesis amounted to around 1 million tonnes nitrogen (1mt N), which accounted for around half of the world's inorganic nitrogen and volumes and proportions rose sharply, especially after 1945. In 1950, fertiliser applications amounted to 3.6mt N; by 1960 to 9.2mt N; in 1970 to 32mt; doubling to 61mt by 1980 then 100mt by 2000. Critically, the greater part of this is used in low- and

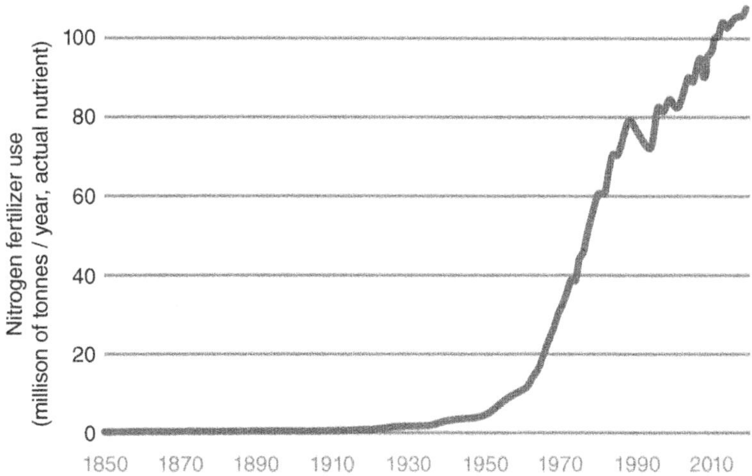

Figure 3.1 Global nitrogen fertiliser use, 1850–2019.

Source: Figure taken from Darrin Qualman's website (https://www.darrinqualman.com/nitrogen-crisis/) under the terms of Creative Commons with thanks to Darrin Qualman.

middle-income countries and for this we have to understand the role of the Green Revolution in driving chemical fertiliser use (Figure 3.1).

While the post-1945 period saw the rapid recovery from conflict and development of the metropolitan industrialised countries, food insecurity continued to blight large parts of the emerging post-colonial world. Here, plant-breeding efforts – principally in rice, maize, and wheat – got underway within a modernising paradigm that placed considerable emphasis upon technological progress and efficiencies. As raising food output was the primary goal, then new methods were introduced and the development of high-yielding crop varieties (often labelled "miracle seeds") were at the centre of this "revolution". Designed to be fertiliser-responsive, these plants demanded optimal conditions for successful performance including adequate moisture (requiring irrigation), weed, and pest suppression (requiring often-frequent applications of pesticide), as well as abundant macro-nutrients (N, P, and potassium, K). The Green Revolution is most usually represented as a hugely significant breakthrough with one of its leading scientists, Norman Borlaug, awarded the Nobel Peace Prize in 1970. In his acceptance speech, Borlaug noted:

> If the high-yielding dwarf wheat and rice varieties are the catalysts that have ignited the green revolution, then chemical fertilizer is the fuel that has powered its forward thrust. The responsiveness of the high-yielding varieties has greatly increased fertilizer consumption.
>
> (Borlaug, 1970)

There is little doubt that the combination of high-yielding crop varieties together with high levels of fertiliser application produced by the Haber-Bosch process has been critical to enabling population growth. Erisman et al. (2008) estimate that the number of humans supported per hectare of arable land increased from 1.9 to 4.3 persons between 1908 and 2008 and that between 30 to 50 per cent of crop yield increase has been due to nitrogen fertiliser application. Indeed, their calculations are consistent with Smil estimating,

> that by 2000, nitrogen fertilisers were responsible for feeding 44 percent of the world's population. Our updated estimate for 2008 is 48 percent – so the lives of around half of humanity are made possible by Haber-Bosch nitrogen.
> (Erisman et al., 2008: 837)

It is important to recall that nitrogen is essential for plants to synthesise amino acids, which are the building blocks for protein, a vital component of our diets. Cropping systems consequently convert nitrogen inputs to the soil into plant matter that is harvested and utilised for human consumption. Note that nitrogen can be supplied as chemical fertilisers, but also as manure or fixed through biological symbiosis using legumes. A lesser amount is deposited through other environmental pathways arising from fossil fuel burning or other industrial processes[5]. The ratio of this total supply of nitrogen to the amount contained in the harvested plant matter is known as nitrogen-use efficiency (NUE). This varies enormously across cropping systems, between regions but above all reflects fertilisation practices. Billen et al. (2014) report an overall global NUE of 43 per cent for cropping systems, which implies that 57 per cent of all supplied nitrogen is lost to the environment. They suggest this amounts to 95Tg N year^{-1}, three-quarters of which is emitted in China, India, Europe, and North America. We look briefly at these regional patterns in the next section, but to close this discussion, we need to consider this total load of nitrogen emissions in relation to planetary boundaries.

In the original formulation of the PB framework, Rockström et al. (2009) suggest that the nitrogen boundary initially be set at approximately 25% of the estimated value for anthropogenic fixation at about 35 Mt N yr^{-1}, which they themselves refer to as "a first guess". Subsequently, de Vries et al. (2013) dismiss this estimate as unnecessarily low and develop a more sophisticated analysis to derive a planetary boundary for nitrogen. They discount the contribution of nitrogen oxides (NOx) derived from industrial emissions and focus upon: atmospheric releases of NH_3 and N_2O; groundwater concentrations of NO_3; and surface water concentrations of dissolved inorganic N. Given the different levels of sensitivity across ecosystems to nitrogen contamination, as well as its different pathways through the environment, they emphasise the difficulty of deriving a singular planetary boundary for N. Moreover, regional variations that arise from the presence of intensive agriculture as well as the food security needs of the current global population – thereby establishing a "social floor" alongside an "environmental ceiling" – further complicate the derivation of singular measures. However, using the most stringent criteria – which Steffen et al. (2015a) do – their planetary boundary at which the eutrophication of aquatic ecosystems would be risked falls at 62 Tg N year^{-1} while for atmospheric ammonia the boundary would be at 89 Tg N year^{-1}. Zhang et al. (2015) conclude that the maximum allowable amount of anthropogenic newly fixed N in agriculture that can be introduced into the earth system is consequently in the order of 62–82 Tg N yr^{-1} while surplus to the environment is currently in the order of 100 Tg N yr^{-1}. For them, the critical challenge is to find ways to achieve ambitious rates of NUE such that better technologies and farm management practices will convert greater proportions of applied N to plant matter leaving less for dispersion into the environment. The question we now have to examine is whether the application of more science and technology to the current food system will deliver sustainable food security for all in the decades ahead.

Nitrogen and the Global Food System

The development of the global food system has utterly transformed the volume and variety of foodstuffs made available for purchase to consumers in many regions of the world. As we have seen, this has been underpinned by the application of science and technology to seeds, other agricultural inputs, and farming practices that have resulted in extraordinary productivity in primary production. Yet the supply of cheap primary food commodities, their processing, manufacturing into final foods, and distribution has resulted in significant environmental burdens (Sage 2012). Agriculture and food production together account for around 30 per cent of greenhouse gas emissions, cause pollution of freshwater and marine aquatic ecosystems, and have been most responsible for land use change and biodiversity loss (Campbell et al., 2017). But alongside this biophysical degradation, one that we might regard as a broader environmental rupture, the global food system has also been responsible for leading a dietary transition resulting in stressed human metabolic states. Exploring the implications of changes in dietary practices is becoming increasingly necessary if we are to recover a safe operating space for humankind. Doing so makes clear that it is no longer sufficient simply to focus upon improving levels of technical efficiency such as NUE in order to achieve the illusion of "sustainable intensification". Rather, a focus on diets establishes the critical link between environmental and human health (Tilman and Clark, 2014).

Horton (2017) itemises a number of deep-seated flaws that characterise a deeply dysfunctional food system that fails to deliver global food security yet imposes a high level of impact on the environment. Among his 13 "root causes" of agri-food system failure lies a clear sense that the current growth-driven model is intrinsically unsustainable due to:

- Intensive fertilisation practices using fossil fuel produced chemicals that are heavily responsible for the transgression of biogeochemical planetary boundaries for N, P, and carbon.
- The production of surpluses that leads, inevitably, to high volumes of food – and packaging – waste while reducing system resilience and leaving it vulnerable to shocks and disruption.
- The dominance of a "small number of large global businesses who relentlessly pursue growth and monopoly" that reinforce the industrial model.

Their promotion of unhealthy, ultra-processed, high-calorie foods leads to poor dietary health outcomes.

- Globalisation of food results in many perverse outcomes, including the diversion of land to export crops (over domestic staples); externalising (and thereby excluding) many of the hidden costs to environmental and human health; and compartmentalisation of research and policy.
- Aggressive marketing together with human omnivorousness has led to an expansion of meat production that is "resource inefficient, demands large lands areas and adds significantly to greenhouse gas emissions" while contributing to "global epidemics of non-communicable disease". (Horton, 2017: 1324)

It is this last point that will provide a final focus for this chapter. While the global expansion of the meat and livestock industry is often represented as a solution to improve dietary intake, particularly in low-income countries where significant numbers of people display nutrient deficiencies, the associated growth of the global feed industry has become a matter of concern. This has largely arisen over recent decades from a process that has seen livestock become progressively decoupled from land (Lassaletta et al., 2014a). The rise of intensive animal rearing – in units often known as confined animal feeding operations (CAFOs) or "factory farms" – has become the dominant mode of production for pork, eggs, and poultry and regionally important for dairy and beef. Such operations necessitate the supply of feeds often requiring long-distance trade. For example, the EU livestock sectors annually use around 500 million tonnes of animal feed, about 40 per cent of which is grass – mostly from intensively fertilised pastures – and 30 per cent cereals. Around 60 per cent of EU cereal production is used for animal feed, while 75 per cent of the protein rich feed – in excess of 35 million tonnes per year – is imported as soybean meal, mainly from Brazil and Argentina (Westhoek et al., 2011).

It is clear that the global ecological footprint for the livestock industry is both large and set to grow as projections of rising demand for meat require new areas to produce feeds. As Havlík et al. (2014) observe, 30 per cent of the global land area is used for livestock rearing; livestock contribute around 80 per cent of all agricultural non-CO_2 emissions; and the sector has been a major driver of land use change, especially in the tropics. This has been especially evident in the expansion of the soybean frontier in South America. Moreover, as Billen et al. (2014) report using 2009 data, at the global scale 74 per cent of cropland protein production is devoted to livestock feeding and only 26 per cent is directly consumed by humans.

In an effort to connect these developments back to the notion of metabolic rift, it is useful to draw upon the distinction made by Billen et al. (2014) between the main types of animal farming practices. The first, extensive pastoral systems, are not generally regarded as commercially significant – though vital to support livelihoods in marginal environments – and so are disregarded here. The second, mixed farming systems where there is a close connection between livestock and crops, including the use of crop residues and manure, were once dominant but have been increasingly displaced by market-driven forces encouraging intensification and specialisation. Finally, a third type are those livestock systems decoupled from land and local cropping systems, so that they become supported by trade in feedstuffs often from distant regions. It is this third and increasingly dominant form worldwide which exemplifies a contemporary metabolic rift which can be illustrated in the following ways:

1. Animal feeds comprise crops that can be used to feed humans directly and consequently places livestock in direct competition with human need.
2. Concentrated animal production units (CAFOs) prevent the efficient cycling of crop residues or manure, a feature of mixed farming systems.
3. With manure now less a resource for fertilising fields (accounting for less than one-fifth of cropland fertilisation compared to three-fifths for synthetic fertilisers), it has become a more problematic and concentrated waste stream often containing high levels of antibiotics and heavy metals (Heinrich Böll Foundation and Friends of the Earth Europe, 2014).
4. The scaling-up and intensification of animal rearing reduces the unit cost of meat enabling rising levels of consumption that contributes to increased incidence of cardio-vascular disease and other diet-related forms of ill-health.
5. Regional specialisation results in significant amounts of N embedded in agricultural products, including meat, being traded internationally. This can result, depending upon agricultural practices, diet, and the composition of imports and exports, in high nitrogen retention that can cause significant environmental contamination.

That there is a significant difference in the nitrogen balance between major world regions is made clear by Billen et al. (2014). In their analysis of the global agro-food system, they identify 12 regions, seven of which are net protein (N) importing regions, three net exporting and two which are close to a trade balance. Again, it is important to emphasise that differences in N fluxes between regions reflect sharp differences in farming systems – including fertilisation methods – and dietary patterns. Thus, among the seven net nitrogen importing regions is Sub-Saharan Africa, with low per capita protein availability, low levels of fertilisation (both from synthetic fertilisers and manure) but with potential for increasing use of legumes (for biological fixation). In contrast to Africa's net imports of 340 GgN year^{-1} or less than 10 per cent of total crop production as proteins, China imports 2900 GgN year^{-1} or 24 per cent. Levels of synthetic fertiliser application in China are the highest in the world at 35 TgN year^{-1}; however, it scores poorly in NUE (25.7) and together with high levels of soy imports for animal feed, China consequently records considerable environmental losses of N, around one-third for the global agro-food system. On the other side of the equation, Billen et al. identify a regional grouping as the South American Soy Countries (comprising Brazil and Argentina, but also Bolivia, Uruguay, and Paraguay). This group is responsible for the export of 5260 GgN year^{-1} or 58 per cent of its total crop production. Given the importance of soy, it achieves the highest level of symbiotic

N fixation but also exports large volumes of embedded N in soy products that are converted to animal protein in importing countries.

In highly simplified terms, calculating the regional – or national – nitrogen balances involves accounting for anthropogenic N inputs to domestic agriculture (invariably dominated by synthetic fertilisers), estimating the volume of N embedded in imported agricultural products as well as those exported, and that metabolised via domestic consumption. This is an analysis conducted for Spain by Lassaletta et al. (2014b) who note how scarce are such studies. Their longitudinal analysis over the period 1961 to 2009 is certainly highly revealing for during this time Spain underwent transformative change. Agricultural intensification witnessed huge increases in fertiliser application as well as mechanisation, irrigation, and associated environmental consequences. Livestock numbers increased considerably, particularly pigs (+358%) and poultry (+256%), with a consequent jump in the importation of animal feeds, especially soy products (by almost 3000%) to over 6 Mt by 2009. Critically they note how total protein consumption has risen in Spain from 4.6 to 6.5 kg N cap^{-1} $year^{-1}$ and where the share of animal protein in the diet has increased from 37 per cent to 65 per cent over this period. It is worth noting that Billen et al. (2014) report that with WHO dietary recommendations of per capita protein intake of 2.8 and 3.5 kg N $year^{-1}$ for women and men respectively place Spain in the highest category of protein ingestion and where high consumption of meat and dairy products is associated with cardiovascular diseases and colorectal cancer risks.

As a consequence of agricultural intensification, anthropogenic nitrogen inputs in Spain have tripled over the period under study with imports of N growing at a faster rate than in Europe as a whole (1986–2009 by 109% as opposed to 8% in Europe). This is explained by the high import volumes of feed commodities (cereals and soy) accounting for 76 per cent of imported N used to feed livestock, 80 per cent of which is consumed domestically. On the other hand, the main Spanish export crops (citrus, wine, tomatoes, olive oil) have low protein (and hence low N) content[6]. Given the huge net surplus of N within Spanish territory, the question that arises is with respect to its fate. Oxidised compounds of N have increased substantially, and a greater part are atmospherically exported. However, the amount of N exported through river systems is very low (6.5%) meaning that the remainder is retained or eliminated inside the country. This is in sharp contrast to, say the Mississippi River which transports huge volumes of N (and P) into the Gulf of Mexico that is responsible for the anoxic conditions over a wide area around its mouth (van Meter et al., 2018). Given such retention, partly a consequence of traditional Mediterranean water management practices, while there is less leakage to the marine environment there is nevertheless high nitrate loading of important aquifers. Given these supply 80 per cent of drinking water in Spain, this has huge implications for human consumption.

In concluding their study, Lassaletta and colleagues highlight the systemic vulnerability of Spain to dependence upon fossil fuel-based inputs such as N fertilisers, and to a reliance on imported inputs such as feeds to maintain dietary consumption practices that present public health risks. Consequently, they argue that a change towards a more traditional Mediterranean diet with lower animal protein content would be a win-win and result in a reduction in N cycle perturbation within and outside Spain. Their work also demonstrates that it is only through auditing such fluxes of nitrogen at national level can measures then be taken to mitigate their consequences.

Finally, widening the lens further, we should note that the average per-capita EU consumption of animal proteins in the form of meat and dairy produce is about 52 kg year $^{-1}$ (corresponding to 85 kilograms in carcass weight), about twice the global average (Westhoek, et al., 2011). Maintaining European and other rich country dietary practices, while rising levels of meat consumption take off in middle- and low-income countries, can only be supported by the continuing expansion of animal feed cultivation. Unfortunately, the inefficient conversion of vegetal to animal protein means that animal food production contributes almost two-thirds of reactive N losses to the environment. With the amount of N traded between countries dramatically increasing (from 3 to 24 Tg N, 1961–2009) (Lassaletta et al., 2014b), the international trade of food and feed constitutes a significant component of the global N cycle. While mainstream approaches will emphasise innovative agricultural technologies in order to improve nitrogen use efficiency, a more holistic and convincing argument will embrace the need for dietary change in order to bring ourselves back within a safe – and just – operating space (Dearing et al., 2014).

Conclusions

As a scientific framework planetary boundaries draws our attention to the ways in which anthropogenic activity is pressing against biospheric limits. While quantitative measures of boundary thresholds for each of the environmental processes may remain contested and its failure to acknowledge different spatial scales leaves it subject to criticism, it nevertheless provides a framework through which to grasp the urgency of reducing human impacts. One area highlighted by this chapter is the significant environmental burden arising from rising global consumption of livestock products. Moreover, recent work has built upon and extended the PB model in order to draw a tighter relationship between human and planetary health (Raworth, 2017) and especially in rethinking diet through a "Great Food Transformation" (Willett et al., 2019). Consequently, the question is whether the metaphor of PB can serve to chart a course of dietary transition, a transformative change, towards a "safe operating space" characterised by lower levels of meat consumption worldwide.

Larson (2018) alerts us to the ways metaphors may be value-laden and advocate for a particular view, and most generally serve to normalise the existing social order. Yet, if it were possible to "populate" the world with environmental rather than productivist metaphors ("duty of care" over "business as usual") would that have sufficient consequences for purchasing decisions that we might make as "consumers" within the food system? We should note, however, that once we move from the simplest metaphor (e.g. "food as fuel") towards embracing more complex relationships ("eating for planetary health"), it becomes more difficult for a metaphor to find its place in the popular lexicon. Two comparative examples here might be "food miles" and "ecological footprint" for which we could ask: just how effectively have they worked to enrol individuals? Arguably, the first has had more success than the second as it presupposes "local food is good" (though the evidence does not necessarily support this assertion). Even if it is scientifically apt, can a metaphor nurture a positive attitude among people towards the environment (Larson, 2018) and encourage change in behavioural practice? Rising numbers of people choosing plant-based diets (vegetarian and vegan) testify to the willingness of many to change in the interests of personal and/or planetary health although it is unclear to what degree – if any – a knowledge of PB has encouraged this.

Many see in the Sustainable Development Goals a way of translating aspirations for human well-being into a programme for living within planetary boundaries (O'Neill et al., 2018). There may well be a strong case as part of this process for calculating national fair shares for a more equitable redistribution of resources (Häyhä et al., 2016). However, as this chapter has argued in relation to the N cycle, and following a broader argument made elsewhere, tackling dietary practices in the wealthiest countries has to be a priority for it would immediately ease environmental burdens while improving human health (Tilman and Clark, 2014). Healthier diets – comprising more vegetables, fruit, and plant-based oils rather than animal products – require less cropland, less pressure on freshwater resources, more opportunities for co-existence with nature and more potential for nutrient cycling without excessive disruption to the global N cycle. However, it is doubtful that the metaphor of metabolic rift would play a key role in encouraging many rich country meat eaters to better understand the global nitrogen cycle and the disruptive effects caused by diets with high embedded N. In contrast, and rather idealistically, the notion of commensality or conviviality – the sharing of food around a hearth or table – might serve to engender a sense of belonging to a common entity – a family or community – where we partake in the abundance of the Earth but within its clearly defined boundaries and on an equitable basis. Yet the prospect of achieving popular support for such an idea demonstrates that the work of formulating effective metaphors that will help us navigate dietary transitions and transformative change across the complex world of food remains a key challenge.

Notes

1 In the original formulation (Rockström et al., 2009a), the following planetary boundaries were identified: climate change, ocean acidification, stratospheric ozone depletion, interference with the global phosphorus and nitrogen cycles, rate of biodiversity loss, global freshwater use, land-system change, aerosol loading, and chemical pollution. In Steffen et al. (2015a) paper, these are largely reproduced but with novel entities replacing chemical pollution. Further discussion of the PBs follows in a subsequent section.

2 Following convention in the literature, I use teragrams (Tg) which is a unit of mass equal to 10^{12} (one trillion) grams with reference to geophysical flows of nitrogen, though use tonnes in relation to manufactured production.

3 In the interests of brevity, this chapter has avoided a detailed discussion of the metabolic rift as used by Marx and subsequently forensically interrogated by contemporary

materialist theorists. Besides struggles over the 'correct' interpretation of Marx's meaning – which are at least partly rooted in dating the origins of capitalism – many social scientists have utilised metabolic rift in their analysis of agrarian change (Schneider, McMichael, 2010; Wittman, 2009) or with particular reference to food as an urban question (McClintock, 2010; Tornaghi et al., 2016; Dehaene et al., 2016).

4 Sage and Kenny (2017) provide a brief application of points 1, 3, and 4 in the context of recent developments in the Republic of Ireland.

5 I have chosen to keep quantitative evidence to a minimum in the text in order to avoid unnecessary clutter and to focus upon the key concepts and relationships of interest. However, to underpin this paragraph, it is worth citing Galloway et al. (2002) who succinctly summarise the sources of nitrogen loading: "Globally, humans currently ingest c. 20 Tg N yr^{-1} in their food. All of this Nr (reactive nitrogen) enters the environment. The resulting environmental consequences are magnified substantially, however, because an additional c.100 Tg N yr^{-1} that was involved in food production, but never entered human mouths, also was released to the environment. In addition to this total of 120 Tg N yr^{-1}, another c.25 Tg N yr^{-1} of Nr was created by fossil fuel combustion, and still another c.20 Tg N yr^{-1} was created for other uses by the Haber-Bosch process, for a total of c.165 Tg N yr^{-1}. This amount is about twice the amount of reactive N created by biological nitrogen fixation (BNF) in natural terrestrial ecosystems 90 Tg N yr^{-1}" (Galloway et al., 2002: 60).

6 Table 2 in Lassaletta et al. (2014b) provides clear evidence of the contrast in the nitrogen content of key import and export commodities. For example, the top four imports for the period 2005-09, their volume and their percentage N content was: wheat 5.5mt (1.95%); maize 5mt (1.52%); soybean cake 3.4mt (7.36%); soybeans 2.7mt (6.08%). In contrast, the top four exports for the same period were: tangerines 1.5mt (0.18%); wine 1.5mt (0%); oranges 1.3mt (0.18%); tomatoes 0.9mt (0.14%).

References

Allan, T., 2011. *Virtual Water: Tackling the Threat to Our Planet's Most Precious Resource*. London: I.B. Tauris.

Billen, G., Lassaletta, L., Garnier, J., 2014. A biogeochemical view of the global agro-food system: Nitrogen flows associated with protein production, consumption and trade. *Global Food Security* 3, pp. 209–219.

Borlaug, N., 1970. Nobel Lecture, The Green Revolution, Peace, and Humanity, December 11, 1970. https://www.nobelprize.org/nobel_prizes/peace/laureates/1970/borlaug-lecture.html. Accessed 27 February 2018.

Campbell, B., Beare, D., Bennett, E., Hall-Spencer, J., Ingram, J., Jaramillo, F., Ortiz, R., Ramankutty, N., Sayer, J., Shindell, D., 2017. Agriculture production as a major driver of the Earth system exceeding planetary boundaries. *Ecology and Society* 22, 4: 8. https://doi.org/10.5751/ES-09595-220408

Castree, N., 2017. Anthropocene and planetary boundaries. In *The International Encyclopedia of Geography: People, the Earth, Environment, and Technology* (Richardson, D., Castree, N., Goodchild, M., Kobayashi, A., Liu, W., Marston, R., eds.). New York: John Wiley & Sons.

Chapagain, A., Hoekstra, A., 2008. The global component of freshwater demand and supply: an assessment of virtual water flows between nations as a result of trade in agricultural and industrial products, *Water International* 33, 1: 19–32.

Clark, B., York, R., 2005. Carbon metabolism: Global capitalism, climate change, and the biospheric rift. *Theory and Society* 34: 391–428

Cornell, S., 2012. On the system properties of the planetary boundaries. *Ecology and Society* 17, 1: r2. http://dx.doi.org/10.5751/ES-04731-1701r02.

Dearing, J., Wang, R., Zhang, K., Dyke, J., Haberl, H., Hossain, S., Langdon, P., Lenton, T., Raworth, K., Brown, S., Carstensen, J., Cole, M. Cornell, S., Dawson, T., Doncaster, P., Eigenbrodm, F., Flörke, M., Jeffers, E., Mackay, A., Nykvist, B., Poppy, G., 2014. Safe and just operating spaces for regional social-ecological systems. *Global Environmental Change* 28: 227–238.

Dehaene, M., Tornaghi, C., Sage, C., 2016. Mending the metabolic rift: Placing the 'urban' in urban agriculture. In *Urban Agriculture Europe* (Lohrberg, F. Licka, L., Scazzosi, L., Timpe, A., eds). Berlin: Jovis Publishers, pp. 174–177.

De Vries, W., Kros, J., Kroeze, C., Seitzinger, S., 2013. Assessing planetary and regional nitrogen boundaries related to food security and adverse environmental impacts. *Current Opinion in Environmental Sustainability* 5: 392–402.

Dixon, J., Hattersley, L., Isaacs, B., 2014. Transgressing Retail: Supermarkets, liminoid power and the metabolic rift. In *Food Transgressions: Making Sense of Contemporary Food Politics* (Goodman, M., Sage, C., eds). Abingdon, UK: Routledge, pp. 131–153.

Erisman, J., Sutton, M., Galloway, J., Klimont, Z., Winiwarter, W., 2008. How a century of ammonia synthesis changed the world. *Nature Geoscience* 1: 636–639.

Foley, J., Ramankutty, N., Brauman, K., Cassidy, E., Gerber, J., Johnston, M., Mueller, N., O'Connell, C., Ray, D., West, P., Balzer, C., Bennett, E., Carpenter, S., Hill, J., Monfreda, C., Polasky, S., Rockström, J., Sheehan, J., Siebert, S., Tilman, D., Zaks, D., 2011. Solutions for a cultivated planet. *Nature* 478: 337–342.

Foster, J. B., 2013. Marx and the rift in the universal metabolism of nature. *Monthly Review*. Available at monthlyreview.org/2013/12/01/marx-rift-universal-metabolism-nature/. Accessed 3 September 2017.

Foster, J. B., Magdoff, F., 2000. Liebig, Marx, and the depletion of soil fertility: Relevance for today's agriculture. In *Hungry for Profit: The Agribusiness Threat to Farmers, Food and the Environment* (Magdoff, F., Foster, J.B., Buttel, F., eds). New York: Monthly Review Press, pp. 43–60.

Galloway, J., Cowling, E., 2002. Reactive nitrogen and the world: 200 years of change. *Ambio* 31, 2: 64–71.

Galloway, J., Cowling, E., Seitzinger, S., Socolow, R., 2002. Reactive nitrogen: Too much of a good thing? *Ambio* 31, 2: 60–63.

Hajer, M., Nilsson, M., Raworth, K., Bakker, P., Berkhout, F., de Boer, Y., Rockström, J., Ludwig, K., Kok, M., 2015. Beyond cockpit-ism: Four insights to enhance the transformative potential of the sustainable development goals. *Sustainability* 2015, 7: 1651–1660; http://dx.doi.org/10.3390/su7021651.

Havlík, P., Valina, H., Herrero, M., Obersteiner, M., Schmid, E., Rufino, M., Mosnier, A., Thornton, P., Böttcher, H., Conant, R., Frank, S., Fritz, S., Fuss, S., Kraxner, F., Notenbaert, A., 2014. Climate change mitigation through livestock system transitions. *PNAS* 111, 10: 3709–3714.

Häyhä, T., Lucas, P., van Vuuren, D., Cornell, S., Hoff, H., 2016. From planetary boundaries to national fair shares of the global safe operating space: How can the scales be bridged? *Global Environmental Change* 40: 60–72.

Heinrich Böll Foundation and Friends of the Earth Europe, 2014. *Meat Atlas: Facts and figures about the animals we eat*. Berlin: Heinrich Böll Foundation.

Horton, P., 2017. We need radical change in how we produce and consume food. *Food Security* 9: 1323–1327.

International Food Policy Research Institute (IFPRI), 2016. *Global Nutrition Report*. Washington, DC: International Food Policy Research Institute.

Jaramillo, F., Destouni, G., 2015. Comment on "Planetary boundaries: Guiding human development on a changing planet". *Science* 348, (6240): 1217.

Knutti, R. Rogelj, J., Sedláček, J., Fischer, E., 2016. A scientific critique of the two-degree climate change target. *Nature Geoscience* 9: 13–19.

Larson, B., 2018. Environmental metaphor. In *Companion to Environmental Studies* (Castree, N., Hulme, M., Proctor, J., eds).). Abingdon, Oxon: Routledge, pp. 645–648.

Lassaletta, L., Billen, G., Grizzetti, B., Garnier, J., Leach, A., Galloway, J., 2014a. Food and feed trade as a driver in the global nitrogen cycle: 50-year trends. *Biogeochemistry* 118: 225–241.

Lassaletta, L., Billen, G., Romero, E., Garnier, J., Aguilera, E., 2014b. How changes in diet and trade patterns have shaped the N cycle at the national scale: Spain (1961–2009). *Regional Environmental Change* 14: 785–797.

Lelieveld, J., Evans, J., Fnais, M., Giannadaki, D., Pozzer, A., 2015. The contribution of outdoor air pollution sources to premature mortality on a global scale. *Nature* 525: 367–371.

McClintock, N., 2010. Why Farm the City? Theorizing Urban Agriculture Through a Lens of Metabolic Rift. *Cambridge Journal of Regions, Economy and Society* 3, 2: 191–207.

Meadows, D.H., Meadows, D.L., Randers, J., Behrens, W., 1972. *The Limits to Growth: A Report for the Club of Rome's Project on the Predicament of Mankind*. New York: Universe Books.

Mew, M., Steiner, G., Geissler, B., 2018. Phosphorus supply chain—Scientific, technical, and economic foundations: A transdisciplinary orientation. *Sustainability* 10, 1087; http://dx.doi.org/10.3390/su10041087.

O'Neill, D., Fanning, A., Lamb, W., Steinberger, J., 2018. A good life for all within planetary boundaries. *Nature Sustainability* 1: 88–95.

Niles, M., Ahuja, R., Eaquivel, J., Gutterman, S., Heller, M., Mango, N., Portner, D., Raimond, R., Tirado, C., Vermeulen, S., 2018. Climate change mitigation beyond agriculture: A review of food system opportunities and implications. *Renewable Agriculture and Food Systems*. https://doi.org/10.1017/S1742170518000029.

Qu, S., Liang, S., Konar, M., Zhu, Z., Chiu, A., Jia, X., Xu, M., 2018. Virtual water scarcity risk to the global trade system. *Environmental Science & Technology* 52, 673–683.

Qualman, D., 2020. *The nitrogen crisis*. https://www.darrinqualman.com/nitrogen-crisis/ (Accessed 1 Aug 2020).

Raworth, K., 2017. A doughnut for the anthropocene: Humanity's compass in the 21st century. *The Lancet Planetary Health*. https://www.thelancet.com/pdfs/journals/lanplh/PIIS2542-5196(17)30028-1.pdf.

Rockström, J., Steffen, W., Noone, K., Persson, A., Chapin, F., Lambin, E., Lenton, T., Scheffer, M., Folke, C., Schellnhuber, H., Nykvist, B., De Wit, C., Hughes, T., Van der Leeuw, S., Rodhe, H., Sörlin, S., Snyder, P., Costanza, R., Svedin, U., Falkenmark, M., Karlberg, L., Corell, R., Fabry, V., Hansen, J., Walker, B., Liverman, D., Richardson, K., Crutzen, P.., Foley, J., 2009a. Planetary boundaries: Exploring the safe operating space for humanity. *Ecology and Society* 14(2). Available at: http://www.ecologyandsociety.org/vol14/iss2/art32/.

Rockström, J., Steffen, W., Noone, K., Persson, A., Chapin, F., Lambin, E., Lenton, T., Scheffer, M., Folke, C., Schellnhuber, H., Nykvist, B., De Wit, C., Hughes, T., Van der Leeuw, S., Rodhe, H., Sörlin, S., Snyder, P., Costanza, R., Svedin, U., Falkenmark, M., Karlberg, L., Corell, R., Fabry, V., Hansen, J., Walker, B., Liverman, D., Richardson, K., Crutzen, P.., Foley, J.., 2009b. A safe operating space for humanity. *Nature* 461(7263): 472–475.

Sage, C., 2012. *Environment and Food*. Abingdon, UK: Routledge.

Sage, C., Kenny, T., 2017. Connecting agri-export productivism, sustainability and domestic food security via the metabolic rift: The case of the Republic of Ireland. In *Advances in Food Security and Sustainability*, vol. 2 (Barling, D., ed). Oxford: Elsevier, pp. 41–67.

Schneider, M., McMichael, P. 2010. Deepening, and repairing, the metabolic rift. *Journal of Peasant Studies* 37(3): 461–484. DOI: 10.1080/03066150.2010.494371

Smil, V., 2000. Phosphorous in the environment: Natural flows and human interferences. *Annual Review Energy and the Environment* 25: 53–88.

Smil, V., 2001. *Enriching the Earth: Fritz Haber, Carl Bosch, and the Transformation of World Food Production*. Cambridge, Mass: MIT Press.

Smil, V., 2002. Nitrogen and food production: proteins for human diets. *Ambio* 31(2): 126–131.

Steffen, W., Persson, A., Deutsch, L., Zalasiewicz, J., Williams, M., Richardson, K., Crumley, C., Crutzen, P., Folke, C., Gordon, L., Molina, M., Ramanathan, V., Rockström, J., Scheffer, M., Schellnhuber, H., Svedin, U., 2011. The anthropocene: From global change to planetary stewardship. *Ambio* 40:739–761.

Steffen, W., Richardson, K., Rockström, J., Cornell, S. E., Fetzer, I., Bennett, E. M., Bigs, R., Carpenter, S. R., De Vries, W., De Wit, C. A., Folks, C., Gerten, D., Heinke, J., Mace, G. M., Ramanathan, V., Reyers, B., Sörlin, S., 2015a. Planetary boundaries: Guiding human development on a changing planet. *Science* 347(6223). http://www.sciencemag.org/content/347/6223/1259855.

Steffen, W., Broadgate, W., Deutsch, L., Gaffney, O., Ludwig, C., 2015b. The trajectory of the Anthropocene: The great acceleration. *The Anthropocene Review* 2, 1: 81–98.

Tilman, D., Clark, M., 2014. Global diets link environmental sustainability and human health. *Nature* 515, pp. 518–522.

Tornaghi, C., Sage, C., Dehaene, M., 2016. Metabolism: An introduction. In *Urban Agriculture Europe* (Lohrberg, F. Licka, L., Scazzosi, L., Timpe, A., eds). Berlin: Jovis Publishers pp. 166–169.

Van Meter, K., Van Cappellen, P., Basu, N., 2018. Legacy nitrogen may prevent achievement of water quality goals in the Gulf of Mexico. *Science* 10.1126/science.aar4462

Vermeulen, S., Campbell, B., Ingram, J., 2012. Climate change and food systems. *Annual Review of Environmental Resources* 37:195–222.

Westhoek, H., Rood, T., van de Berg, M., Janse, J., Nijdam, D., Reudink, M., Stehfest, E., 2011. *The Protein Puzzle*. The Hague: PBL Netherlands Environmental Assessment Agency

Willett, W., Rockström, J., Loken, B., Springmann, M., Lang, T., Vermeulen, S., et al., 2019. Food in the Anthropocene: the EAT–Lancet Commission on healthy diets from sustainable food systems. *The Lancet* 393(10170): 447–492.

Wironen, M., Bennett, E., Erickson, J., 2018. Phosphorus flows and legacy accumulation in an animal-dominated agricultural region from 1925 to 2012. *Global Environmental Change* 50: 88–99

Wittman, H., 2009. Re-working the metabolic rift: La Via Campesina, agrarian citizenship and food sovereignty. *Journal of Peasant Studies* 36, 4: 805–826.

Zhang, X. Davidson, E., Mauzerall, D., Searchinger, T., Dumas, P., Shen, Y., 2015. Managing nitrogen for sustainable development. *Nature* 528: 51–59.

4 Alchemical and cyborgian imaginings in technoscientific discourse relating to holistic turns in food processing and personalised nutrition

Shane V. Crowley

Introduction

Everyday language is rich with food metaphors (Lakoff and Johnson, 2003: 46). We *are what we eat* and are given *food for thought*. We are advised not to *sugar-coat* things and are wary of *half-baked* plans. If we are fortunate, our friends and family will consider us the *salt of the earth*, although a *salty* temperament could make one unpopular or, worse, earn one a reputation as an *unsavoury* character. All of these examples function by treating food as a source of metaphors directed towards the target of another concept. Food may also be the target [1] of metaphors, however; for example, since the computer has become the dominant technology in society, and therefore a common metaphor source (Jayasinghe, 2012), we have heard of "*food hacks*", such as "*open source*" nutritional beverages that get software-esque updates ('*v.1*', '*v.2*',..,'*v.n*').

In this chapter, I will use metaphorical concepts to interpret two utopian imaginings in technoscientific discourse. The first is personalised nutrition, a cybernetic system in which biometric and genetic data determine the nutrient intake of the individual. The second is circular economics, a design philosophy for industrial manufacture that aims to transform all waste into a raw material for further consumption. These imaginings will be shown to be generative of metaphors signifying a "holistic turn" in the discourse around sustainable food, thereby influencing epistemological motivations, axiological preferences, and metaphysical worldviews. To this end, personalised nutrition is interpreted through the metaphor of *humans as cyborgs* and circular economics is interpreted through the metaphor *production as alchemy*. Each metaphor signifies the unification of opposing parts. The cyborg unifies consumer and machine, while alchemy unifies food and waste.

In the first section of the chapter, I review some theoretical background that informs the analysis to follow. I draw from concepts in linguistics (structural and cognitive), structuralism (Saussurian), and psychoanalysis (Jungian and Lacanian). Next, I elaborate on the cyborgian and alchemical metaphors as they apply to sustainable food. The chapter is interspersed with historical examples, which are intended to highlight the intimate relation between science and technology, the role of the imaginary in scientific discovery and the defamiliarising capacity of technoscientific discourse. The deployment of cyborgian and

DOI: 10.4324/9781003143567-5

alchemical metaphors reveals a food system that is responding to a time of crisis, fragmentation, and chaos. They function as signifiers of a "holistic turn" (Zwart, 2018, 2019) and an aspiration for food and bodies to be made whole again. Such metaphors can serve as guides for researchers in the discovery phase of science but may also be therapeutic aides for those at the mercy of "impossible professions" (Freud, 1937).

Theoretical background

Structural linguistics and the science of signs

This chapter draws heavily from semiotic [2] approaches to discourse analysis derived from the theories of structural linguists, namely Ferdinand de Saussure (1857–1913) and Roman Jakobsen (1896–1982). In his *Course in General Linguistics* (*CGL*), Saussure (2009) gave an account of language as a system of signs. He presented a dyadic model of the sign, comprising a signifier and signified (Figure 4.1). The signifier was conceived of as a verbal "sound pattern" and signified as a mental "concept" (*CGL*, 65–67). The relation between signifier and signified is arbitrary for a given sign, amply demonstrated by the unique words used by speakers of different languages to signify the same concept (*CGL*, 68). Signs, however, exhibit a temporal fixedness, as once established by convention they become difficult to shift (*CGL*, 67). Language, to Saussure, is a structure built from basic units – the signs – which exist in systematic relation to one another (*CGL*, 121). The value inherent in one sign is that it is related to other signs by difference or association. Some important features of the dyadic model are notable, including the "bar" between signifier and signified, the arrows pointing from each to the other, and their relative positioning on the vertical axis. Respectively, these can be taken to indicate that there is no necessary connection between the parts of a sign, that each part triggers the other part and that Saussure's sign privileges the signified. This last remark was a key insight of the psychoanalyst Jacques Lacan, who critiqued these ordering relations.

For Saussure, meaning arises from differences between signifiers across horizontal (syntagmatic) and vertical (paradigmatic[3]) axes. A syntagm is a concatenation of signifiers, a spatio-temporal succession of elements, rendered intelligible by its adherence to rules (i.e. a grammar). While a syntagm is based on successive relations, a paradigm is based on associative relations. A paradigmatic transformation involves a substitution of one element for another. Importantly

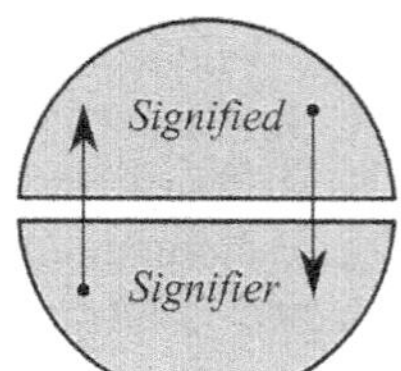

Figure 4.1 Saussurian model of the sign.

Source: Drawing author's own.

for our purposes, metonyms and metaphors are generated by such substitutions. According to Jakobsen (1956), metonymic substitutions function by contiguity while those of a metaphoric nature function by similarity. Consider synecdoche, which we will treat as a species of metonym, where a part is substituted for the whole. The part-whole relation is one of contiguity, a proper part, and its whole necessarily "touch". Metaphor, on the other hand, only requires two objects to be *similar* in some sense for them to be deemed substitutable. Jakobsen used this distinction to differentiate between literary genres with, for example, realism being characterised by metonymy and romanticism being characterised by metaphor.

Structuralism and culinary signification

Saussurian linguistics was later adopted as a model for "structuralist" analysis of a much broader range of cultural signs. Indeed, Saussure himself suggested that language was "only one type" of sign system (*CGL*, 68). Key figures associated with structuralism, such as Claude Lévi-Strauss, devoted considerable attention to the study of food. Characteristic of these structuralist analyses was a preoccupation with oppositions, exemplified in Lévi-Strauss' use of his "culinary triangle" (the raw, cooked, and rotted) to study the nature|culture opposition. The quality of naturalness which language confers upon boiled food is purely metaphorical: the "boiled" is not the "spoiled"; they simply share a conceptual resemblance (Lévi-Strauss, 2008).

Culinary practices in a given society can be interpreted as symbols of cultural differentiation, with raw food having two primary routes of transformational change: raw → rotted and raw → cooked. Lévi-Strauss represented these transformations using the Culinary Triangle (Figure 4.2), on which he positioned various cooking methods and cooked foods on a nature-culture continuum. Roasting is close to the nature, as it involves direct, unmediated heating of food by fire. Boiling is closer to culture, as the food is heated in a medium (water,

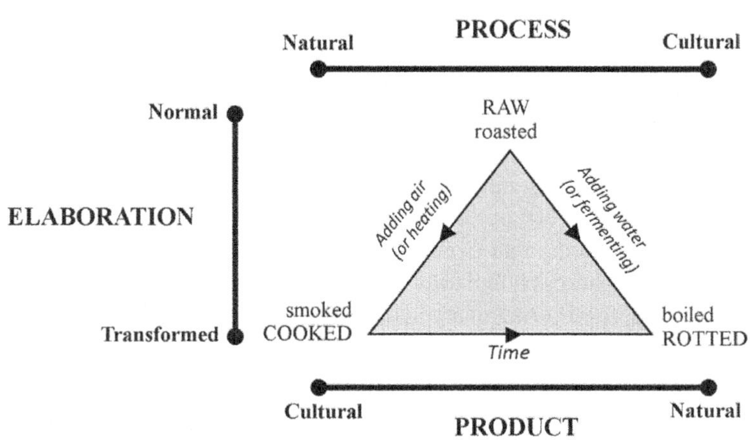

Figure 4.2 The Culinary Triangle.

Source: Drawing is author's own with modifications of elements found in Leach (1970) and Vilgis (2013).

mediating between food and fire) housed in a receptacle (cultural artefact, mediating between man and the world). Smoking is closer to nature than boiling as, like roasting, only air mediates between the food and heat source; however, as the mediating distance between food and heat source is comparatively greater, smoking is considered closer to culture than roasting.

Lévi-Strauss concludes that culinary transformation can be considered closer to nature or culture based on a complex interplay between cultural mediation, degree of elaboration, and the final properties of the end-product. Thus, the highly mediated process of boiling is *cultural*, but the boiled end-product physically resembles the rotted and is therefore *natural*; on the other hand, smoking is *natural* in its low degree of mediation, but *cultural* in its production of a durable and thoroughly cooked product. Roasting is always *natural*, being an unmediated process resulting in a partially cooked (i.e. raw on the inside) end-product. Leach interprets the significance of the Culinary Triangle as follows:

> Cooking is thus universally a means by which Nature is transformed into Culture, and categories of cooking are always peculiarly appropriate for use as symbols of social differentiation.
>
> (Leach, 1970)

Curiously, Lévi-Strauss' structuralist approach to food has had a quite resurgence in more technoscientific writings on food (Knorr and Watzke, 2019; Vilgis, 2013). Vilgis, a soft matter physicist, supplements the culinary triangle with concepts familiar to any food scientist. For example, Lévi-Strauss' opaque assertion that the boiled "resembles" the raw is instead explained through common hydrolysis reactions that soften the texture of both structures. In addition, the raw → cooked transformation is considered to be temperature-driven, while the raw → rotted transformation is considered to be microbiological (Figure 4.2). On this basis, processes based on fermentation can be grouped with those due to rotting, both being "natural" but only the former being acceptable to a consumer. The "cultural and physical structuralism" of Vilgis could be viewed as a productive marriage of cultural theory and technoscientific knowledge. Sophia Roosth, however, in her ethnographic study of a molecular gastronomy laboratory, found that the "afterlives" of structuralism can have problematic effects on technoscientific discourse (Roosth, 2013). Extant "old wives' tales" are conceptualised as being in opposition to the technoscientific imagining of "rational cooking", with the former criticised as "magical thinking" in need of displacement by "reason and experiment" (Roosth, 2013). It is worth noting that even the most revered of scientists, such as Isaac Newton, have also been admonished for their "magical" works (Newman, 2015). This effort to suppress the imaginary in technoscientific discourse, despite its role in the discovery phase *necessary* to all science, is a key theme of this chapter.

Psychoanalysis of technoscience

Lacan's psychoanalytic theory was greatly influenced by Saussurian linguistics and structuralism, and this informed his critical re-reading of Freud. He adopted

Saussure's dyadic model of the sign but in Lacan (2007) formalised it quasi-algebraically as $\frac{S}{s}$, with the signifier (S) in a vertically "privileged" position separated from the signified (*s*) by a bar. That we skillfully manipulate signifiers in everyday communication belies to Lacan our uncertainty about concrete signifieds, resulting in a tendency for "slippage" of signifieds underneath a given signifier. In this respect, Lacanian semiotics has a Kantian aspect in its distinction between the accessible phenomena (signifiers) and inaccessible noumena (signifieds). Lacan considered the Freudian concepts of *condensation* (combination of disparate ideas into a unified dream symbol) and *displacement* (transfer of significance from one idea to an associated idea) to correspond to Jakobsen's metaphoric and metonymic poles, respectively. The metaphorical formation of symbols in the unconscious can yield remarkable insights, as exemplified by the story of how German chemist August Kekulé discovered the structure of benzene.

A quarter century after his discovery Kekulé gave an account (translated in Benfey, 1956) of his discovery to an audience who had gathered in Berlin to honour his work. His speech reveals him to be an advocate of a holistic epistemology in which "something absolutely new has never been thought", where old ideas are not "the ruins" on which new ideas stand but are, rather, "utilised in the later structure, and form with it one harmonious whole". He goes on to recount that while working on the structure of benzene he dreamed of chains of atoms swirling in a "snake-like" motion, and before long "one of the snakes had seized hold of its own tail". Kekulé worked long into the night developing the structure of benzene that was hinted by his reverie. This creative event can be considered as a process of metaphorical construction (or Freudian condensation), in which a signifier of alchemical origin (the *ouroborus*) is substituted for a chemical signifier (the known formula showing the proportions of elements in the compound). The resultant sign is the "flash of lightening" that led to the elucidation of benzene's ring-like structure (Figure 4.3). As Zwart (2019: 30) asserts, scientific

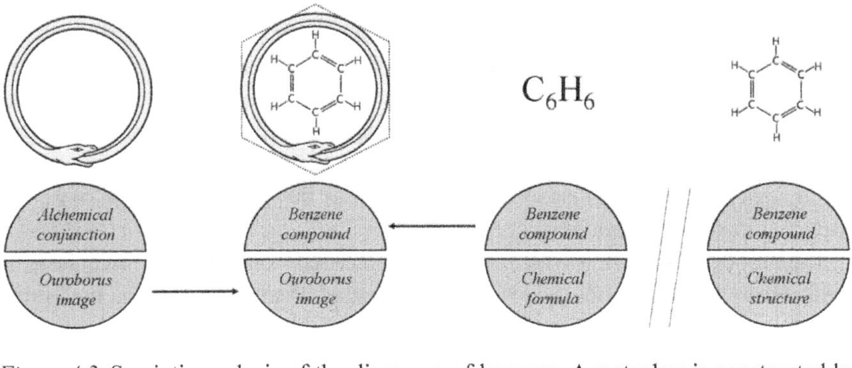

Figure 4.3 Semiotic analysis of the discovery of benzene. A metaphor is constructed by substituting the signified of one sign with the signified of another. The process of discovery is typically barred from normal technoscientific discourse (e.g. in conventional scientific communications).

Source: Drawing is author's own, but an element is included that is licensed under Creative Commons.[4]

discourse privileges the symbolic form and so suppresses the image, leading to the chemical sign being separated from its metaphorical construction (signified by the "//"in Figure 4.3), the terminus of the diachronic process is the fixing of the chemical structure. The imaginary is this way suppressed in histories of science. Kekulé himself urged scientists to "learn to dream" and attributed his discovery, in part, to a talent for drawing and his studies in architecture, which gave him "an irresistible urge to visualise everything". As scientists, we must not only be aware of the possibilities that are found in imaginative work but also for the tendency of scientific discourse to erase imagination from its own history.

Kekulé's story highlights the capacity of technoscientific discourse to be both iconogenetic and iconoclastic. In his early work, the French philosopher Bachelard considered scientific progress to involve the destruction of "pre-scientific" images and metaphors, which he termed "epistemological obstacles" (Bachelard, 2002: 88). Cultivating a scientific mindset required one to clear the path of rational inquiry of such obstacles through a regime of self-surveillance (Zwart, 2019: 16). Technoscience, however, involves both the creation and destruction of images (Zwart, 2019: 18). The ouroborus was a structural template for Kekulé's discovery but "falls away" from the final technoscientific symbolisation (Zwart, 2019: 30). More generally, various metaphors can be thought of as persisting (unconsciously) in the symbols that guide researchers during a "holistic turn". In my discussions of cyborgs and alchemy in this chapter, I do not claim that scientists explicitly desire to create the alchemical "opus" or to manufacture the cyborg, but rather that these "circular" metaphors are used to interpret a structure of entailments from which the template has, as it were, fallen away.

Cognitive linguistics and conceptual metaphor

The approach to metaphor adopted in this chapter is based on the work of George Lakoff and Mark Johnson (L&J) in their book *Metaphors We Live By*. At a basic level, metaphor is when one thing is described in terms of something else. To L&J, the very fact that metaphorical expressions are possible indicates that metaphors are a feature of cognitive processing (L&J, 2003/1980: 6). Once a metaphor is adopted in conventional usage, it can form a system of related metaphors through entailment relations. Once we come to conceive of *food as fuel* then, by entailment, we may also come to conceive of *food as a resource* and the *body as machine* (Van Der Weele, 2006). Hence, a single system of inter-related metaphorical concepts is established that ensures the intelligibility of linguistic expressions (e.g. "I need to refuel").

The association of food with fuel is not a necessary one, yet it is ubiquitous in modern industrialised societies. However, a metaphorical expression is not equivalent to a one-to-one mapping, in the mathematical sense. If x is a domain (source) and y its co-domain (target), it is not necessarily the case that an element in x maps to an element in y. It is also worth noting that a given concept, such as food, can be both a source or target in a metaphorical system (Korthals, 2008). A metaphor hence functions in at least three ways: 1) *correspondence* (highlighting known similarities between source and target); 2) *creation* (creating new similarities between source and target); and 3) *masking* (hiding properties of a domain not captured in the mapping) (L&J, 10–11, 252–253). While metaphors can aid our understanding of experience and inform our actions, they may also constrict our thinking by blocking us from aspects of reality (ibid., 139–146). The value of reflecting on metaphor use is therefore not just a matter of discovering correspondences that are absolute, or of breaking free of conventions by generating new metaphors, it can simply be the elaboration of the conceptual weightings we have inherited. In this sense we may inquire into which aspects of our present discourse are emphasised and which are obscured by the prevailing metaphorical concepts and their systematic entailments.

Literary analysis and defamiliarisation

Jakobsen (1958) found that literary and artistic genres could be interpreted based on their metonymic or metaphoric character. In this sense, realist works (e.g. Tolstoy) were opposed to romantic works (e.g. Wordsworth). Realism is not to be understood as a technique of correspondence, where wholes are represented to the reader *as they are* phenomenologically. Rather, realist fiction is populated by salient parts (shoulders, doorways) standing in place for obscured wholes (people, houses). Similarly, the fragmented figures of the cubists deconstruct wholes into contiguous sets of parts. Realism, thus, is essentially metonymic. Romanticism, on the other hand, is metaphoric, appealing often to images of nature (flowers, thunder) to represent human emotions (love, anger). Consider how in food discourse, a *war* has emerged between activists who wish to *defend* "whole food" from realist/metynomic science (see Pollen's popular 2008 book '*In Defense of Food*) and those who wish to *defend* food science from the romantic/metaphoric activists (see McClements et al., 2011. '*In Defense of Food Science*'). It can be seen in Table 4.1 that metaphor and metonym can thus be considered to correspond with different literary, psychoanalytic, and technoscientific modes.

Jakobsen quotes a contemporary critic who wrote of metonymic works where "the reader is crushed by the multiplicity of detail [...] and is physically unable to grasp the whole [...]" (ibid., 80). We might say that the whole is made strange to the reader through its deconstruction into synecdoches. Yet, the metaphoric mode also makes the whole strange, by substituting the familiar object for the unfamiliar. Thus, to borrow a phrase from the Russian formalist Viktor Shklovsky, both metonymy and metaphor cause a degree of defamiliarisation in the reader. In his "Art as Device", Shklovsky (1988) described how routine phenomenological encounters with objects lead to the subject's "over automisation". Art then, to

Table 4.1 Alignment between tropes and concepts from literary analysis, psychoanalysis, and food technoscientific discourse.

Trope	*Metaphor*	*Metonym*
Mechanism	Similarity	Contiguity
Literary genre	Romantic	Realist
Unconscious activity	Condensation	Displacement
Discourse	Activist	Technoscience

Shklovsky, is uniquely powerful in rendering what is familiar to the subject in an aspect that is unfamiliar, thereby restoring or even enhancing subjective experience of the once automatic. Consider the writing of Franz Kafka:

> [...] his food soon stopped affording him the least pleasure, and so, to divert himself, he got into the habit of crawling all over the walls and ceiling.
>
> (Kafka, 2007: 117)

In Kafka's famous short story, the *Metamorphosis*, a man awakes to realise he has been transformed into an insect. Gregor, the man/insect, narrates the story. To the reader, the role "narrator" is defamiliarised through its voicing by an insect, while the animal "insect" is defamiliarised through its having a voice. Gregor's sense of embodiment becomes defamiliarised through his physical transformation, while to his family, this metamorphosis defamiliarises Gregor as a son and brother. Food also becomes defamiliarised to Gregor, with a substitution of the pleasant for the disgusting. Gregor begins to reject food he once enjoyed ("the milk did not taste at all nice"), coming instead to prefer "half-rotten vegetables", "congealed white sauce" and cheese that was "unfit for human consumption" (Kafka, 2007: 109).

The influential social scientist Claude Fischler (2011) has written of the "nutritional cacophony", a metonymic noise generated by the food industry that results in a generalised anxiety about food choice. Barthes (1983) made similar comments about the rhetoric of fashion "neomania", which obscures the collective memory of past fashions and imbues the new with mythical qualities.

> In short, the system is drowned under literature, the Fashion consumer is plunged into a disorder which is soon an oblivion [...].
>
> (Barthes, 1983: 300)

Fischler (1992) suggests that the transition of Western society from a state of "lipophilia" (fat loving) to "lipophobia" (fat hating) was caused by a number of socio-cultural factors, including reduced energy expenditure, the demonisation of individual nutrients, and an altered feminine ideal. This transition resulted in tremendous investment in butter-replacement technology (e.g. margarine, hydrogenated fats, cholesterol-reducing spreads). Butter has recently become "good" again, as evidenced by TIME magazine's cautious claims that "butter may, in fact, be back" (Sifferlin, 2016) in the same year that the New York Times heralded 'the return of the '90s' in a fashion commentary (Fury, 2016). Everyday speech is abundant with metonymic descriptions that bypass the phenomenological aspect of food. Someone rushes to buy a coffee uttering 'I want my caffeine', or another who eats a chicken salad thinking "I need my protein", these people are conceptualising food and drink metonymically, substituting wholes for their physico-chemical parts. These "realist" expressions are related to the prevailing technoscientific discourse surrounding food, which is opposed to Fischler's more "romantic" substitution of meals for their socio-psychological effects (e.g. identity formation and social cohesion). The result: both sides of the argument are read as "making food strange" by the other, the natural scientists by being downwardly reductive (to the molecular level) and the social scientist by being upwardly reductive (to the social level); as "up" is generally privileged relative to "down" as an orientational metaphor (L&J, 14–15), it is no surprise that the criticism of the molecular-orientation has received widespread traction (as in the best-selling works of Michael Pollen).

The intimate connection of science and technology is signified in this chapter by the portmanteau "technoscience" (Zwart, 2019: 9). The history of how protein was discovered as a nutrient, described in Carpenter (2003), highlights this intimacy in the context of the historic "molecularisation" of food. The "animal substance", as it was then known, was discovered in 1785 after Claude Berthollet observed that rotting meat gives off ammonia, indicating that animal matter contained nitrogen. In 1816 François Magendie designed a (flawed) experiment in which a series of dogs were fed diets of sugar, oil, gum, or butter (i.e. materials lacking nitrogen). All of the dogs died and, in the absence of a positive control, the animal substance began to come into focus as "the only true nutrient" for members of the scientific community; as Carpenter states, it was both "the machinery of the body and the fuel for its work" (Carpenter, 2003). Several years later, Magendie set about testing whether gelatin from boiled bones could be a complete source of nutrition, as it contained the necessary animal substance; however, the experiment once again had fatal consequences for the dogs involved. It was hoped at the time that a successful result could be the basis for gelatin technology to be applied to cheaply feed hospital patients. The scientific experiments and technological applications evolved in parallel. Although Magendie's experiments did not bear technological fruit, it was only the beginning of a "protein obsession" that was to grip the famous chemist Justus Liebig and his followers for many years (Carpenter, 2003), an obsession that remains influential in the "proteinophilic" discourse of today.

Cyborg imaginings for sustainable diets

The factory as rational ideal

Western philosophical and religious thought has traditionally been dominated by a dualistic metaphysics that divides the body (material, machine-like, non-rational) and mind (immaterial, rational). Plato (1997[5]) considered the body as an earthly structure that imprisons and grounds the soul. The philosopher can release her soul from this prison by avoiding "so-called pleasures" like food, drink, and sex, so that she "turns away from the body" (*Phaedo*, 64b-e). In the tripartite soul of Plato's most famous student, Aristotle (1995[6]), the nutritive soul was the lowest, being possessed by "everything that must grow, reach maturity and decay" (*On the Soul*, III.12, 434a23–29; Shields, 2016: 75–77). Descartes contrasted rational humans with non-rational animals, the latter being considered mere *automata*, lacking the immaterial substance that imparts the faculty of reason (Descartes, 1985: 30–31); hence, according to his substance dualism, the body itself is a non-rational automaton. In defending utilitarianism from critics who pronounced the ethical theory "a doctrine worthy only of swine", John Stuart Mill argued that humanity is capable of experiencing pleasures that are qualitatively superior to the "animal appetites", with these "higher pleasures" being largely intellectual (Mill, 2016: 375–376). Given the legacy of mainstream Western thought, it is not surprising that perceived non-rational food behaviours are framed using animalistic metaphors (e.g. 'I *grazed* on snacks, 's/he *wolfed* down the meal), which López-Rodríguez (2016) found to be frequently used in negative depictions of women in mass media. The *body as machine* metaphor and the *food as fuel* metaphor it entails have been influential in shaping technoscientific discourse (Van der Weele, 2006). In a recent study of an online blogging community, it was shown that contributors viewed *food as fuel* that enabled exercise and that, in turn, exercise was a justification for eating pleasurable foods (Lynch, 2010).

In Figure 4.4, a Taylorist vision of the human body as a "scientifically managed" factory is shown. The visual metaphor is attributed to Fritz Kahn[7], a popular science writer in pre-WW2 Germany. The various functions of the body are imagined as consisting of tanks, conveyor belts, hoppers, and so on, while offices, switchboards, and control rooms are found in the brain. Each area is staffed with homunculi; in the famous wall poster version "Man as Industrial Palace", each homunculus can be seen wearing attire befitting their role (e.g. business suit, lab coat, work clothes). Borck (2007) considers the body-factory illustration characteristic of an idealised representation of a "rational and clean, technological and sanitized modernity" typical of the Weimer Republic in which Fritz did most of his work. It has been suggested that the "mechanistic biomedical model" of health, which emphasises Cartesian and Newtonian approaches to the body, has its origin in such factory metaphors (Jayasinghe, 2012). Kahn's image shows how the technoscientific ideals of an era inform our conceptions of the body.

The *body as factory* metaphor is alive and well in the writings of some food scientists. An interesting development in the field is the gastrointestinal digestion model – an apparatus mimicking the human digestive system, which simulates the breakdown, transformation, and absorption of foods in the body. To explain how this model corresponds to the body, food scientists have described the human gastrointestinal tract as a *series of unit operations*, engineering terminology for an integrated series of machines in a factory:

> All foods pass through a common *unit operation*, the gastrointestinal tract, yet it is the least studied and *least understood of all food processes*. To design the foods of the future, *we need to understand what happens inside people in the same way as understanding any other process.*
>
> (Norton et al., 2006; italics added for emphasis)

In the review of Boland (2016), some of the following factory metaphors are used in descriptions of human digestion: *unit operation*, *reactor*, *control point*, *batch operation*, *input/output,* and *stirred storage vessel.* Bornhorst et al. (2016) state that understanding this digestion process is important for "functional foods", which the food industry have been "prompted" to develop due to links between

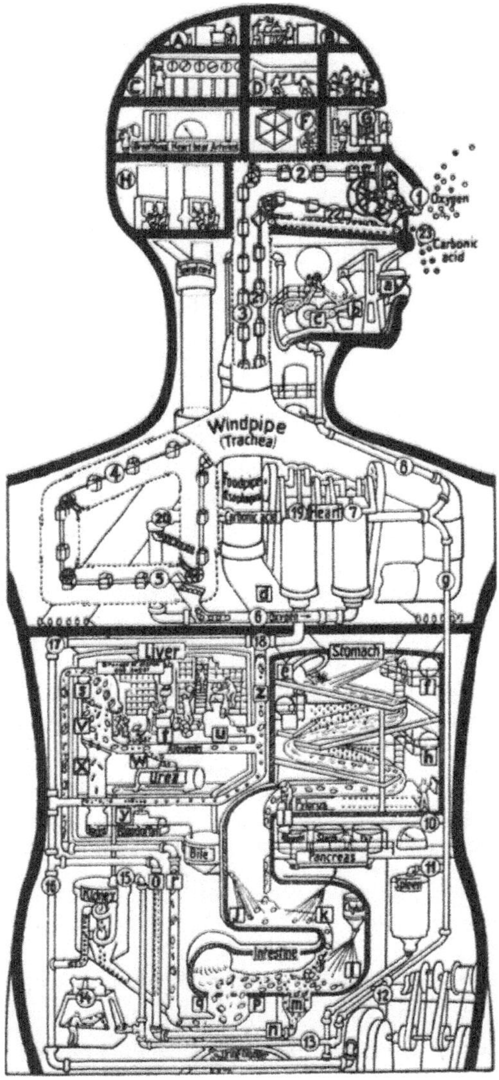

Figure 4.4 "The Human Factory" (by Fritz Kahn).

Source: Courtesy of Leo Baeck Institute New York.

diet and a range of diseases (malnutrition, diabetes, cardiovascular disease, cancers) and "consumer awareness" of same. Whether *body as factory* metaphors are useful scientific models and/or textbook cases of *déformation professionnelle* may warrant some reflection. It is undeniable that this type of apparatus is increasing our understanding of how foods are digested, yet we must be cautious about its metaphorical entailments, which may include the idea that food choice is a *process* for *optimisation* to maximise *efficiency*, an idea of "the new man" that Kahn himself believed in (von Debschitz and von Debschitz, 2017).

As metaphors of the body generally reflect the dominant technology in society, we might expect that metaphors based on computers, information theory, and cybernetics have started displacing factory metaphors (Jayasinghe, 2012). The transition from factory metaphors to cyborg metaphors can be conceived as a shift from institutional (panoptic) to automated (cybernetic) modes of dietary discipline.

Panoptic diets and molecularised discourse

Michel Foucault believed that a shift from the disciplining of diet to the disciplining of sex occurred after the seventeenth century (Taylor, 2010). The ancient Greeks were acutely concerned with the regulation of diet. References abound to food in Platonic texts. It is stated that provision of food to sustain life is the "first and greatest need" (*Republic*, 369d), and that a diet mostly comprised of plain fare ("they'll knead and cook the flour and meal they've made") and a few luxuries ("salt, olives, cheese [...]") will help citizens live in "good health" and die "at a ripe old age" (ibid., 372d)' on the other hand, luxury foods like "sweet desserts" and "pastries", are associated with "illness" (ibid., 404b-e). The need for medical intervention in cases of illness was considered a sign of societal decline by Plato (ibid., 404b–e; Skiadas and Lascaratos, 2001) who was skeptical about the prescription of "strange drink[s]" as remedies (ibid., 405d–e). Today, concerns persist over the "discipline" of consumers in their food choices, and the public health issues linked with overconsumption (e.g. obesity, diabetes), as do the "strange drinks" that are put forward as remedies. Although Foucault's treatment of food was limited, some of his conceptual tools, namely the panopticon metaphor, are useful in the analysis of diet.

Jeremy Bentham developed an architectural design (Figure 4.5) for a prison (the panopticon) that theoretically maintained control over inmates through the imposition of self-discipline:

> It is obvious that, in all these instances, the more constantly the persons to be inspected are under the eyes of the persons who should inspect them, the more perfectly will the purpose of the establishment have been attained. Ideal perfection, if that were the object, would require that each person should actually be in that predicament, during every instant of time.
>
> (Bentham, 1843: 40)

Bentham designed the panopticon as an annular multi-levelled structure where the inmates were housed in their cells; at the centre of the ring was an observation tower in which the guards were located; although the guards were greatly outnumbered by the cell occupants, they could observe the activities of each while they themselves were obscured; although all prisoners could not be subject to surveillance at once, a given prisoner could not determine if he was being watched or not; the panopticon was intended to impose self-discipline through the *possibility* of surveillance alone. Foucault adopted the structure as a metaphor to describe a *panoptic society* in which the few could observe the many, a transformation from the "civilisation of spectacle" that was antiquity, where the many observed the few (Foucault, 1991: 216) at events such as public executions (ibid., 3–7). Foucault considered state-run institutions (e.g. schools, hospitals, prisons) as exerting their power on private individuals through a *panoptic* mechanism. In such a structure, the need for "spectacular manifestations of power" (ibid., 217) is obviated. Panoptic and post-panoptic metaphors of dietary surveillance are visually represented in Figure 4.5.

Panoptic diets can be considered to consist of foods designed by the few for the many in order to discipline, control, or shape the bodies of the larger group. This food can function directly as a punitive measure in a panoptic institution (e.g. in prison), as a method to ensure survival (e.g. in hospital) or as nutritional solution for the public (e.g. meal-replacement beverage). *Food as punishment*, a strange distortion of the role of food as "the first need" (*Republic*, 369d), may be employed in cases of individual or collective failings of self-discipline. An example of this

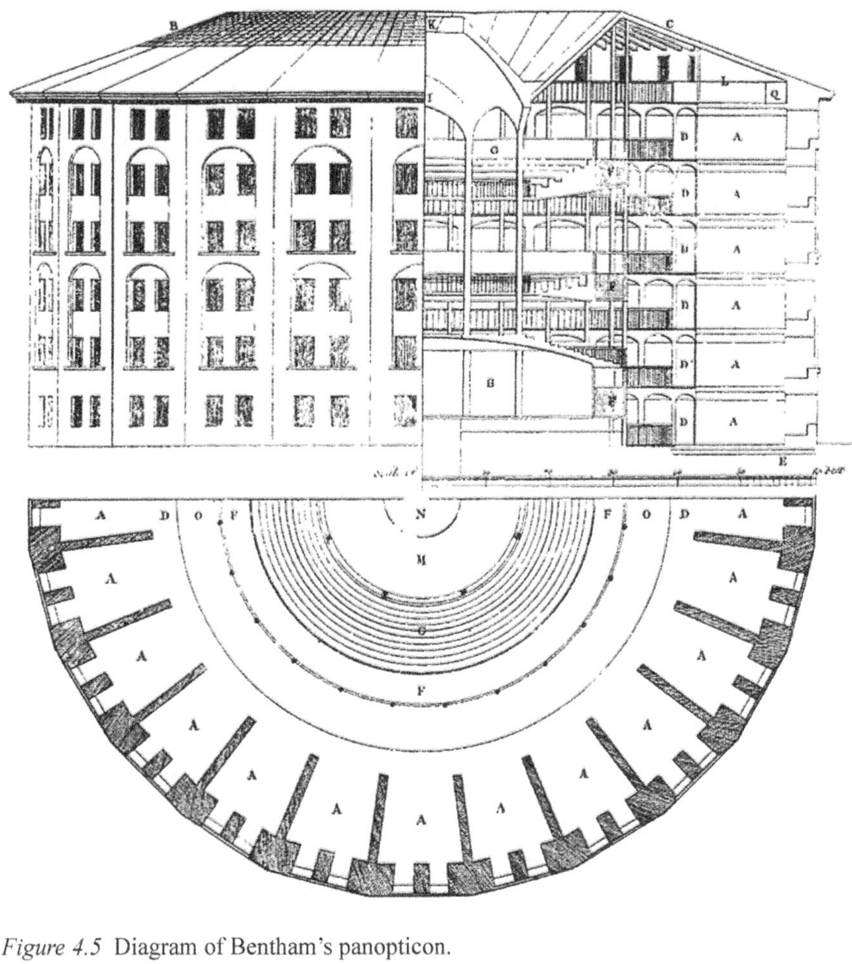

Figure 4.5 Diagram of Bentham's panopticon.

Source: Image is Public Domain content from Bentham (1843).

type of punishment food is "nutraloaf" (discussed below). Foucault considered the prison as an example of a "heterotopia", a space within society that is distinct in its otherness (Foucault, 1986). More specifically, Foucault considered the prison to be a heterotopia of deviation, in which subjects that have not complied with societal norms are placed. In this heterotopia, punitive restrictions on autonomy extend to food:

> In outside society dietary habits serve to establish and symbolise control over one's body. In prison, that control is taken away as the prisoner and their body become the objects of external forces.
>
> (Smith, 2002)

Food as punishment has long been a feature of prison life. In the now distant past, new prisoners were restricted to bread-and-water diets until they "earned the right" to eat other foods. Nutraloaf (also known as "Food loaf" and, earlier "Grue") is a more recent controversy in prison food and has been the subject of over 20 lawsuits in the US between 2012 and 2016, in which the question was not whether the food was a punishment but whether the punishment was cruel and unusual (Zoukis, 2016). In prisons where it is used, nutraloaf is an accepted punishment for food-related infractions (e.g. throwing food) of institutional disciplinary codes. In the requirements for use of nutraloaf (referred to as "Special Management Meal") provided by the state of Florida, responsibilities of those who prepare the "meal" are limited to aspects such as meeting "recommended dietary allowances", "religious and medical needs", and "sanitary" standards (Florida Administrative Code and Florida Administrative Register, 2013). Nutraloaf is a composite food with much inter-institutional variability, consisting generally of combinations of canned vegetables, minced meats, condiments, milk powders, beans, oil, water, and bread-crumbs[8]. It has also been reported that nutraloaf may be prepared from the collected leftovers of fellow inmates (Spanos, 2013). It is always formed into a block and baked before serving hot or cold. Legal arguments that Nutraloaf violates the 8th Amendment of the US constitution have been repeatedly defeated, largely off the strength of arguments in favour of its nutritional adequacy (Spanos, 2013).

Food products often develop initially for specific heterotopic contexts (military camps, prisons, hospitals, space travel) before crossing into our everyday lives (Weaver et al., 2014). *Panoptic foods* are dispensed based on the few (health professionals, commercial marketers) observing the many (populations, markets) through the relevant datasets (Couch et al., 2015). Meal-replacers, a panoptic food, can be useful in heterotopic circumstances where specific contexts (e.g. spatio-temporal restrictions) or conditions (e.g. swallowing difficulties) limit food choice. In the case of Soylent, a liquid meal-replacer aimed at healthy adults, the claim is that "both food making and the body" are "problematic aspects of life" (Miles and Smith, 2016). A vision of food as such a nutrient vehicle, strictly for bodily maintenance, is a recurring feature of utopian (or dystopian) literature. Thomas More (1478–1535) wrote that the inhabitants of his imaginary "Utopia" believed food to be worthy of enjoyment, although "purely in the interests of health" and "only as methods of resisting the stealthy onset of disease" (More, 2009: 86). The novel *Brave New World* (Huxley, 2014) abounds with references to nutrient-centric products such as "carotine sandwiches" and "'vitamin A pâté" (ibid; 152), "vitaminized beef surrogate" (217), and "magnesium-salted almonds" (226). These diets serve the *machine body*, maintaining it in *proper working order*. Visions for sustainable diets guiding technoscientific research are no longer strictly panoptic but, instead, are cyborgian. The cyborg signifies the obviation of dietary choice, an escape from the "nutritional cacophony" characteristic of the metonymic chaos of molecularised discourse. The molecules are still real, but choice is bypassed in favour of automation. This cyborg imagining symbolises a "holistic turn" for sustainable diets, where the gap between objective data and dietary intake is closed by cybernetics.

The cyborg as a (strange) holistic turn

The cyborg[9] concept originated in a 1960 article by Clynes and Kline, where the authors argued that the best way to adapt astronauts to space travel may be cybernetic enhancement of the human rather than modifications of the extra-terrestrial environment. Walmsley (2014: 181–182) suggests that humans may already be cyborgs, arguing that devices such as pacemakers exhibit the transparency (unconscious operation of cybernetic enhancement), poise (smooth and natural information exchange), and trust (knowledge of reliability of exogenous technology) befitting of true organism-machine symbiosis. So, what do cyborgs eat? This is the question Margaret Morse attempted to answer in a 1994 article. The cyborgian "smart foods" discussed by Morse, however, are classic utopian (or dystopian) fare, being highly molecularised ("reduced to its byte-sized chemical constituents like decontextualized data") and medicalised ("food-ceuticals as longevity enhancers"). Morse's vision is explicitly cyberpunk in its aesthetic and concerns the provision of corporate products to the masses in the form of dematerialised code. Miles and Smith (2016) suggest that metaphors borrowed from information theory, cybernetics and computer science have had a major influence on what they refer to as "*digital* food", citing the example of Soylent:

> Soylent, however, aims to *disconnect* the body from physical space and control it as if it were an *information processing machine.*
>
> (Miles and Smith, 2016, my italics)

Soylent is formulated from various ingredients (e.g. protein isolates, oils, minerals, vitamins) and is claimed to fully meet the nutritional needs of an adult. Interestingly, Soylent is given versions (e.g. *v1.1*, *v1.2*…) for new products, much like software updates (http://blog.soylent.com/). Soylent, and related "smart drinks" like Huel (i.e. "human fuel"), are only weakly cyborgian, however. They aim to displace '4D' tasks (Dangerous, Dirty, Dull, or Difficult), such as cooking and chewing, that could be accomplished by a machine and are a *waste* of human intellect (Walmsley, 2014: 17). In this case, the 4D tasks are outsourced to the workers in the factory, the few with access to the necessary data manufacturing the product for the many seeking technoscientific discipline. The vision for "personalised nutrition" is strongly cyborgian, a unification of bodily output/input data in a cybernetic feedback loop that obviates the need for panoptic surveillance. There is much interest in technology that can achieve so-called "personalisation" of food. Personalised food is a new and ill-defined concept. The experimental psychologist, Charles Spence, refers to the "Share a Coke" campaign, where individuals purchased bottles of Coke with their names on the label, as an example of "superficial personalisation" (Spence, 2017: 177). What I will term "personalised nutrition" has deeper practical and moral implications for how people interact with food. The vision for this type of personalisation is one where unique nutrigenomic and/or biometric feedback from each individual determines the nutritional profile of their food (Komdur et al., 2009; Ronteltap et al., 2013; Berezowska et al., 2015). A simple model of personalised nutrition is shown in Figure 4.6.

As a mechanism of dietary discipline, the cybernetic apparatus is distinct from the panopticon, although the two are not incompatible. Firstly, the actual or apparent surveillance of the subject is not necessary to ensure control; instead, the subject must simply "opt in" to the system, which will proceed with a feeding procedure based on any (mis)alignments between personal data and databases of nutritional requirements. Secondly, the mechanism is explicitly cybernetic, in that deviations from an ideal "healthy" state are corrected by a feedback loop. It is an exogenous technology that optimises homeostasis:

> Illness is entropic, irregular, an error in the living system, while healing is cybernetic, restoring the body to its original state, correcting the error.
>
> (Campbell, 1984: 23)

The system is, however, compatible with the panopticon, in that "big data" from cyborg populations can be collected, if such a practice is permitted by the users, for analysis by the few. The context of this cybernetic enhancement deviates from that imagined by the originators of the cyborg concept. Clynes and Kline (1960) were concerned with human survival during space travel and considered the development of the cyborg as optimal by virtue of its "adapting man to his environment, rather than vice versa". In our case, the "extreme" environment is a world where the impossibility of food choice places individuals at risk of contributing to personal illness ("obesogenic space") or environmental devastation ("anthropocenic space"). For example, the "*nightmare*" of increasing numbers of overweight and obese individuals, deviants from the norm, is described in mass

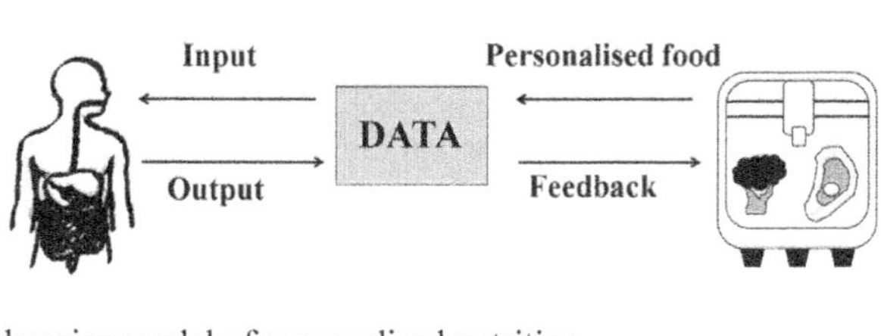

Figure 4.6 Cyborgian model of personalised nutrition.

Source: Diagram is author's own.

media using disaster metaphors (*"time-bomb"*, *"epidemic"*) characteristic of a moral panic (Couch et al., 2015). Roland Barthes, writing in 1960s France, a time of tension between gastronomic orthodoxy and the new sciences of nutrition, observed that "food has a constant tendency to transform itself into situation", not necessarily as a result of scientific rationalism but rather "mythical notions" orientated "towards *an adaption of man to the modern* world" (Barthes, 2008). Nestec S.A., a research arm and subsidiary of Nestle S.A. – the world's largest food company, have recently filed a patent under the name "System, Process and Device for Producing a Nutritional Composition with Personalized Nutrient Content", which includes the following cybernetic description:

> [The] system comprises a feedback loop after delivery of the personalized nutritional composition back to the measurement device to monitor the Subject's [10] nutrient status to ensure that the target value and the actual value of the level of the one or more nutrients converge.
>
> (US 2017/0156386 A1)

Patents for similar devices have been filed by Coca Cola Company and PepsiCo Inc., among others.

Food and non-food

The cybernetic diet promises to free the mind from dietary choice, permitting survival in a hostile space characterised by data overload ("the nutritional cacophony") and medical risk ("the obesogenic environment"). Physiological output determines nutritional input for the cyborg. The subject forms a closed loop with the object, protected from outside interference and achieving wholeness. Does this, however, negate the very idea of food, rendering it as "non-food" (Morse, 1994)? An objection to the cyborg ideal is that it omits the symbolic act of choosing an object one wishes to incorporate into the body. Humans seek food for nourishment and pleasure, yet the act of eating results in food becoming part of us, shaping our identity as well as our bodies (Mol, 2008). This has been referred to as "the principle of incorporation":

> To incorporate a food is, in both real and imaginary terms, to incorporate all or some of its properties: we become what we eat. Incorporation is a foundation for identity
>
> (Fischler, 1998)

To live we must consume (eat and drink) food, which necessitates a series of embodied transformations: outside to inside (oral intake of food); whole to deformed (chewing into bolus); conscious to unconscious (swallowing and oesophageal passage); whole to digested (formation of chyme in stomach); discreet to assimilated (intestinal absorption and tissue repair); food to waste (excretion); and from hungry to fed (satiation). This "internalisation" (the subject incorporating the object) and "destruction" (the dematerialisation of the object) is the foundation for innumerable metaphorical constructs (Newman, 1997). In its potential as external and internal, incorporated and destroyed, food can represent subject or object, us or other, and the liminal states in between (Lupton, 1996: 17), infusing food with much power and versatility as a metaphorical concept.

Van der Weele (2006) has argued that machine metaphors (e.g. machine, cyborg), while not without value, should not be allowed to hide the hedonic aspects of food and diet. Komdur et al. (2009) evaluated the "*script*" of nutrigenomics (a data source for personalisation), identifying implicit normative premises such as "The healthy life is a life without disease" and "Health is quantified risk minimisation". Citing duty- and virtue-based ethical theories, the authors suggest that 'health should be in balance with other duties and values or should only be a means to leading a virtuous life', arguing that in pursuing the 'healthist' vision of personalised diets "we may lose a rich and diverse society with a variety of interpretations of the good life". The integration of social and cultural aspects into the framework of personalised nutrition was considered by Nordström et al. (2013) to be a key factor influencing the future success of personalisation. The risk with personalised nutrition is no longer the disorientation caused by metonymic technoscientific discourse, as in Morse's smart foods, rather it is the potential ceding of dietary autonomy and the closure of the subject to the symbolic order through the displacement of choice by automation.

Alchemical imaginings for sustainable processing

Alchemy and modern science

In normal technoscientific discourse, alchemy is generally viewed negatively as an occult practice, steeped in a mysticism that was displaced by the more rigorous and successful methodologies adopted during the scientific revolution. In this sense, Gaston Bachelard considered modern chemists to be in "radical opposition" with their "pre-scientific" counterparts, the "chymists" (Bachelard, 2002: 55). Yet it is undeniable that techniques and equipment (e.g. distillation) developed by the alchemists would later find use in science. In addition, some of the general principles of the alchemical work, namely the centrality of separation and synthesis, remain a major focus of technoscientific inquiries. Perhaps more controversially, it can be argued that alchemical symbols still hold some power for practising scientists (at least unconsciously, though occasionally explicitly). The ouroborus signified the *opus* in the alchemist's reveries, yet it also signified the structure of benzene in Kekulé's dream. The physicist Martin Rees (2000) chose the image of the ouroborus to explain the relative scales of different objects in the universe, suggesting that the "gastronomic" symbol of the serpent's tail (the quantum) meeting its mouth (the cosmos) is the "ultimate synthesis that still eludes us". In his book *The Beginnings of Infinity*, another physicist, David Deutsch (2012; 425), regrets that the alchemists lacked sufficient knowledge to achieve transmutation; this goal, Deutsch argues, will inevitably be achieved by humans, who he identifies as 'universal constructors' (ibid., 157). In just these two examples, we witness scientists directly appealing to alchemical symbols and metaphors when envisioning some ultimate synthesis.

Bachelard characterises the alchemists as repeatedly finding "their precious matter in the 'belly of corruption', just as miners seek it in the impure belly of the earth" (Bachelard 2002; 200). The alchemists subscribed to a metaphysics in which "the highest is in the lowest and the lowest in the highest" (Agrippa, 1510/2002), hence, materials of the sun (gold) and moon (silver) could be found in the earth. Paracelsus, like many alchemists, believed that God had designed a world of chaotic elements, and that skilful alchemy could order those elements into remedies for the physician's practice (Paracelsus, 2002a/1520–1530s). Aspects of classical Greek metaphysics has a strong presence in alchemical texts, particularly the division of matter into four elements (earth, water, air, fire) following Plato (*Timaeus*, 33a–d), and the doctrine of four causes (material, efficient, formal, final) following Aristotle (*Physics*, II, 7, 198a14–32). Paracelsus (2002b/1520–1530s) conceived the world as *prima materia* (Aristotle's material cause), in which virtues remain concealed until the alchemist discloses its hidden secrets; by subjecting the *prima materia* to heating (efficient cause), it was possible to generate the *quinta essentia* or "spirit" (formal cause), thereby carrying "to its *end* something that has not yet been completed" (the final cause). The great physicist Isaac Newton echoed many of these ideas in his alchemical work (Newman, 2015), which occupied him for large periods of his life (Rickey, 1987), yet barely a single trace of his alchemical thought remained in the first edition of the *Principia*, his masterwork, which included a hypothesis regarding the possibility of transmuting one kind of matter into any other kind that was removed from subsequent editions (Cohen and Whitman, 1999: 59–61). Newton's deliberate cleavage of his scientific work, which focussed on matter and motion (material and efficient causality), from his metaphysical work, which focussed on essences and purpose (formal and final causes), is characteristic of the "split" nature of modern science. Newton's contemporary, Francis Bacon had advocated for such an epistemological division in in his *Novum Organon* (1902/1620: 119) and *Advancement of Learning* (1930/1605: 90–94). For Bacon, physics (material and efficient) and metaphysics (formal and final) dealt with different causes and the two should not be mixed. Although Karl Popper's falsificationism upheld the Baconian demarcation of science from metaphysics he did not argue for a Humean casting of metaphysics "to the flames" (Hume, 1998: 120); rather, Popper asserted that metaphysical claims can indeed inform scientific inquiry in the discovery phase, though not the justification phase, as in the case of Greek atomistic theories that pre-dated evidence of the atom by centuries (Popper, 2002: 16).

Finding unity in chaos

In circular economics waste generated during one process is "up-cycled" in a subsequent process. Advocates of this approach, also known as "cradle to cradle", have promoted the idea that "waste is food" through appeals to metabolic metaphors (Braungart and McDonough, 2008). In their *Manifesto for a Resource-Efficient Europe*[11] the European Commission envisions a transition to a circular economy involving "use of waste as raw materials" and "moving waste up the value chain". One of the earliest proponents of circular-style model of economics was Kenneth Boulding. In his *The Economics of the Coming Spaceship Earth*, he wrote that humans for much of their history have had an image of a frontier, beyond which existed a limitless plane (Boulding, 1966: 1). Gradually, it was recognised that the biosphere is a closed system, but the psychological, moral, and political transition to this reality has been slow and remains incomplete (ibid., 1–2). He uses the metaphor of the spaceship – a closed system of finite resources – a place necessitating 'a system [where] all outputs from consumption would constantly be recycled to become inputs for production' (ibid., 5). This 'spaceman economy' is opposed to the 'cowboy economy' (ibid., 7), in that humans must establish themselves in a 'cyclical ecological system' (ibid., 8), rather than chasing the *"gold"* that lies beyond the frontier. Yet Boulding was critical of economic models that privileged production and consumption, and called for measures of economic success that considered "... the nature, extent, quality, and complexity of the total capital stock, including in this the state of the human bodies and minds included in the system" (ibid., 8). Circular economics is typically signified using a loop where the beginning is linked to the end (Figure 4.7). Notably, this *ouroborus* represents a partial unification, with an arrow off to the side accounting for inevitable imperfections.

Circular economics dictates that waste signifies *goldmine*, *resource,* or *biorefinery* to scientists and engineers tasked with its systematic reordering. Such top-down attempts to valorise (perceived) waste are not unprecedented. In WWII-era USA during a time of rationing, the Department of Defence and National Research Council established the Committee on Food Habits to identify dietary sources of protein that could replace meat for malnourished citizens (National Research Council (US) Committee on Food Habits, 1943). Their solution was "organ meats", also known as "offal", the etymology of the latter, incidentally, traces back to waste or "that which *falls* or is thrown *off*" (off+al[l] – OED, 2018). As "... some people perceived organ meats as useless parts of livestock to be discarded... " plans were drawn up to "restructure social norms" by, for example, making the act of offal consumption signify a patriotic commitment to one's country (Wansink, 2002). Braungart and McDonough (2008) urge us to "imitate nature's highly

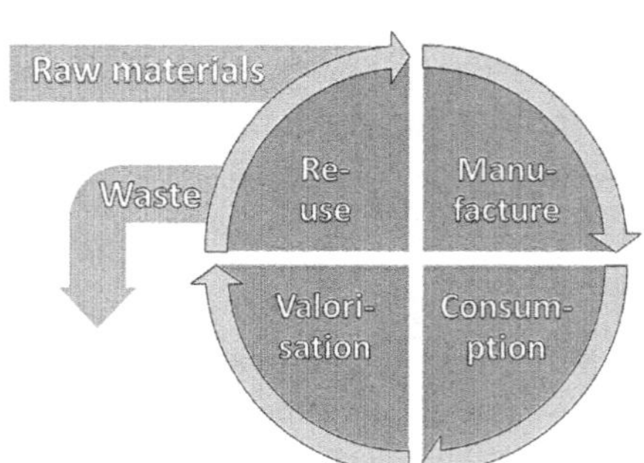

Figure 4.7 The circular economy concept.

Source: Diagram is author's own.

effective cradle-to-cradle system of nutrient flow and metabolism, in which the very concept of waste does not exist" (Braungart and McDonough, 2008: 90). This is an image of nature that extends back to at least Plato, in which reference is made to a world that "uses its own waste for its food" (*Timaeus*, 1238). When circular economics is conceived of alchemically, however, a different picture emerges. As Bachelard suggests, the alchemist must value waste, not seek its erasure. For Jung (1967, 1968, 1970), alchemy is a therapy for its practitioner, who experiences an internal chaos. Following Zwart, such a chaos can have the demands of an "impossible profession" as its cause. The alchemical metaphor bears insights of a metaphysical (the potentiality of oppositions), technological (the orderability of chaos), and psychological (the individuation of the self) nature.

In the rhetoric of sustainable production, the circular economy is an image signifying a work of separation and synthesis. It is an alchemical unification of opposites (waste and food), a *coniunctio oppisitorum*, that guides researchers towards a "holistic turn" and a resolution of contradictions (Zwart, 2018, 2019: 33). This can be conceived as both a technical and therapeutic exercise for those involved. The Swiss psychiatrist Carl Jung studied alchemy as a source of spiritual and psychological symbolism, rather than simply as a precursor to modern chemistry (Jung, 1968: 23). Alchemical procedures consisted of two principal stages, in which materials would be subject to separation and synthesis "for only separate things can unite". (Jung, 1970: 471). To Jung, alchemy was not simply the attempt to transmute matter (e.g. from base metal into gold), it involved the projection of the alchemist's unconscious into the matter that was to be transformed (Jaffé, 1964: 292; Jung, 1968: 67). Jung considered the alchemical process to represent psychological "individuation", in which opposing elements of the self become integrated into a unified whole. Jung identifies the mandala (Sanskrit for "magic circle") as the symbol for this (re)turn to wholeness, an archetypal symbol recurring in the collective unconscious via dreams, imaginative works, and alchemy. Such circular images abound in alchemical texts; for example, gold was represented symbolically as an open circle (signifying that it was "of the sun"). The individuation process and the symbol of the mandala has been identified in "holistic turns" in science and technology studies where "after fragmentation and specialisation, time has now come to put the pieces back together again" (Zwart, 2018).

Technoscientific discourse in food processing often features a separation of the subtle from the gross. A highly cited article in the scientific literature reviews the extraction of valuable components from whey, formerly a waste product of cheese-making, is titled "From Gutter to *Gold*" (Smithers, 2008). Another review article in the field of dairy science is titled the "*Mining* of Milk" (Mills et al., 2011). Elsewhere trade magazines for the food industry refer to whey as a potential *goldmine* [12] . The alchemical preoccupation with "the seed of unity which lies hidden in the chaos" (Jung, 1968: 25) can thus be found in current discourse on the circular economy. Our bifurcation of matter has led to a distressing chaos of waste, a frustrating crisis that challenges technoscience and leads to a resurgence of the imaginary (Zwart, 2019: 48). In alchemy, such chaos is the *prima materia* from which the whole is constructed. In describing the *opus* (the work of alchemy), Jung says the following:

> Time and time again the alchemists reiterate that the *opus* proceeds from the one and leads back to the one, that it is a sort of circle like a dragon biting its own tail [...] For this reason the *opus* was often called circulare (circular)
> (Jung, 1968: 293)

The dragon Jung mentions is the ouroborus, one of the oldest and best known of the alchemical symbols (see Figure 4.3). The circular economy as an ouroborus is a whole constructed from the object (food) and the abject (waste). Waste becomes pure potential.

Waste-free but wasteful

One aspect that remains hidden in the alchemical imaginings of the circular economy is the vice of wastefulness. Consumption replaces the alchemists' God as the generator of *prima materia*. Some authors have suggested that circularity leads to a "market valuation of the wasted", in which excessive wastefulness becomes positivised in the growth-driven capitalist framework (Valenzuela and Böhm, 2017). In his Nicomachean Ethics (NE), Aristotle discusses wastefulness in the context of getting and using money, but he makes clear that this should be understood as applying to anything the value of which can be measured monetarily (*NE*, IV.1.1195b, 27–28). According to his "doctrine of the mean", two types of wastefulness (prodigality and vulgarity) may be considered vices of excess (Table 4.2). Here, we will concern ourselves with the prodigal person.

A prodigal person is guilty of "wasting his substance; because a prodigal man is one who is ruined by himself, and the wasting of a man's property is considered to be a sort of self-destruction [...]" (*NE*, IV.1. 1120a, 1–4). Following Aristotle, Aquinas in his *Summa Theologica* considers *consumptio* (Latin for "waste") the vice of excess that corresponds to the opposing vice of deficiency, "meanness". He also uses the Greek term *apyrokalia* (i.e. "lacking good fire"), as "like fire it [the vice of wastefulness] consumes all, but not for a good purpose" (*Aquinas*, 2018, 43–44). When considering waste, we are apt to consider its generation, but, on the Aristotelian-Thomist account, wastefulness can arise in the act and manner of consumption [13]. If waste is generated due to our consumption habits, then perhaps we need to reflect on those habits rather than render waste another

Table 4.2 Vices of excess corresponding to wastefulness, their opposite vices of deficiency, and their respective means of virtue. Minor/major indicates lesser/greater quantities of material at the disposal of the agent.

Sphere of Action or Feeling	*Excess*	*Mean*	*Deficiency*
Getting/using (minor)	prodigality *asōtia*	liberality *eleutheriotēs*	illiberality *aneleutheria*
Getting/using (major)	vulgarity *apeirokalia* or *banuasia*	magnificence *megaloprepeia*	pettiness *mikroprepeia*

consumable. Augustine (2016) recounts a personally shameful episode when he and his "accomplices" stole pears from an orchard, most of which they later fed to pigs (*Confessions*, II, 24). In this story, the teenage Augustine is wasteful in that he both consumes what he does not need and discards what could have been needed by others. Augustine reflects that he stole not out of a desire for pears, of which an abundance of superior kinds were available in his father's garden, but for the thrill of the sin itself. This episode can be interpreted as an allegory with allusions to The Fall (stealing of fruit) and The Prodigal Son [14] (feeding of pigs). Later in the *Confessions,* Augustine treats hunger and thirst as "pains" that "burn and kill" unless relieved by a "medicine of nourishments" (ibid., X, 179). He confesses that in seeking nourishment his body has occasioned to deceive him into an activity of gratification disguised as the pursuit of health. As per Aristotle's account of restorative pleasure, the need to consume (food) is cyclical but self-limiting (Smith, 2001: 239–240), and to go beyond the limit is wasteful and self-destructive. From an Aristotelian perspective, it is not a matter of limiting waste but of limiting wastefulness. If the food scientist transforms a waste material into a protein supplement, and this merely leads to affluent consumers exceeding their protein requirements further, have we really become less wasteful on the whole?

Mary Douglas borrowed from William James the notion that dirt can be defined simply as "matter out of place". Waste, similarly, can be categorised in this way. Douglas asserted that this designation was a feature of a larger (symbolic) system, stating that "Dirt is the by-product of a systematic ordering and classification of matter, in so far as ordering involves rejecting inappropriate elements". (Douglas, 2001: 36). Hence, our concepts of clean and dirty are fundamentally relational. We are alerted to a danger when the dirty object threatens to cross a symbolic threshold to the clean subject. Julia Kristeva calls such a "jettisoned object" the *abject*, citing biological fluids and cadavers as examples of "what disturbs identity, system, order…" (*Powers of Horror: An Essay on Objection*, 1982: 2). Natural matter can also exist "out of place", such as the weed that "invades" the cultivated garden. Morse summarises Kristeva's response to physical expression of "food loathing", her spitting out milk with a surface skin that she found nauseating, as an act that "… establishes a limit, a not-self, but at the cost of expelling a substance, food waste, that is ambiguously self and not-self". Such abjection is no longer commonplace, due to "modern networks of disposal" that render waste invisible and the "difficulty people have in dealing with life-process issues" (Ewen, 2009: 401). The circular economy, rather than masking material waste in an invisible disposal network, seeks to mask our wastefulness through the valorisation of waste. Just as the cyborg confuses the boundary between human and machine, alchemy confuses the boundary between the object and abject.

The dynamics of sustainable food discourse

Metaphors as roadmaps for discovery

In a Jungian-Bachelardian investigation of research in synthetic biology, Zwart (2018) observed an initial drive to reduce the living organism to a *code* being superseded by efforts to (re)build the organism from that code into a synthetic *cell*; here, Zwart sees a dialectic at play in which a reductionist approach initially fragments the whole into code before a "holistic turn" orientates scientists towards a recovery of the whole. An analogous dialectic is witnessed in sustainable food: food is first negated (the initial whole is fragmented), which leads to the negation being negated (components are combined into a sustainable "whole"). Metaphorical representations of the cell as a *mandala* in the field of synthetic biology are viewed by Zwart as useful "*roadmaps*" for research in synthetic biology to converge in its complex "*journey*" to wholeness (or individuation). Just as Zwart considers the mandala as an "ideal end state of convergence" in synthetic biology, circular images can be seen to guide food researchers back to some idea of the whole (Figures 4.3, 4.6, 4.7). Such "ideal end states" are here found in the form of the cyborg (signifying cybernetic nutrition) and the spaceship (signifying circular economics). The noise of nutritional information is reduced through the formation of a cyborgian whole, while the chaos of food waste is reduced through its alchemical synthesis with food. These holistic metaphors point to a better future and are a response to the reductionist metonymies of the past. In what follows I will attempt to use a Lacanian approach to analyse this discourse, with a view to presenting a unified picture of how it evolves through time under the (partial) influence of the imaginary and metaphorical.

Introduction to Lacan's four discourses

In Lacan's Seminar XVII (*The Other Side of Psychoanalysis*), he outlines a novel quasi-algebraic structure within which four principal discourses can be formed. In a recent monograph, Zwart applied this method to analyse technoscientific discourse (Zwart, 2019). To close this chapter, I will apply the method in a diachronic analysis of technoscientific discourse in food, with the aim of probing the tension between the scientific ideal and the imaginary.

Lacan's method replaces the disinterested inquirer of Enlightenment science with the desiring subject ($), an individual "paralysed by uncertainties" and "suffering from a basic inability of lack" (Zwart. 2016). In place of Kant's split object (both as *phenomenon* and *noumenon*) we get the desired object (*a*), which is both the 'target and 'cause' of human desire. The relationship between subject and object is symbolised as the 'matheme of desire' – *$ ◊ a* – where a two-placed relation (◊) signifies that the object induces a desire to which the subject submits. Scientists are thus drawn to an impossible object or panacea, such as meals-in-a-pill (panopticon), personalised nutrition (cyborg), or circular economy (alchemical). Each of four basic symbols, representing the Master (S_1), the University (S_2), the Hysteric ($), and the analyst (*a*), can be inserted into the following structure to generate four discourses, named after signifier occupying the Agent position.

Agent	(Desired) Other
(Disavowed) Truth	By-product

In pursuing the Other, the Agent is driven by some unconscious Truth, indicated by the position of the disavowed Truth under the bar. The pursuit itself often leads to the generation of a By-product, which, being unexpected is also positioned under the bar. Lacan outlines four discourses, that of the Master, University, Analyst, and Hysteric. The basic four-part structure remains fixed, but the internal arrangement undergoes shifts. During such shifts, there is a change in the position of the four signifiers, S_1 (master signifier), S_2 (expert knowledge), \$ (divided subject), and *a* (object of desire), signifying a change in their discursive relations. Lacan referred to each transition as a "quarter turn", in which the signifiers shift one place in the clockwise direction. The diachronic structure is represented in Figure 4.8, fitting in its circularity, in which each discourse is symbolised quasi-algebraically.

The Master's discourse is based on the words of some master teacher, such as Aristotle or Hippocrates (Zwart, 2016), and is symbolised as follows:

$$\uparrow \frac{S_1}{\$} \rightarrow // \frac{S_2}{a} \downarrow$$

The authority (S_1) can be a figure (e.g. the philosopher, the messiah) or their words (e.g. the Aristotelian Corpus, The Bible). By exploring such works, one may become a scholar (S_2), competent in the art of studying the truths imparted by S_1. A by-product (*a*) of this activity is the generation of joy for the scholar, in the form of hard-won insights (Zwart, 2016). What the scholar ignores is that the authority itself was a divided subject (\$), with doubts and uncertainties, though this may not be readily apparent from their writings. Thus, the master is misconceived as a signifier of pure objective truth.

The Master's Discourse may give way to University Discourse when the scholars themselves become the generators of knowledge. The projects of Descartes and Kant were fuelled by an impassioned rejection of the Master's Discourse. Descarte's effort to establish an indubitable foundation for rational inquiry stemmed from his doubting of scholastic Aristotelianism, while Kant's answer to the question 'What is Enlightenment?' advocated for an emancipation from

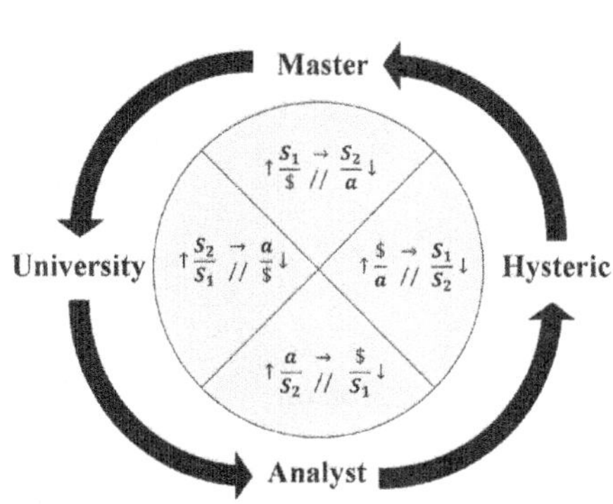

Figure 4.8 Lacan's diachronic model of the four discourses.

Source: Diagram is author's own.

institutional dogma in favour of intellectual autonomy. In University Discourse, the expert (S_2) becomes the agent:

$$\uparrow \frac{S_2}{S_1} \rightarrow \frac{a}{\$} \downarrow \quad (S_1 \,//\, \$)$$

The expert (S_2) applies robust techniques to study their object of interest (a). The master no longer initiates knowledge generation, signified by the master's (suppressed) position below the bar ($\frac{S_2}{S_1}$). The difficulty of the inquiry into the object of desire may result in the subject becoming divided ($). The scientist may feel that s/he is incompetent, or that their profession is impossible. The master signifier (S_1) may still exert an influence, in the form of an extant worldview or private imagination. For example, Isaac Newton made his historical achievements in physics while secretly practising alchemy, an activity suppressed by the Baconian institution of science. Newton kept his alchemical works (informed by Aristotle's metaphysics) secret while making his physics (informed by Bacon's epistemology) public.

The authority that has emerged in University Discourse may itself come to be challenged, by a divided subject ($) dissatisfied with technoscientific paradigms (e.g. dualism, reductionism). The resultant Hysteric's[15] Discourse may be rejected by the embattled authority (S_1), who finds the voice of protest to be irrational and based on baseless fantasy. What drives the protester is an archetypal image (a), a fear of some monster or great catastrophe. If, however, the protesting voices are heard, then academic discourse may absorb its concerns, leading to reflection and the potential for reform (signified, perhaps, by holistic imagery).

$$\uparrow \frac{\$}{a} \rightarrow \frac{S_1}{S_2} \downarrow \quad (a \,//\, S_2)$$

The final discourse, that of the Analyst, allows one to investigate why the subject ($) is provoked by the object of inquiry (a), which has developed a monstrous aspect. The priority of expertise (S_2) is suspended and in favour of a focus on the relationship between Agent and Other. By engaging with the subject paradoxes and contradictions may be revealed, but rather than this being cause to dismiss their concerns, an epistemologically productive *aporia* is reached. A by-product of the resultant dialectic may be the insertion of a new Master (S_1), instigating a return to the Master's discourse.

$$\uparrow \frac{a}{S_2} \rightarrow \frac{\$}{S_1} \downarrow \quad (S_2 \,//\, S_1)$$

Discourse analysis: The return to Holism

Holistic (re)turns in science can be traced diachronically through a series of structural changes in the nature of technoscientific discourse. After the scientific revolution, heroes of the scientific ideal became epistemic exemplars (S_1). Newtonian and Cartesian scholars (S_2) followed their heroes, each considered to have presented important truths in a *magnum opus* (the *Principia* and *Meditations*, respectively). The scholars, however, would come to suppress the split nature ($) of these famous thinkers, both being hybrids of the natural scientist and the metaphysician; for example, Newton was a practising alchemist and Descarte's *cogito* argument was premised partially on theological grounds. The byproduct of their followers' scholarship was the revelation of useful tools of scientific inquiry, such as Newton's calculus and Descarte's algebraic geometry.

S_1 (Scientific heroes – Newton/Descartes)	S_2 (Followers of the reductionist tradition)
$ (Suppression of their natural philosophy)	*a* (Useful tools – laws, methods, apparatus)

Having harnessed the tools of their masters, scientists gained an unprecedented ability to symbolise and quantify natural objects and forces. This led to a dislodging of the master as oracle of truth, with university scientists themselves becoming generators of knowledge (S_2). Metaphysical aspects (e.g. formal and final causality) of the former masters' thought were partially suppressed in modern science. The image of humans as machines/factories remained (following Descartes), a background metaphor that profoundly affected developments in physiology, medicine, and nutrition. The unconscious influence of alchemy persisted (after Newton), exemplified by the ouroborus' unlocking the mystery structure of benzene for August Kekulé (Benfey, 1956). The success of modern technoscience led to a proliferation of technological artefacts and associated waste (*a*), while yielding economic (commercial profit) and academic (peer-reviewed publications) as "surplus value". Ultimately, both promising technologies (e.g. GM foods, nutraceuticals) and industrial by-products (e.g. waste) were rejected by the public, leading to conflicted experts ($) uncertain about how their discipline can progress.

Discourse of the university

S_2 (The scientific discipline)	*a* (Scientific objects, technologies, waste)
S_1 (Hidden imperative – rejection of images)	$ (Public rejection of scientific knowledge)

For a scientist to reflect on the situation, an oblique perspective must be adopted, in which their expertise (S_2) is placed in a subordinate position (Zwart, 2019: 71). What is it about their "vision" of the impossible object, *a* (i.e. the rational cyborg, the waste-food synthesis) which leads to their desperation ($)? The scientists can be prompted to such reflections by identifying underlying conceptual metaphors that might structure their thinking. Such a reflection can prompt the scientist to consider alternative sources of truth, such as those found in great philosophers (S_1), who deal directly with categories (e.g. the whole, the natural, the beautiful, the good) disavowed by science.

Discourse of the analyst

a (The impossible research object)	$ (Desperate scientists, seeking a new way)
S_2 (Suppressed scientific orthodoxy)	S_1 (Return to axiology and metaphysics)

The scientists may be prompted to such reflection by protesting voices ($) from a technophobic public, activist journalists, or whistle-blowers, who are driven by an intangible object of fear (*a*) that is difficult to articulate. Criticisms may also emerge from the human sciences, with insightful remarks from scholars problematising scientific discourse on topics such as automation and (un)sustainability. The scientific establishment (S_1) may reject these criticisms, dismissing "irrational" protests driven by pre-scientific imagery (of Frankenstein's monster) or "magical thinking" (Roosth, 2013). Frankenstein is a typical monster archetype (Zwart, 2019: 33), emerging in an age of technophobic fears of synthetic threats. Alternatively, scientists can acknowledge the fears people express and reflect on their practice; for example, the franken-food criticism suggests a mereological anxiety, in which aberrant wholes are created from disparate parts. Such reflections may assist in the developments of new types of expertise (S_2), a more holistic inquiry, whatever that might be. Maybe it is a purposeful revisiting of classical hylomorphic theories, as designers often do (see Buchanen, 1999), to attend to both the metaphysics of form/function and the science of matter/motion. Though these were firmly demarcated by Bacon, and suppressed in the Cartesian/Newtonian worldviews, they re-emerge united in the metaphors and images that guide scientific discovery in a time of crisis.

Discourse of the hysteric

$ (Voices of dissent)	S_1 (Scientific authority under attack)
a (Fear of monster archetype)	S_2 (Emergence of "holistic" technoscience)

Conclusion

Bachelard was scathing in his criticism of metaphors, which he considered "epistemological obstacles" to scientific understanding. Once adopted, he claimed only an 'autonomous kind of thought' that came to 'completion in the realm of images' could result (Bachelard, 2002: 88). The imaginary is not the preserve of non-scientists, however, though it is frequently suppressed in technoscientific discourse. Historically, the imaginative aspects of normal science are displaced by mathematical symbolism. In the current epoch, characterised by complexity and fragmentation, images are returning to the forefront of technoscientific discourse. Two prominent utopian imaginings in the area of sustainable food are personalised nutrition and circular economics. In this chapter, I have suggested that both can be understood using metaphors that signify a "holistic turn". Personalised nutrition imagines *humans as cyborgs*, while the circular economy imagines *food production as alchemy*. Considered diachronically, both can be conceived of as response to a metonymic chaos, characteristic of reductionist, and molecularised representations of food. The cyborgian consumer escapes the nutritional cacophony of technoscientific jargon while the alchemical scientist escapes the chaos of industrial waste. Both metonymic discourse and metaphoric discourse have the power to defamiliarise and make us *see* what has become familiar and automatic in a strange aspect. By studying these conceptual metaphors, we can identify problematic entailments: human-machine and waste-food hybridisation challenge our dualistic notions of self|other and object|abject; cyborgian futures privilege the mind over the body and potentially neglect the role of food choice in identity formation. The work of alchemists necessitates a valorisation of waste (psychologically) rather than simply the elimination of waste (materially). The oblique, defamiliarising power of these metaphors helps to reveal aspects of technoscientific discourse that silently influence our efforts to transition to a more sustainable society.

Notes

1 That food is a prominent source and target of metaphors in everyday language and scientific discourse has been the subject of excellent articles by Korthals (1998) and van der Weele (2006).
2 The term 'semiology' is typically associated with the Saussurian approach, but here I use 'semiotics', which is effectively synonymous and common in academic usage.
3 Saussure actually used the term 'associative', but Jakobsen replaced this with 'paradigmatic', and this remains more common in academic usage.
4 Note: figure contains an element "File:Benzene Ouroboros.png" by DMGualtieri, which is licensed with CC BY-SA 3.0. To view a copy of this license, visit https://creativecommons.org/licenses/by-sa/3.0.
5 All citations of Plato refer to classical writings (book names indicated with italics) that appear in his Complete Works (Plato, 1997). The citations are written in the conventional style using Stephanus numbers, which is the standard form of citation for Plato's corpus
6 All citations of Aristotle refer to classical writings (book names indicated with italics) that appear in his Complete Works (Aristotle, 1995). The citations are written in the conventional style using Bekker numbers, which is the standard form of citation for Aristotle's corpus
7 Further commentary on the nature of Fritz's precise role in this and other illustrations is provided in Borck (2007).
8 In court filings, the Illinois Department of Corrections provides the following ingredients for 'Meal loaf': Ground Beef; Canned, Chopped Spinach; Canned Carrots, Diced; Vegetarian Beans; Applesauce; Tomato Paste; Potato Flakes; Bread Crumbs; Dry Milk Powder; Garlic Powder; or Flakes.
9 **Cyb**[ernetic]**org**[anism]
10 'The subjects may comprise infants, children, adults, and elderly people. The subject may be healthy or suffering from a disease, e.g. the subject may be a normal healthy subject, or a nursing home resident and/or a bed-ridden person. The term may also comprise animals, in particular companion animals such as a cat or dog' (taken from US 2017/0156386 A1).
11 Available at: http://europa.eu/rapid/press-release_MEMO-12-989_en.html
12 https://www.foodnavigator-usa.com/Product-innovations/Acid-whey-isn-t-waste-it-s-a-goldmine
13 The etymology of 'consumption' traces its earliest meaning as referring to the wastage or decay of bodies (OED, 2009).
14 The parable of the prodigal son (Luke 15:11-32) describes a wayward son who squanders his inheritance before getting a job feeding pigs where 'He longed to fill his

stomach with the pods that the pigs were eating, but no one gave him anything' before being reconciled with his family.

15 Not to be interpreted in a pejorative sense.

References

Agrippa, C. (1510/2002). De Occulta Philosophia, in, *Science in Europe, 1500–1800 – A Primary Sources Reader*. M. Olster (ed.). Hampshire: Palgrave MacMillan, p. 94.

Aquinas, T. (2018). *Summa Theologica: Third Part of the Second Part*. Translated by Father of the English Dominican Province. Ontario, Canada: Devoted Publishing.

Aristotle. (1995). *The Complete Works: Revised Oxford Translation*. Translated by J. A. K. Thomson. J. Barnes (Ed. and trans.). Princeton: Princeton University Press.

Augustine (2016). *Confessions*. Translated by E. B. Pusey, Hertfordshire: Wordsworth Editions Limited.

Bachelard, G. (2002). *The Formation of the Scientific Mind: A Contribution to a Psychoanalysis of Objective Knowledge*. M. McAllester Jones (trns.). Manchester: Clinamen Press Ltd.

Bacon, F. (1902). *Novum Organum*. New York: P. F. Collier and Son.

Bacon, F. (1930). *On the Advancement of Learning*. London: J. M. Dent & Sons.

Barthes, R. (1983). *The Fashion System*. Translated by M. Ward and R. Howard. California, USA: University of California Press Berkeley and Los Angeles.

Barthes, R. (2008). Towards a pyschosociology of contemporary food consumption. In *Food and Culture: A Reader*, 2nd edition, by C. Counihan and P. Van Esterik (eds.). New York: Routledge, 2008, 28–35.

Benfey, O. T. (1956). August Kekulé and the birth of the structural theory of organic chemistry in 1858. *Journal of Chemical Education*, 35, 21–23.

Bentham, J. (1843). *The Works of Jeremy Bentham Vol 4*. Edinburgh, UK: William Tait.

Berezowska, A., Fischer, A. R. H., Ronteltap, A., van der Lans, I. A. and van Trijp, H. C. M. (2015). Consumer adoption of personalised nutrition services from the perspective of a risk-benefit trade-off. *Genes and Nutrition*, 10, 42.

Boland, M. (2016). Human digestion - a processing perspective. *Journal of the Science of Food and Agriculture*, 96, 2275–2283.

Borck, C. (2007). Communicating the modern body: Fritz Kahn's popular images of human physiology as an industrialised world. *Canadian Journal of Communication*, 32, 495–520.

Bornhorst, G. M., Gouseti, O., Wickham, M. S. J., and Bakalis, S. (2016). Engineering digestion: Multiscale processes of food digestion. *Concise Reviews in Food Science*, 81, 534–543.

Boulding, K. E. (1966). The economics of the coming spaceship earth. In *Environmental Quality in a Growing Economy*, H. Jarrett (ed.). Baltimore, MD, USA: Resources for the Future. Maryland: Johns Hopkins University Press.

Braungart, M., & McDonough, W. (2008). *Cradle to Cradle*. London: Penguin

Buchanen, R. (1999). Design research and the new learning. *Design Issues*, 17, 3–23.

Campbell, J. (1984). *Grammatical Man: Information, Entropy, Language and Life*. London, UK: Pelican Books.

Carpenter, K. J. (2003). A short history of nutritional science: Part 1 (1785–1885). *The Journal of Nutrition*, 133, 638–645.

Clynes, M. E., and Kline, N. S. (1960). Cyborgs and space. *Astronautics*, September, 26–27 and 74–76.

Cohen, B., and Whitman, A. (1999). *Some General Aspects of the Principia, In, Isaac Newton's The Principia: Mathematical Principles of Natural Philosophy – translated by B Cohen and A. Whitman*. Berkley: University of California Press.

Couch, D., Han, G.-S., Robinson, P., and Komesaroff, P. (2015). Public health surveillance and the media: A dyad of panoptic and synoptic social control. *Health Psychology and Behavioural Medicine*, 3, 128–141.

Descartes, R. (1985). *Discourse on Method and Meditations on First Philosophy*. Translated by D. A. Cress. Indianapolis, USA: Hackett Publishing Company.

Deutsch, D. (2012). *The Beginnings of Infinity*. London: Penguin.

Douglas, M. (2001). *Purity and Danger: An Analysis of Concepts of Pollution and Taboo*. New York: Taylor & Francis.

Ewen, S. (2009). Form follows waste. In *Earthcare: An Anthology in Environmental Ethics*, D. Clowney & P. Mosto (Eds.). Lanham, MD, USA: Rowman & Littlefield Publishers, Inc. pp. 399–405.

Fischler, C. (1988). Food, Self and Identity. *Social Science Information*, 27, 275–293.

Fischler, C. (1992). From lipophilia to lipophobia. Changing attitudes and behaviours toward fat – a sociohistorical approach. In *Dietary Fats*, pp. 103–115, D. Mela (Ed.). London, UK: Elsevier.

Fischler. C. (2011). The nutritional cacophony may be detrimental to your health. *Progress in Nutrition*, 13, 217–221.

Florida Administrative Code and Florida Administrative Register (2013). *Rule 33–602.223, Special Management Meal*. Retrieved on 9/3/18 from https://www.flrules.org/gateway/RuleNo.asp?ID=33-602.223

Foucault, M. (1986). Of other spaces. Translated by J. Miskowiec. *Diacritics*, 16, 22–27.

Foucault, M. (1991). *Discipline and Punish: The Birth of the Prison*. Translated by A. Sheridan. London, UK: Penguin Books.

Freud, S. (1937). Analysis Terminable and Interminable. *International Journal of Psychoanalysis*, 18, 373–405.

Fury, A. (2016). The return off the '90s. *New York Times*. Retrieved online on 9/3/18: https://www.nytimes.com/2016/07/13/t-magazine/fashion/90s-fashion-revival.html

Hume, D. (1998). *An Enquiry Concerning Human Understanding*. Oxford, UK: Oxford University Press.

Huxley (2014). *Brave New World*. London: Vintage.

Jaffé, A. (1964). Symbolism in the visual arts. In C. G. Jung, M.-L. von Franz and J. Freeman (Eds.) *Man and His Symbols*. 255–322. New York: Dell Publishing.

Jakobsen, R. (1956). Two aspects of language and two types of aphasic disturbance, in, *Fundamentals of Language* (eds., R. Jakobsen and M. Halle). The Hague: Mouton & Co.

Jayasinghe, S. (2012). Complexity science to conceptualize health and disease: Is it relevant to clinical medicine. *Mayo Clinic Proceedings*, 87, 314–319.

Jung, C. G. (1967). Alchemical Studies, 2nd ed. In H. Read, M. Fordham, G. Adler, and W. McGuire (Eds.), R. F. C. Hull (Trans.) *The Collected Works of C. G. Jung*, Vol. 13. London and Henley: Routledge & Kegan Paul.

Jung, C. G. (1968). Psychology and Alchemy, 2nd ed. In H. Read, M. Fordham, G. Adler, and W. McGuire (Eds.), R. F. C. Hull (Trans.) *The Collected Works of C. G. Jung*, Vol. 12. London: Routledge & Kegan Paul.

Jung, C. G. (1970). Mysterium Coniunctionis: An Inquiry into the Separation and Synthesis, 2nd ed. In H. Read, M. Fordham, G. Adler, and W. McGuire (Eds.), R. F. C. Hull (Trans.) *The Collected Works of C. G. Jung*, Vol. 14. London: Routledge & Kegan Paul.

Kafka, F. (2007). *Metamorphosis & Other Stories*. M. Hofmann (Trans.). London, UK: Penguin Books.

Knorr, D., and Watzke, H. (2019). Food processing at a crossroad. *Frontiers in Nutrition*, 6, 85.

Komdur, R. H., Korthals, M., and te Molder, H. (2009). The good life: Living for health and a life without risks? On a prominent script of nutrigenomics. *British Journal of Nutrition*, 101, 307–316.

Korthals, M. (2008). Food as a source and target of metaphors: Inclusion and exclusion of foodstuffs and persons through metaphors. *Configurations*, 16, 77–92.

Kristeva, J. (1982). *Powers of Horror: An Essay on Objection*. L. S. Roudiez (Trans.). New York, USA: Columbia University Press.

Lacan, J. (2007). *The Seminar of Jacques Lacan Book XVII: The Other Side of Psychoanalysis*. R. Grigg (Trans.). New York: WW. Norton & Company.

Lakoff, G., and Johnson, M. (2003). *Metaphors We Live By*. Chicago: The University of Chicago Press.

Leach, E. (1970). *Lévi-Strauss*. London: Wm. Collins & Co Ltd.

Lévi-Strauss, C. (2008). The culinary triangle. *In Food and Culture: A Reader*, 2nd ed., pp. 36–43, C. Counihan and P. Van Esterik (Eds.). New York: Routledge.

López-Rodríguez, I. (2016). Feeding women with animal metaphors that promote eating disorders in the written media. *Linguistik Online*, 75, 71–101.

Lupton, D. (1996). *Food, the Body and the Self*. London: SAGE Publications.

Lynch, M. (2010). From food to fuel: Perceptions of exercise and food in a community of food bloggers. *Health Education Journal*, 7, 72–79.

McClements, D. J., Vega, C., McBride, A. E., and Decker, E. A. (2011). In defense of food science. *Gastronomica*, 11, 76–84.

Miles, C., and Smith, N. (2016). What grows in silicon valley? The emerging ideology of food technology. In *The Ecopolitics of Consumption: The Food Trade*, H. L. Davis; K. Pilgrim and M. Sinha (Eds.). Maryland, USA: Rowman & Littlefield, 119–138

Mill, J. S. (2016). Utilitarianism. In *On Liberty and Utilitarianism and Other Works*, T. Griffith (Ed.). Hertfordshire: Wordsworth Editions Limited. Ch. 2 ('what utilitarianism is'), 374–394.

Mills, S., Ross, R. P., Hill, C., Fitzgerald, G. F., and Stanton, C. (2011). Milk intelligence: Mining milk for bioactive substances associated with human health. *International Dairy Journal*, 21, 377–401.

Mol, A. (2008). I eat an apple. Theorizing subjectivities. *Subjectivity*, 22, 28–37.

More, T. (2009). *Utopia*. P. Turner (Trans.). London: Penguin Books Ltd.

Morse, M. (1994). What do cyborgs eat? Oral logic in an information society. *Discourse*, 16, 86–123.

National Research Council (US) Committee on Food Habits. (1943). *The Problem of Changing Food Habits: Report of the Committee on Food Habits 1941–1943*. Washington DC, USA.

Newman, J. (1997). Eating and drinking as sources of metaphor in english. *Cuadernos de Filología Inglesa*, 612, 213–231.

Nordström, K., Coff, C., Jönsson, H., Nordenfelt, L., and Görman, U. (2013). Food and health: Individual, cultural, or scientific matters? *Genes & Nutrition*, 8, 357–363.

Norton, I., Fryer, P., and Moore, S. (2006). Product/process integration in food manufacture: Engineering sustained health. *AIChE Journal*, 52, 1632–1640.

Paracelscus. (1520s-1530s/2002a). The physician's remedies. In *Science in Europe, 1500–1800 – A Primary Sources Reader*. Hampshire: Palgrave MacMillan, p. 99.

Paracelscus. (1520s-1530s/2002b). Alchemy, art of transformations. In *Science in Europe, 1500–1800 – A Primary Sources Reader*. Hampshire: Palgrave MacMillan, p. 100.

Plato. (1997). *Complete Works*. Translated by various. J. M. Cooper & D. S. Hutchinson (Eds.). Indianapolis: Hackett.

Pollen, M. (2008). *In Defense of Food*. London: Penguin.

Popper, K. (2002). *The Logic of Scientific Discovery*. Oxon, NY, USA: Routledge.

Rees, M. (2000). *Just Six Numbers: The Deep Forces That Shape the Universe*. New York, Basic Books.

Rickey, V. F. (1987). Man, myth, and mathematics. *College Mathematics Journal*, 18, 362–389.

Ronteltap, A., van Trijp, H., Berezowska, A., and Goossens, J. (2013). Nutrigenomics-based personalised nutritional advice: In search of a business model? *Genes and Nutrition*, 8, 153–163.

Roosth, S. (2013). Of foams and formalisms: Scientific expertise and craft practice in molecular gastronomy. *American Anthropologist*, 115, 4–16.

Saussure, F. de (2009) *Course in General Linguistics*. R. Harris (trans.), Chicago, IL: Open Court.

Shklovsky, V. (1988). Art as device. In *Modern Criticism and Theory: A Reader*. L. T. Lemon and M. J. Reis (Trans.), David Lodge (Ed.), London: Longmans. pp. 16–30.

Sifferlin, A. (2016). The case for eating butter just got stronger. *TIME*. Retrieved online on 9/3/18: http://time.com/4386248/fat-butter-nutrition-health/

Skiadas, P. K., and Lascaratos, J. G. (2001). Dietetics in ancient Greek philosophy: Plato's concepts of healthy diet. *European Journal of Clinical Nutrition*, 55, 532–537.

Smith, T. W. (2001). *Revaluing Ethics: Aristotle's Dielectical Pedagogy*. New York: State University New York Press.

Smith, C. (2002). Punishment and pleasure: women, food and the imprisoned body. *The Sociological Review*, 50, 197–214.

Smithers, G. W. (2008). Whey and whey proteins—From 'gutter-to-gold'. *International Dairy Journal*, 18, 695–704.

Spanos, A. J. (2013). The eighth amendment and nutraloaf: A recipe for disaster. *Journal of Contemporary Health Law and Policy*, 30, 222–248.

Spence, C. (2017). *Gastrophysics: The New Science of Eating*. London, UK: Penguin Random House.

Taylor, C. (2010). Foucault and the ethics of eating. *Foucault Studies* 9, 71–88.

Valenzuela, F., and Böhm, S. (2017). Against wasted politics: A critique of the circular economy. *Ephemera Journal*, 17, 23–60.

Van der Weele, C. (2006). Food metaphors and ethics: Towards more attention for bodily experience. *Journal of Agriculture and Environmental Ethics*, 19, 313–324.

Vilgis, T. A. (2013). Texture, taste and aroma: Multi-scale materialsand the gastrophysics of food. *Flavour*, 2, 12.

Von Debschitz, U., and von Debschitz, T. (2017). *Fritz Kahn: Infographics Pioneer*. Cologne: TASCHEN.

Walmsley, J. (2014). *Mind and Machine*. Hampshire: Palgrave MacMillan.

Wansink, B. (2002). Changing Eating Habits on the Home Front: Lost Lessons from World War II Research. *Journal of Public Policy & Marketing*, 21, 90–99.

Weaver, C. M., Dwyer, J., Fulgoni III, V. L., King, J. C., Leveille, G. A., MacDonald, R. S., Ordovas, J., and Schnakenberg, D. (2014). Processed foods: Contributions to nutrition. *The American Journal of Clinical Nutrition*, 99, 1525–1542.

Zoukis, C. (2016). Use of nutraloaf on the decline in U.S. prisons. *Prison Legal News*, Retrieved on 9/3/18 at https://www.prisonlegalnews.org/news/2016/mar/31/use-nutraloaf-decline-us-prisons/

Zwart, H. (2016). Psychoanalysis and bioethics: A Lacanian approach to bioethical discourse. *Medicine, Health Care and Philosophy*. DOI: 10.1007/s11019-016-9698-1

Zwart, H. (2018). Scientific iconoclasm and active imagination: synthetic cells as techno-scientific mandalas. *Life Sciences, Society and Policy*. 14, 10.

Zwart, H. (2019). *Psychoanalysis of Technoscience*. Zürich: LIT Verlag.

5 Carbon budgets

A metaphor to bridge the science–policy interface on climate change action

James Glynn

Introduction

The group of gases that are warming our planet are referred to as greenhouse gases or GHGs for short. Carbon dioxide (CO_2) is one such GHG which makes up the majority of anthropogenic GHG emissions by volume and by warming impact on our home, Earth. Other GHGs such as methane (CH_4) and nitrous oxide (N_2O) are both smaller in their annual volume emitted but have more potent global warming potentials (GWP*) per year over their shorter lifespan in the atmosphere (Allen et al., 2018). The greenhouse gas effect, global warming, climate change, climate chaos, and climate breakdown are all iterations of metaphors used in communicating the same phenomena now impacting the earths biosphere (Harré, Brockmeier, & Mühlhäuser, 1999).

To stop warming the climate, we know that we must reduce our carbon dioxide (CO_2) emissions, but by how much and by when? The latest analysis feeding into the Intergovernmental Panel on Climate Change (IPCC) special report on 1.5°C points to the reality that, by 2050, the average person's CO_2 emissions need to be zero or even negative if we are to achieve the goals set out in the Paris Agreement of limiting post-industrial temperature increase to 2°C and pursue efforts to limit warming to 1.5°C (Grubler et al., 2018; Kriegler et al., 2018; Luderer et al., 2018; Rogelj et al., 2018; Strefler et al., 2018; UNFCCC, 2015; van Vuuren et al., 2018). Climate change is not our grandchildren's problem as is so often espoused by elder states-people typically double the age of the general population average age. It is important to understand intergenerational impacts from variations in perceptions of time from the old to the very young, and resultant variations in reality brought about by the time lag in myopic policy and climate inaction. Scientists understand that to stop man-made temperature increase, humankind must stop emitting CO_2. More accurately, humankind must balance the greenhouse gases emitted by our energy, industry, consumption, and food choices, with ways of capturing greenhouse gases before they get to the earth's atmospheric commons[1] (Gardiner, 2010; Hardin, 1968). To stop temperature increase, we must stop (net) CO_2 emissions. The time at which the temperature increase stops, and temperature stabilises at its new level above the pre-industrial average temperature, depends on the cumulative amount of CO_2 released into the atmosphere. This cumulative amount of CO_2 emitted is called a "*carbon budget*".

DOI: 10.4324/9781003143567-6

A carbon budget is the cumulative amount of carbon dioxide emitted from a specific start year to an undefined time in the future that would push the global mean temperature to a specified temperature limit with a given probability (Rogelj et al., 2016). Typically the IPCC use a starting date of 1870 to signify a baseline for measuring post-industrial warming of 2°C when quantifying a carbon budget. A carbon budget is the sum of annual CO_2 emissions, or the area (coloured black) under the annual CO_2 emissions curve such as outlined in Figure 5.1. From an 1860–1890 baseline, the carbon budget to keep global mean temperature below a 2°C threshold with 66% probability is about 3,200 $GtCO_2$, acknowledging that historical global emissions from fossil fuels and industry are approximately 2,200 $GtCO_2$ and the global mean temperature has risen over +1°C during this period. There is approximately a 1 trillion tonne CO_2 carbon budget remaining from 2015 to stay below 2°C with 66% probability, which is being depleted by approximately 41$GtCO_2$ per year. This rate of CO_2 emissions is accelerating, and so the remaining carbon budget is being consumed faster and faster. These types of carbon budgets are referred to as *avoidance* carbon budgets in keeping with the goal of avoiding a future temperature limit. Carbon budgets that quantify the volume of CO_2 that will lead to the *Exceedance* of a temperature threshold, or a *Peak* carbon budget when temperature peaks as a result of net zero emissions being reached are referred to as *Exceedance Budgets and Peak Budgets*. A carbon budget whereby there is a temperature overshoot and return below a temperature limit is called an *overshoot and return* budget.

This chapter focusses on avoidance *Carbon Budgets*, the metaphor that is bridging both scientific and policy lexicons (Peters, 2018). The hope is that this metaphor can be interpreted further from global policy to personal perspectives as a

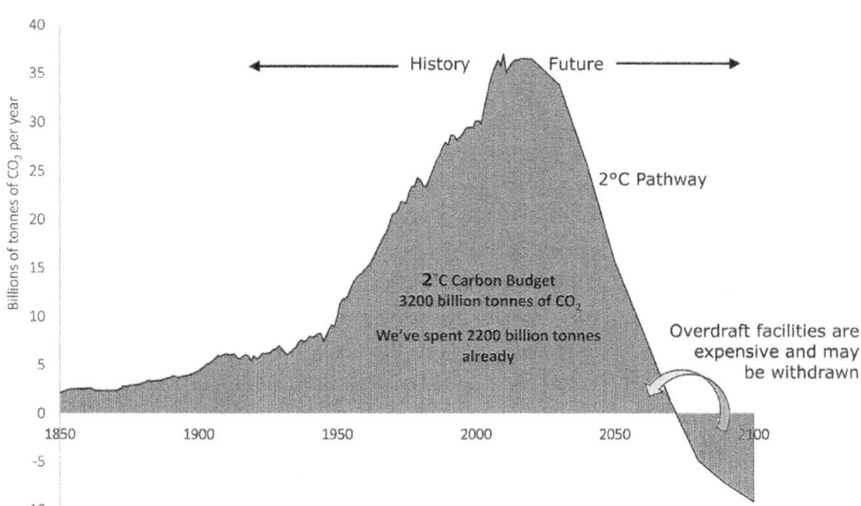

Figure 5.1 The post-industrial 2°C avoidance carbon budget of 3,200$GtCO_2$ with less than 1,000$GtCO_2$ remaining.

Source: Author created chart using the following data sources: Boden, Andres, & Marland, 2017; Quéré et al., 2018.

useful means to understanding the personal scale of response required to minimise climate change impacts in the context of inter-national fairness, inter-generational fairness, and intra-national fairness across wealth classes (Skeie et al., 2017).

Carbon budgets as a metaphor is primarily useful in that it serves a dual purpose. It is *quantitatively* scientifically robust, is simple, and succinctly provides enough information to policy makers and citizens of how their individual and local actions add up to the global target of temperature stabilisation. The carbon budget is algebraic, mathematically simple, and no jargon or scientific background is required to understand the implications. If you can add, subtract, and divide, carbon budgets enable an intuitive method to understand how much of the complexity of climate science and climate diplomacy comes down to this single number. Once understood that there is less than 1,000GtCO_2 remaining in the global 2°C carbon budget, and that we are producing 41GtCO_2 per year, it is simple to divide the carbon budget by the speed at which we are consuming the carbon budget to work out that breaching the 2°C carbon budget is less than 25 years away on current trends. Similarly, breaching the 1.5°C carbon budget is easily calculable to occur in about 12 years give or take a year.

Secondly carbon budgets are intuitively analogous to familiar constraints we all deal with on a day-to-day basis. Carbon budgets, like a household cashflow budget, is a metaphor that no longer focusses on communicating the phenomenon's problem, or scientific complexity, but describes a solution space which is scientifically quantifiable, useful from global, to local and personal scales because of its linear additive nature, while still remaining intuitively familiar to national policy makers, national negotiation delegations, and perhaps at a personal level for the public (Cohen, 2011). You can chop the carbon budget per country to simply explore what might be diplomatically considered a fair share of the carbon budget the same way you'd slice up a birthday cake. Why should Uncle Sam get such a large slice of the cake, leaving crumbs for your brothers and sisters? You can even chop up the carbon budget into a slice for each person on the planet leaving about a 100 tonne slice for each person for their lifetime.

Carbon budgets as a metaphor communicate the outcomes of a scientifically complex integrated climate-energy-economy-environment system in a single simple number, which metaphorically is analogous to many familiar and intuitive household activities. A carbon budget is most analogous to our mortgage on our homes, or our cumulative lifetime income. Like a household budget ensuring that we have enough savings to pay our mortgage and bills into the future, we can ensure we budget our carbon emissions so that our household stays hospitable in the future with a stable climate. It also makes the scale and pace of climate action required for global temperature stabilisation tangible, tractable, and quantifiable at the personal level.

Global context

The evidence is unequivocal; global anthropogenic emissions are leading to an average warming of the climate (IPCC, 2015). Decarbonisation targets need to be ambitious to stabilise temperature and equitable to enable global collective action, while acknowledging common but differentiated responsibilities and respective capabilities to mitigate and adapt (Schellnhuber, Rahmstorf, & Winkelmann, 2016; UNFCCC, 2015). The nationally determined contributions (NDCs) which built the consensus underpinning the 2015 Paris Agreement have reduced the probability of global temperature increase below the previous projections of 3.7°C to 4.8°C during the twenty-first century (Edenhofer et al., 2014), but only marginally so to approximately 3°C by 2100. On their current trajectory, the NDCs will not put global GHG emissions on a trajectory consistent with the Paris Agreement goals of limiting temperature increase well below 2°C and pursuing efforts towards 1.5°C (International Energy Agency, 2015; UNEP, 2016; UNFCCC, 2016). Global CO_2 emissions are still rising at 2% per year (Quéré et al., 2018).

Cumulative emissions of anthropogenic CO_2 is the primary driver of post-industrial anthropogenic temperature increase (Allen et al., 2009; IPCC, 2013; Meinshausen et al., 2009; Millar et al., 2017; Rogelj, Reisinger, et al., 2015a) (see Figure 5.2). The remaining cumulative CO_2 emissions that would result in a 1.5°C or 2°C temperature increase with a given probability can be ascribed to a total carbon budget, measured in tonnes of CO_2 (tCO_2) over a time horizon (Friedlingstein et al., 2014; Rogelj et al., 2016). The time horizon can span from the present day to the end of this century, 2100, to a peak in global mean temperature, or a peak in global GHG emissions. van Vuuren et al. (2016) point to the simple strength of this near linear relationship in that, (1) long-term temperature stabilisation does not depend on CO_2 emissions at a specific time, (2) near-term emissions are important as they also [most rapidly] exhaust the carbon budget, and finally (3) CO_2 emissions will need to be phased out to net zero eventually to achieve temperature stabilisation (Rogelj, Schaeffer, et al., 2015b; van Vuuren et al., 2016). The remaining global carbon budgets for a 66% probability of limiting temperature rise to 2°C are estimated between 590 and 1240 billion tCO_2 ($GtCO_2$) (Friedlingstein et al., 2014; Rogelj et al., 2016). Carbon budget uncertainty is largely driven by the decadal time scale dynamics of non-CO_2 GHGs, such as methane and

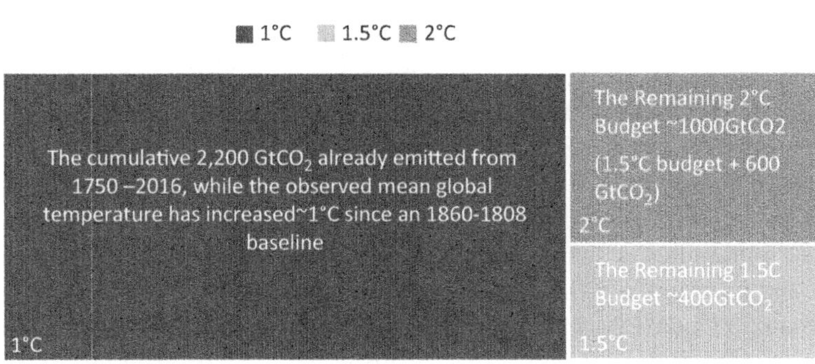

Figure 5.2 "The Carbon Budget Cake" – Historical cumulative global CO_2 emissions that have raised the mean global temperature 1°C, and the remaining 1.5°C carbon budget, and remaining 2°C carbon budget.

Source: Author's work.

nitrous oxide from agricultural emissions (IPCC, 2014; Rogelj et al., 2016). The carbon budget to return below 1.5°C by 2100 with greater than 50% probability is estimated at between 200GtCO_2 and 700GtCO_2 (IPCC, 2015; Millar et al., 2017; Rogelj et al., 2018). Annual CO_2 emissions from fossil fuel and industry for 2017 are estimated at 36.8 ± 2 GtCO_2, accelerating year on year, 2% higher than they were in 2016, with additional highly uncertain land use change emissions (deforestation) in the order of 5 GtCO_2 in 2016 (Quéré et al., 2018).

The algebra of how much time is left before the carbon budget for a peak 2°C or 1.5°C scenario is spent is relatively simple; dividing the remaining carbon budget by the future annual CO_2 emissions gives an estimate of the remaining time before this temperature target is breached. The concept of a carbon budget is increasingly relied upon in IPCC assessment reports, special reports, and within the integrated assessment modelling (IAM) community, which analyse the impacts of the energy system upon the economy, environment, and climate, while others warn that this over-reliance on carbon budgets as a communication tool over simplifies the problem (Peters, 2018) and that using the range of shared socio-economic pathways remains the best analysis method (Bauer et al., 2017; Calvin et al., 2017; Fricko et al., 2017; Fricko et al., 2016; Fujimori et al., 2017; Kriegler et al., 2017; McCollum et al., 2018; O'Neill et al., 2014; Riahi et al., 2017; Rozenberg, Guivarch, Lempert, & Hallegatte, 2013; van Vuuren et al., 2018). The relationship between temperature increase and cumulative CO_2 emissions, or the carbon budget, was most succinctly presented in the 5th assessment report of the Intergovernmental Panel on Climate Change (IPCC, 2015). They show the simple linear relationship between cumulative anthropogenic CO_2 emissions along the horizontal X axis, and resultant temperature change along the vertical Y axis (IPCC, 2015). Given this straight light relationship one can read the expected temperature increase for a given cumulative CO_2 emissions – a carbon budget. RCP stands for representative concentration pathways and are climate model scenarios where the radiative forcing (additional heat) on the planet by the end of the century ranges from 2.6 watts per square metre, up to 6 watts per square metre. RCP 6.0 is a scenario equivalent to a world where temperatures have risen on average by 4°C by 2100, whereas RCP 2.6 is equivalent to a high probability of staying below 2°C where concentrations of CO_2 in the atmosphere stay below 450 parts per million (PPM). Mankind has already emitted 2,200 GtCO_2 since the industrial revolution, concentrations of CO_2 in the atmosphere have increased beyond 410PPM and the planet has warmed on average about 1°C with more frequent record-breaking high temperatures, and less frequent minimum low temperatures globally (Millar et al., 2017; Quéré et al., 2018). Limiting future CO_2 emissions below 1,000 GtCO_2 from 2015 onwards gives mankind a 66% probability of keeping total average temperature increase below 2°C (Friedlingstein et al., 2014). This is already a difficult task. Keeping temperatures below 1.5°C, as is the goal of the Paris Agreement, requires the remaining carbon budget to be cut in half, and is essentially a doubling of decarbonisation ambition and speed of required action when compared to the previous 2°C increase limit.

Imagine arriving at a (COP) party expecting to have to share a cake of a certain size among all the guests, then the nutritionists say half that cake is too bad for your health. You can only share half that cake, and the rest must not be eaten for the sake of our health and increasing waistlines (rising sea level).

Some party delegates still want cake, especially those that are bakers and patissiers, even though we all know it's not good for us.

This problem is further exasperated by the fact that a few party goers arrived at the party before everyone else and have already eaten over two thirds of the cake that the nutritionist's thought was ok to share without causing too many health impacts. In reality, there isn't a fair share of the cake (carbon budget) remaining for each guest at the party.

Carbon budgets as an instrument of equitable climate action

Given that cumulative carbon emissions are a strong linear indicator of temperature increase, historical territorial cumulative emissions are one simple measure of proportional responsibility for anthropogenic temperature increase which occurred in the past. Equitable mitigation can be enabled through fair share carbon budgets via simple fair effort sharing allocation rules based on historical responsibility as a function of cumulative emissions. (Bows & Anderson, 2008; Kartha et al., 2018; Kober, Van Der Zwaan, & Rösler, 2014; Raupach et al., 2014; Robinson & Shine, 2018; Robiou du Pont et al., 2017).

Ethically attributing responsibility and capability is far from politically simple, with complex national development circumstances to be accounted for and balanced during climate negotiations (CSO Equity Review, 2018; Gardiner, 2010; Holz, Kartha, & Athanasiou, 2018; Skeie et al., 2017). Responsibility and capability to mitigate are two guiding principles of the United Nations Framework Convention on Climate Change (UNFCCC). Estimation of equitable national carbon budgets typically includes weighting methods to balance historical responsibility for warming, and capability to mitigate, based on a start date of responsibility (1750, 1850, 1890, 1900, 1950, 1990… etc.), cumulative emissions from this start date, cumulative embodied carbon in traded energy and goods, and gross domestic product (GDP) per capita as an indicator of wealth and capacity to mitigate, or to fund other regions to mitigate more cost effectively.

There are multiple potential effort sharing rules that incorporate carbon budgets to account for historical warming and to attribute responsibility for prior warming. One such burden sharing rule was the "Brazilian proposal" in 1997 where countries would take responsibility for national proportion of warming based on historical emissions (UNFCCC, 1997). This rule is simple and attractive in that cumulative historical CO_2 emissions are directly relatable to warming, and attributable on a regional boundary basis. If a country has already emitted more than its fair share carbon budget, then it is proposed that they should give aid and reparations to those countries that will not emit their fair share of the carbon budget. However, this does not consider whether the emissions within a country were emitted for the benefit of that country or exported to another country embodied in the goods exported, or further in fossil fuels exported from one country to another. Therefore, assigning responsibility for historic warming needs to take into account whether the CO_2 is accounted for using consumption-based, production-based, or extraction-based accounting (Skeie et al., 2017). Also, the date at which we take responsibility has significant impact on the level of responsibility; some nations argue that countries should not take responsibility for emissions before climate change was scientifically understood as a problem, giving rise to the debate over the start years in calculating an attributable carbon budget for historical responsibility.

This simplest and strictest version of the Brazilian carbon budget proposal is outlined here. The total carbon budget that would warm the planet 2°C post industrialisation is 3,200GtCO$_{2.}$ In an equitable scenario, each person, past, present, and future, would have an equal share of that 3,200GtCO$_2$ carbon budget, and the development potential it represents. Given some countries industrialised much earlier than other countries, fossil-fuelled industrialised nations have a higher CO_2 per capita than developing countries. If we were to share the all-time 2°C carbon budget – 3,200GtCO$_2$ – on a per capita production basis within territorial boarders, Figure 5.3 outlines the per capita weighted share of the carbon budget sorted with low-income regions on the left and wealthy regions on the right. Africa's (AFR), India's (IND), and developing Asia's (ODA) fair share carbon budget is considerably larger than their historical emissions, or the regions projected CO_2 emissions in a future least cost[2] 2°C scenario. The fair share of the carbon budget is sensitive to the projected rapid population growth in Africa and south east Asia. Conversely the wealthier countries of the world, western Europe (WEU), Canada (CAN), Japan (JPN), and the USA, have typically produced considerably more CO_2 than their fair share carbon budget and are projected to

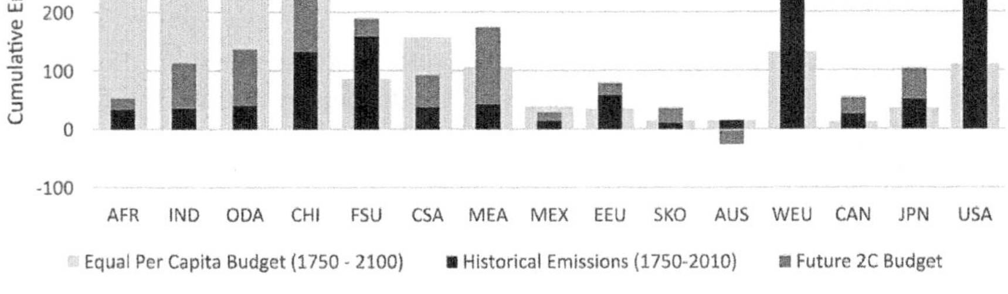

Figure 5.3 Regional distribution of the equal per capita share of the 2°C carbon budget (3,200GtCO$_2$), with historical emissions per region and potential future least cost CO_2 emissions given socio-economic trends for a single energy system pathway to a 2°C limit. Sorted in order of GDP per capita in 2015 from low income to high income per capita.

Source: Author's work with data adapted from Kypreos et al., 2018.

continue to do so in all scenarios from pathways to 6°C or 2°C. Therefore, there are calls for financial contributions above national mitigation efforts for countries who have emitted beyond their fair share carbon budgets to contribute in development aid, innovation, and knowledge transfer towards an equitable and efficient transition to a zero carbon society (Kypreos, Glynn, Panos, Giannakidis, & Ó Gallachóir, 2018).

However, there are other equity and fairness issues to consider such as intergenerational equitable distribution of the carbon budget as well as fair distribution of the carbon budget across income groups with progressive policy instruments. Figure 5.4 outlines the CO_2 emissions per capita for a selection of developed and developing countries, showing that while historically there have been high emissions per capita in developing countries, all countries in the future need to produce net-zero CO_2 per capita. Therefore, to stabilise global temperatures, generation Z will need to have less carbon-intensive lifestyles than generation Y, and my generation Y cannot expect to have the same carbon-intensive lifestyle as my parents' generation, generation X. While inter-regional equity is discussed by generation X and Y within the UNFCCC and conference of party meetings, inter-generational equity between generation X, Y, and Z is generally not addressed, and only raised in emotive speeches and pleas for climate action from young adults and future leaders of generation Z. More recently generation Z has started suing[3] the political institutions of generation X to protect their fundamental human rights (Robinson & Shine, 2018).

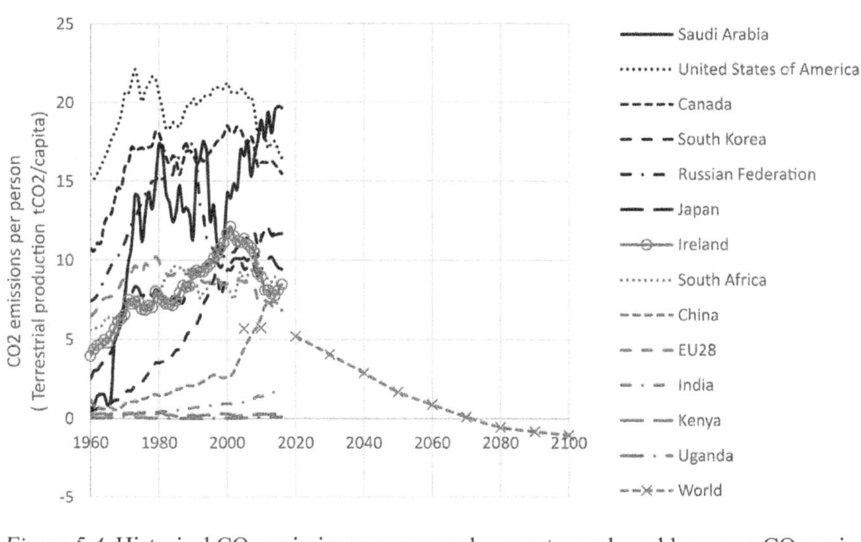

Figure 5.4 Historical CO_2 emissions per person by country and world average CO_2 emissions per person for a 2°C scenario from the next generation IPCC IAM scenario database.

Source: Author's work. Data sources: Global Carbon Project (2018), CDIAC (Boden et al., 2017; Quéré et al., 2018), and SSP2-2.6 marker model: MESSAGE-GLOBIOM, IIASA (O. Fricko et al., 2017).

Irish context

In Ireland, we produce about 8.5 tonnes of carbon dioxide (tCO_2) per person per year. These gases are emitted by the fuels we choose to provide our heat, transport, and electricity, as well as the fuels and processes used to manufacture and transport the goods and services produced in Ireland (Quéré et al., 2018). The Irish produce half the CO_2 per person of the average American, 25 % more than the average European, double the world average, 4 times the average Indian and over 25 times the average Kenyan.

In comparison to many other developed nations that industrialised earlier in the 18th and 19th centuries (Peters, Andrew, Solomon, & Friedlingstein, 2015), Irish historical CO_2 emissions from fossil fuel combustion and cement production (1751–2015) are not significantly larger than their per capita fair share of a 2°C carbon budget of 3,200$GtCO_2$, at approximately 2.04$GtCO_2$ (Quéré et al., 2018) and thus Ireland is not in significant carbon debt (Gignac & Matthews, 2015; Matthews, 2015). Ireland has however consumed more than its per capita fair share of a 1.5°C carbon budget.

In countries, or demographic groups within countries, where cumulative emissions per person are considerably larger than their equitable fair share of the total carbon budget, it can be reasonably argued that these groups should ambitiously decarbonise towards a net-zero carbon society and enable developing countries and low carbon intensity demographic groups to decarbonise by providing financial aid, access to financial capital, knowledge transfer, and technology transfer (CSO Equity Review, 2018). A grandfathering rule considers current per capita emissions and aims to contract high per capita emissions countries and converge with low per capita emissions countries towards an average global emission per capita that is in line with a temperature stabilisation trajectory. The problem with this method is that it enables countries and individuals to continue to emit CO_2 levels above fair share levels and results in an inequitable distribution of the remaining carbon budget (Kartha et al., 2018). Effort sharing rules in this regard can represent the authors normative bias against developing countries when using grandfathering type pathways to a zero carbon energy system (Kartha et al., 2018). Figure 5.5 shows the CO_2 emissions pathway from the Irish energy system where fair shares of the remaining carbon budget are applied as a constraint for equal per capita shares for the remaining global carbon budgets for 2°C and 1.5°C – the rate of reduction of CO_2 emissions required is higher than EU emission reduction targets and significantly higher than the current GHG emissions projected trajectories by the Irish Environmental Protection Agency (EPA, 2018; Glynn et al., 2018).

The average Irish person from 1850 to 1950, or a current average Indian person, has similar carbon emissions of less than 1.2 tCO_2 per year. The average farmers in Ireland in the nineteenth century, and in twentieth century India, relied on horses, oxen, and water buffalo for energy and power, and both following British innovation and knowledge transfer, began to use diesel tractors to provide the energy and automation to supplant their manual labour energy in agricultural production. Automation and mechanisation of production brings with it improvements in quality of life, health, and education enabled by cheap abundant energy

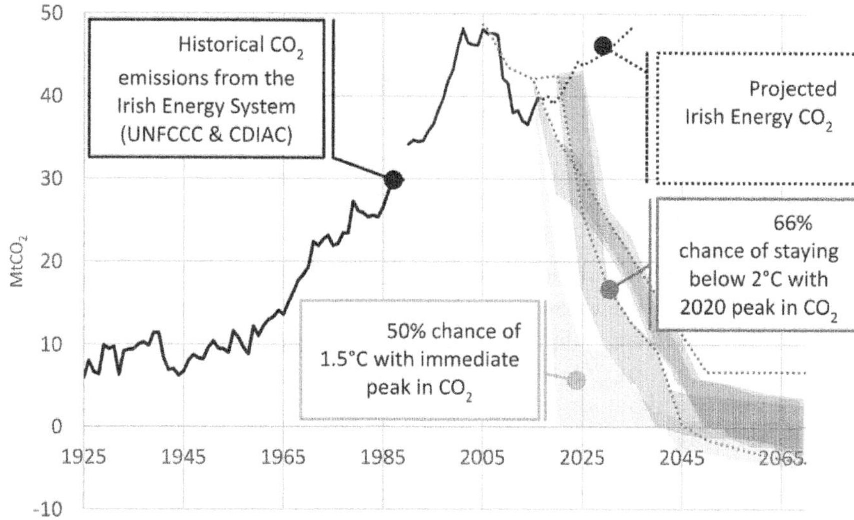

Figure 5.5 Irish historical CO_2 emissions and future pathways with carbon budgets consistent with the Paris Agreement.

Source: Adapted from Glynn et al., 2018.

by liberating labourers from labour and allowing the pursuit of higher education and more leisure time. The average person can output 800 watt hours (Wh) of energy per day. One litre of diesel contains 10,000 Wh of energy. A diesel powered tractor can carry out the labour of 10–100 labourers at the fraction of the time and cost. The average Irish person today currently emits 8 times the CO_2 that their grandparents emitted at the same age in the 1950s from the consumption of larger volumes of fossil fuels and imported globally transported goods. This trend is not sustainable, in that the biosphere system that supports life cannot sustainably (indefinitely) absorb our collective pollution.

Lastly, the carbon budget and instruments such as carbon taxes to ration the carbon budget need to be progressive within all countries, developed and developing, across young and old, as within each of these demographic groups there are rich and poor distributions of families already with low carbon intensity lives due in many cases to low income and poverty. In Ireland for example, the richest income decile produces double the CO_2 per person than the poorest income decile, largely from indirect emissions embodied in the consumption of luxury goods and services. Single parent families, or traditional families, have a quarter the carbon intensity per person than single occupant households (Lyons, Pentecost, & Tol, 2012). Carbon taxes levied to ration and distribute the carbon budget should be progressive, given the unequal income and carbon intensity distribution across society, with particular attention given to the difficulty for families in energy poverty to switch their fuels for heating and transport, which make up a considerable proportion of the cohort's disposable income (Farrell, 2017).

The charitable organisation, Saint Vincent de Paul, now spends 75% of their cash budget on supplementary purchasing of fuel for those in need of heat in

Ireland. This is on top of the €400 million per year that the Irish government spend on financial support to provide heat for households in energy poverty. Unfortunately, oftentimes this support purchases the most carbon intensive and health damaging of fuels, coal and peat, for heating-inefficient poor-quality housing.

What can I do? Figure out your carbon budget!

When scientists add up all the cumulative GHG emissions from all the sources, mostly fossil fuel combustion, industry, and deforestation, over all the years of available emissions data estimates (Boden et al., 2017), they can work out that for every trillion tonnes of CO_2 humans emit, the temperature will go up about half a degree Celsius (Simmons & Matthews, 2016; Tachiiri, Hajima, & Kawamiya, 2015). One trillion tonnes of CO_2 emitted causes half a degree temperature increase. Given that the planet has already warmed on average by 1°C since 1860–1890, if we want to limit future warming to 2°C with a high probability of success, we can only emit around 1 trillion tonnes more of CO_2. To limit warming to well below 2°C and towards 1.5°C, as is the goal of the Paris Agreement, we can only emit between 200 $GtCO_2$ and 700 $GtCO_2$. Remember, that global CO_2 emissions in 2017 are estimated at 41 $GtCO_2$ and are rising. Tick tock.

Friends and family justifiably bamboozled, often ask me the question, "What can I do?", when confronted with the seemingly insurmountable task of reducing our personal carbon footprint as part of humanities' collective environmental impacts which have long pushed passed sustainable planetary boundaries (O'Neill, et al., 2018; Rockström et al., 2009). There are no easy fixes and there are no single technological silver bullets to solve this problem. Analysis shows we already have the suite of technology options required to stabilise temperatures if we are also behaviourally and politically willing to change lifestyles towards ambitious climate action. We are however running out of time to mobilise behavioural and political change. Globally we need to reduce our carbon emissions in the order of 5% per year and in the order of 10% per year in the wealthier carbon-intensive countries. Global CO_2 emissions are growing at 2% per year from 2015 to 2017.

Environmental campaigns during the 1990's advocated that if we all did a "little bit", it would add up to significant collective change. This does not make mathematical sense. Unfortunately, many small individual changes still add up to a small collective change – we are far beyond the time for this sort of naïve thinking. Our collective individual carbon budgets need to – literally – add up to the global carbon budget which corresponds to the temperature limit we collectively wish to stay below. Individually, the first task is to measure and understand the scale of the problem. Understanding carbon budgets gives a meaningful and simple method to understand and communicate the scale of our required actions on the atmospheric commons so that they add up to a realistic transition towards our Paris Agreement goals (UNFCCC, 2015). So, how do we share out the remaining 1 trillion tonnes of CO_2 in a way that is fair and takes responsibility for the past emissions, the resultant climate impacts, and accounts for those countries that have capacity to change?

How do we share out the carbon budget cake? Who gets a big slice and who has eaten enough already? Who is suffering from carbon obesity and needs a crash diet? Is anyone not allowed a slice of carbon cake?

If you were to give every person on the planet their equal slice of the remaining 2°C carbon budget, it would be about 100 tonnes of CO_2 per person, for their lifetime. This is a rather simplified approach and ignores historical emissions mentioned previously, as well as responsibility, capacity, equity, and justice, but is useful to illustrate a point.

If we were to have personal carbon budgets, the same way we budget our family income, what would that be like? You know how much money you hope to earn over your life, and you take a bet that you will earn that income by getting a mortgage for your home in the hope that you'll achieve your projected cumulative earnings before you retire, enough to pay for your house and interest to the bank. How much are you willing to bet on your carbon budget, or taking an overdraft on your carbon budget? Will your children inherit your carbon debt in the form of runaway climate change? Or will your pension be vulnerable to your carbon debt through financial insurance markets volatility and uninsurable risks? Perhaps your pension fund will be spent on critical flooding adaption measures in a "bail out" of a different type – will we need to ensure our flood defences really are "too big to fail".

If you have a 100 tCO_2 carbon budget for your lifetime, how would you spread your fair share of the global carbon budget over your 80-year life expectancy? You could spread your carbon budget thinly over your whole life, that would roughly equate to 1.2 tCO_2 per year. For scale, the average European family fossil fuel car emits 2 tCO_2 per year, as does one transatlantic flight or one cow.

Would the unsustainable expectation of a more materially and carbon-intensive lifestyle be like an enjoyable Irish night out, prioritising today over tomorrow, where you spend your weekly budget on whiskey and beer today, and then you're left with a hangover tomorrow and nothing for dinner for the rest of the week? Or would you plant a biologically diverse forest to capture CO_2 to try to grow your personal carbon budget?

Carbon dioxide is as fundamental to life as is food. Like the familiar food industry, there are inequities spanning obesity and health impacts in developed countries from nutritionally poor food, to a calorie and a nutritional deficit of famine victims in some developing countries. Global charitable organisations provide food aid to famine-stricken countries, while food regulators in developed countries enforce information requirements on ingredients, nutritional value, and calorie content of foods and meals to encourage informed food choices. A similar approach could be taken with carbon dioxide, alongside lists of ingredients and calorie content of a food item, the embodied carbon dioxide content of a meal could be displayed on the menu or on the packaging of general items which would enable individuals to understand the impacts of their choices on their carbon budget and the global carbon budget.

I was born in 1981, and in my lifetime, the average Irish person has emitted 340 tCO_2 each. Remembering that Irish per person CO_2 emissions are currently 8.5 tCO_2 per year. All else remaining equal, this would result in 646tCO_2 per Irish lifespan into the future, and if each person on earth had this same lifestyle, the global collective CO_2 emissions over the next three generations (82 years) would be approximately 6,500 $GtCO_2$ – this is equivalent to greater than 3°C warming from CO_2 alone and likely greater than 4°C warming with associated other GHGs from agricultural methane and nitrous oxide.

This raises the obvious question of urgency. Making financial comparisons, I don't like taking loans, or owing money, or being over drawn at the bank especially if I don't see a means of commencing immediate repayment. The simple realisation is that different nations, generations, and income groups owe a carbon debt, to those who have not benefited from the external damages imposed on the climate commons, to be paid from the wealth that freedom from manual labour that burning fossil fuels in machines has enabled. Further, our individual solo efforts won't add up to much of a change in the national carbon budget without our neighbours, family, and friends doing the same – we'll only succeed by collective global action, by immediately pursuing the radical provision of zero carbon infrastructure, community supports, and a just transition (Obama, 2017; Robinson & Shine, 2018). A just transition can have health, development, and well-being benefits in the form of warmer homes, cleaner air, energy access, eradication of energy poverty, cheaper energy bills, and meaningful employment.

Like the IPCC climate stabilisation scenario analysis points to, there is no technological single silver bullet solution, and there is no single person or group who single-handedly cause climate change. Anyone who has benefited from industrialisation of civilisation has some responsibility for changing the climate and thus a responsibility to understand their remaining carbon budget. However, the strength of carbon budgets is that they clearly convey three simple messages:

1. Long-term temperature change does not depend on CO_2 emission at a specific moment in time, but on the cumulative CO_2 emissions over a long period of time,
2. Therefore, CO_2 emissions emitted now and in the near term are important as they, too [most rapidly] exhaust the remaining available carbon budget, and
3. Finally, to stabilise temperature and stop temperature increase, CO_2 emissions will need to be phased out to net zero.

Considering carbon budgets as a metaphor, can remove the scientific jargon commonly used in climate change conversations and can enable better understanding of the scale and urgency of climate action required to stabilise temperature. Carbon budgets imbue a familiar analogy to household budgeting to demystify the scale of the climate challenge, enabling planning pathways to destinations worth saving for – a family holiday or a functional climate – and drawing roadmaps for climate action commensurate with the timeline before the climate breaks down. A carbon budget could be seen as analogous to our cumulative savings for that holiday we want or our cumulative income for a mortgage for our home that protects us. The carbon budget metaphor is also analogous to institutional and governmental budgeting practices accounting for cumulative carbon flows instead of cumulative cashflows and capital stocks or investments and savings bonds.

Carbon budgets as a metaphor are useful also as it is independent of individual perspectives and is flexible to individual, international, intergenerational, institutional, and governmental perspectives and norms without losing its descriptive succinctness. Carbon budgets are relative to all stakeholders and agencies.

A carbon budget does not inform how you could choose to spend your carbon budget. However, given how much carbon we produce each year, it does inform the remaining timeline to breaching the safe temperature limit, and the required speed of the collective transition to a zero carbon society linking your individual local choices to national policy targets and to the global temperature stablisation target. It all has to add up.

Irrespective of whether a person from the developed north or the developing south produced an individual tonne of CO_2, each tonne of carbon dioxide going into the atmospheric commons reduces the remaining global carbon budget. Each tonne of carbon dioxide each Irish citizen produces reduces the remaining fair share national Irish carbon budget. To make your fair share carbon budget last as long as possible, you could consider the two action steps that you can do today, one focussed on how you vote politically and the second the action I can take.

1. Vote for political leadership who are familiar with carbon budgets. When your local elected representative asks for your vote, ask them what is their carbon budget? Demand climate action commensurate with the Paris Agreement. Demand zero carbon electricity, zero carbon heat and zero carbon transport from your local elected representative.
2. Reduce superfluous carbon-intensive habits and consumption from your day-to-day life. The carbon reductions you achieve and maintain over your lifetime have a greater cumulative impact on your carbon budget and temperature stabilisation, than once off momentary efforts you achieve.

All our climate action efforts need to add up, so that we don't produce more carbon dioxide than the global carbon budget. It all has to add up.

Notes

1 The *commons* typically refers to land or resources belonging to or affecting the whole of a community. The tragedy of the agricultural grazing commons in medieval England in Hardin's work is the typical example (Hardin, 1968). Commons is used here in reference to a specific type of environmental economic problem of free riding and abuse of a commons resource where without sufficient social capital and collective norms it is in homoeconomicus's individual interest to marginally increase their use of the commons, to the detriment to the commons and thus the community as a whole. (Gardner & Ostrom, 1991; Ostrom, 1990, 2000; Ostrom & Field, 1999).

2 Integrated Assessment Modelling of the energy, economy, environment, and climate system generally calculate pathways towards the goal of limiting the global rise in temperature while minimising the social cost of achieving this goal by investments in new technologies weighted against the relative increased costs to society. This is referred to as "least cost" modelling in the Integrated Assessment Modelling community and in the climate policy community.

3 US Lawsuit "Juliana vs. US.https://static1.squarespace.com/static/571d109b04426270152febe0/t/57a35ac5ebbd1ac03847eece/1470323398409/YouthAmendedComplaintAgainstUS.pdf . See also: Urgenda. Dutch Lawsuit against Netherlands 2030 climate targets: http://www.urgenda.nl/en/themas/climate-case/ and: The Peoples Climate Case. Swedish family court case against the level of ambition of EU climate targets.

References

Allen, M. R., Frame, D. J., Huntingford, C., Jones, C. D., Lowe, J. A., Meinshausen, M., & Meinshausen, N. (2009). Warming caused by cumulative carbon emissions towards the trillionth tonne. *Nature*, 458(7242), 1163–1166. https://doi.org/10.1038/nature08019

Allen, M. R., Shine, K. P., Fuglestvedt, J. S., Millar, R. J., Cain, M., Frame, D. J., & Macey, A. H. (2018). A solution to the misrepresentations of CO_2 -equivalent emissions of short-lived climate pollutants under ambitious mitigation. *Npj Climate and Atmospheric Science*, 1(1), 16. https://doi.org/10.1038/s41612-018-0026-8

Bauer, N., Calvin, K., Emmerling, J., Fricko, O., Fujimori, S., Hilaire, J., … van Vuuren, D. P. (2017). Shared socio-economic pathways of the energy sector – Quantifying the narratives. *Global Environmental Change*, 42(Supplement C), 316–330. https://doi.org/10.1016/j.gloenvcha.2016.07.006

Boden, T., Andres, R., & Marland, G. (2017). *Global, Regional, and National Fossil-Fuel CO2 Emissions (1751 - 2014) (V. 2017)*. Oak Ridge, TN (United States): Carbon Dioxide Information Analysis Center (CDIAC), Oak Ridge National Laboratory (ORNL). https://doi.org/10.3334/CDIAC/00001_V2017

Bows, A., & Anderson, K. (2008). Contraction and convergence: An assessment of the CCOptions model. *Climatic Change*, 91(3–4), 275–290. https://doi.org/10.1007/s10584-008-9468-z

Calvin, K., Bond-Lamberty, B., Clarke, L., Edmonds, J., Eom, J., Hartin, C., … Wise, M. (2017). The SSP4: A world of deepening inequality. *Global Environmental Change*, 42, 284–296. https://doi.org/10.1016/j.gloenvcha.2016.06.010

Cohen, M. J. (2011). Is the UK preparing for "war"? Military metaphors, personal carbon allowances, and consumption rationing in historical perspective. *Climatic Change*, 104(2), 199–222. https://doi.org/10.1007/s10584-009-9785-x

CSO Equity Review. (2018). *Equity and the Ambition Ratchet: Towards a Meaningful 2018 Facilitative Dialogue*. Figshare. https://doi.org/10.6084/m9.figshare.5917408

Edenhofer, O., Pichs-Madruga, R., Sokona, Y., Kadner, S., Minx, J. C., Brunner, S., … Zwickel, T. (2014). Technical summary. In *Climate Change 2014: Mitigation of Climate Change. Contribution of Working Group III to the Fifth Assessment Report of the Intergovernmental Panel on Climate Change* [Edenhofer, O., R. Pichs-Madruga, Y. Sokona, E. Farahani, S. Kadner, K. Seyboth, A. Adler, I. Baum, S. Brunner, P. Eickemeier, B. Kriemann, J. Savolainen, S. Schlömer, C. von Stechow, T. Zwickel and J.C. Minx (eds.)]. Cambridge, United Kingdom and New York, NY, USA: Cambridge University Press.

EPA. (2018). *Ireland's Greenhouse Gas Emissions Projectons 2017 - 2035*. Dublin: Environmental Projection Agency. Retrieved from http://www.epa.ie/pubs/reports/air/airemissions/ghgprojections2017-2035/

Farrell, N. (2017). What factors drive inequalities in carbon tax incidence? decomposing socioeconomic inequalities in carbon tax incidence in Ireland. *Ecological Economics*, 142, 31–45. https://doi.org/10.1016/j.ecolecon.2017.04.004

Fricko, O., Havlik, P., Rogelj, J., Klimont, Z., Gusti, M., Johnson, N., … Riahi, K. (2017). The marker quantification of the Shared Socioeconomic Pathway 2: A middle-of-the-road scenario for the 21st century. *Global Environmental Change*, 42, 251–267. https://doi.org/10.1016/j.gloenvcha.2016.06.004

Fricko, Oliver, Parkinson, S. C., Johnson, N., Strubegger, M., van Vliet, M. T., & Riahi, K. (2016). Energy sector water use implications of a 2°C climate policy. *Environmental Research Letters*, 11(3), 034011. https://doi.org/10.1088/1748-9326/11/3/034011

Friedlingstein, P., Andrew, R. M., Rogelj, J., Peters, G. P., Canadell, J. G., Knutti, R., ... Le Quéré, C. (2014). Persistent growth of CO_2 emissions and implications for reaching climate targets. *Nature Geoscience*, 7(10), 709–715. https://doi.org/10.1038/ngeo2248

Fujimori, S., Hasegawa, T., Masui, T., Takahashi, K., Herran, D. S., Dai, H., ... Kainuma, M. (2017). SSP3: AIM implementation of Shared Socioeconomic Pathways. *Global Environmental Change*, 42, 268–283. https://doi.org/10.1016/j.gloenvcha.2016.06.009

Gardiner, S. M. (Ed.). (2010). *Climate Ethics: Essential Readings*. Oxford; New York: Oxford University Press.

Gardner, R., & Ostrom, E. (1991). Rules and games. *Public Choice*, 70(2), 121–149.

Gignac, R., & Matthews, H. D. (2015). Allocating a 2°C cumulative carbon budget to countries. *Environmental Research Letters*, 10(7), 075004. https://doi.org/10.1088/1748-9326/10/7/075004

Glynn, J., Gargiulo, M., Chiodi, A., Deane, P., Rogan, F., & Gallachóir, B. Ó. (2018). Zero carbon energy system pathways for Ireland consistent with the Paris Agreement. *Climate Policy*, 0(0), 1–13. https://doi.org/10.1080/14693062.2018.1464893

Grubler, A., Wilson, C., Bento, N., Boza-Kiss, B., Krey, V., McCollum, D. L., ... Valin, H. (2018). A low energy demand scenario for meeting the 1.5°C target and sustainable development goals without negative emission technologies. *Nature Energy*, 3(6), 515–527. https://doi.org/10.1038/s41560-018-0172-6

Hardin, G. (1968). The tragedy of the commons. *Science*, 162(3859), 1243–1248. https://doi.org/10.1126/science.162.3859.1243

Harré, R., Brockmeier, J., & Mühlhäuser, P. (1999). *Greenspeak: A Study of Environmental Discourse*. Thousand Oaks, CA: Sage Publications.

Holz, C., Kartha, S., & Athanasiou, T. (2018). Fairly sharing 1.5: National fair shares of a 1.5°C-compliant global mitigation effort. *International Environmental Agreements: Politics, Law and Economics*, 18(1), 117–134. https://doi.org/10.1007/s10784-017-9371-z

International Energy Agency. (2015). *Energy and Climate Change: World Energy Outlook Special Briefing for COP21*. Paris, France: International Energy Agency. Retrieved from https://www.iea.org/media/news/WEO_INDC_Paper_Final_WEB.PDF

IPCC. (2013). *Climate Change 2013: The Physical Science Basis*. New York, USA: Cambridge University Press. Retrieved from http://www.ipcc.ch/report/ar5/wg1/

IPCC (Ed.). (2014). Long-term climate change: Projections, commitments and irreversibility pages 1029 to 1076. In *Climate Change 2013 - The Physical Science Basis* (pp. 1029–1136). Cambridge: Cambridge University Press. Retrieved from http://ebooks.cambridge.org/ref/id/CBO9781107415324A032

IPCC. (2015). *Climate change 2014: synthesis report*. Retrieved from http://www.ipcc.ch/pdf/assessment-report/ar5/syr/SYR_AR5_FINAL_full.pdf

Kartha, S., Athanasiou, T., Caney, S., Cripps, E., Dooley, K., Dubash, N. K., ... Winkler, H. (2018). Cascading biases against poorer countries. *Nature Climate Change*, 8(5), 348–349. https://doi.org/10.1038/s41558-018-0152-7

Kober, T., Van Der Zwaan, B. C. C., & RöSler, H. (2014). Emission certificate trade and costs under regional burden-sharing regimes for a 2C climate change Control target. *Climate Change Economics*, 05(01), 1440001. https://doi.org/10.1142/S2010007814400016

Kriegler, E., Bauer, N., Popp, A., Humpenöder, F., Leimbach, M., Strefler, J., ... Edenhofer, O. (2017). Fossil-fueled development (SSP5): An energy and resource intensive scenario for the 21st century. *Global Environmental Change*, 42, 297–315. https://doi.org/10.1016/j.gloenvcha.2016.05.015

Kriegler, E., Luderer, G., Bauer, N., Baumstark, L., Fujimori, S., Popp, A., … van Vuuren, D. P.. (2018). Pathways limiting warming to 1.5°C: A tale of turning around in no time? *Philosophical Transactions of the Royal Society A*, 376(2119), 20160457. https://doi.org/10.1098/rsta.2016.0457

Kypreos, S., Glynn, J., Panos, E., Giannakidis, G., & Ó Gallachóir, B. (2018). Efficient and equitable climate change policies. *Systems*, 6(2), 10. https://doi.org/10.3390/systems6020010

Luderer, G., Vrontisi, Z., Bertram, C., Edelenbosch, O. Y., Pietzcker, R. C., Rogelj, J., … Kriegler, E. (2018). Residual fossil CO 2 emissions in 1.5–2°C pathways. *Nature Climate Change*, 1. https://doi.org/10.1038/s41558-018-0198-6

Lyons, S., Pentecost, A., & Tol, R. S. J. (2012). Socioeconomic distribution of emissions and resource use in Ireland. *Journal of Environmental Management*, 112, 186–198. https://doi.org/10.1016/j.jenvman.2012.07.019

Matthews, H. D. (2015). Quantifying historical carbon and climate debts among nations. *Nature Climate Change, advance online publication*. https://doi.org/10.1038/nclimate2774

McCollum, D. L., Zhou, W., Bertram, C., de Boer, H.-S., Bosetti, V., Busch, S., … Riahi, K. (2018). Energy investment needs for fulfilling the Paris Agreement and achieving the Sustainable Development Goals. *Nature Energy*, 3(7), 589–599. https://doi.org/10.1038/s41560-018-0179-z

Meinshausen, M., Meinshausen, N., Hare, W., Raper, S. C. B., Frieler, K., Knutti, R., … Allen, M. R. (2009). Greenhouse-gas emission targets for limiting global warming to 2 °C. *Nature*, 458(7242), 1158–1162. https://doi.org/10.1038/nature08017

Millar, R. J., Fuglestvedt, J. S., Friedlingstein, P., Rogelj, J., Grubb, M. J., Matthews, H. D., … Allen, M. R. (2017). Emission budgets and pathways consistent with limiting warming to 1.5°C. *Nature Geoscience, Advance Online Publication*. https://doi.org/10.1038/ngeo3031

Obama, B. (2017). The irreversible momentum of clean energy. *Science*, aam6284. https://doi.org/10.1126/science.aam6284.

O'Neill, B. C., Kriegler, E., Riahi, K., Ebi, K. L., Hallegatte, S., Carter, T. R., … van Vuuren, D. P.. (2014). A new scenario framework for climate change research: The concept of shared socioeconomic pathways. *Climatic Change*, 122(3), 387–400. https://doi.org/10.1007/s10584-013-0905-2

O'Neill, D. W., Fanning, A. L., Lamb, W. F., & Steinberger, J. K. (2018). A good life for all within planetary boundaries. *Nature Sustainability*, 1(2), 88–95. https://doi.org/10.1038/s41893-018-0021-4

Ostrom, E. (1990). *Governing the Commons: The Evolution of Institutions for Collective Action*. Cambridge: Cambridge University Press.

Ostrom, E. (2000). Collective action and the evolution of social norms. *The Journal of Economic Perspectives*, 14(3), 137–158.

Ostrom, E., & Field, C. B. (1999). Revisiting the commons: Local lessons, global challenges. *Science*, 284(5412), 278.

Peters, G. P. (2018). Beyond carbon budgets. *Nature Geoscience*, 11(6), 378–380. https://doi.org/10.1038/s41561-018-0142-4

Peters, G. P., Andrew, R. M., Solomon, S., & Friedlingstein, P. (2015). Measuring a fair and ambitious climate agreement using cumulative emissions. *Environmental Research Letters*, 10(10), 105004. https://doi.org/10.1088/1748-9326/10/10/105004

Quéré, C. L., Andrew, R. M., Friedlingstein, P., Sitch, S., Pongratz, J., Manning, A. C., … Zhu, D. (2018). Global carbon budget 2017. *Earth System Science Data*, 10(1), 405–448. https://doi.org/10.5194/essd-10-405-2018

Raupach, M. R., Davis, S. J., Peters, G. P., Andrew, R. M., Canadell, J. G., Ciais, P., … Le Quéré, C. (2014). Sharing a quota on cumulative carbon emissions. *Nature Climate Change*, 4(10), 873–879. https://doi.org/10.1038/nclimate2384

Riahi, K., van Vuuren, D. P., Kriegler, E., Edmonds, J., O'Neill, B. C., Fujimori, S., … Tavoni, M. (2017). The Shared Socioeconomic Pathways and their energy, land use, and greenhouse gas emissions implications: An overview. *Global Environmental Change*, 42, 153–168. https://doi.org/10.1016/j.gloenvcha.2016.05.009

Robinson, M., & Shine, T. (2018). Achieving a climate justice pathway to 1.5°C. *Nature Climate Change*, 8(7), 564–569. https://doi.org/10.1038/s41558-018-0189-7

Robiou du Pont, Y., Jeffery, M. L., Gütschow, J., Rogelj, J., Christoff, P., & Meinshausen, M. (2017). Equitable mitigation to achieve the Paris Agreement goals. *Nature Climate Change*, 7(1), 38–43. https://doi.org/10.1038/nclimate3186

Rockström, J., Steffen, W., Noone, K., Persson, Å., Chapin, F. S., Lambin, E. F., … Foley, J. A. (2009). A safe operating space for humanity. *Nature*, 461(7263), 472–475. https://doi.org/10.1038/461472a

Rogelj, J., Popp, A., Calvin, K. V., Luderer, G., Emmerling, J., Gernaat, D., … Tavoni, M. (2018). Scenarios towards limiting global mean temperature increase below 1.5°C. *Nature Climate Change*, 1. https://doi.org/10.1038/s41558-018-0091-3

Rogelj, J., Reisinger, A., McCollum, D. L., Knutti, R., Riahi, K., & Meinshausen, M. (2015a). Mitigation choices impact carbon budget size compatible with low temperature goals. *Environmental Research Letters*, 10(7), 075003. https://doi.org/10.1088/1748-9326/10/7/075003

Rogelj, J., Schaeffer, M., Friedlingstein, P., Gillett, N. P., van Vuuren, D. P., Riahi, K., … Knutti, R. (2016). Differences between carbon budget estimates unravelled. *Nature Climate Change*, 6(3), 245–252. https://doi.org/10.1038/nclimate2868

Rogelj, J., Schaeffer, M., Meinshausen, M., Knutti, R., Alcamo, J., Riahi, K., & Hare, W. (2015b). Zero emission targets as long-term global goals for climate protection. *Environmental Research Letters*, 10(10), 105007. https://doi.org/10.1088/1748-9326/10/10/105007

Rozenberg, J., Guivarch, C., Lempert, R., & Hallegatte, S. (2013). Building SSPs for climate policy analysis: A scenario elicitation methodology to map the space of possible future challenges to mitigation and adaptation. *Climatic Change*, 122(3), 509–522. https://doi.org/10.1007/s10584-013-0904-3

Schellnhuber, H. J., Rahmstorf, S., & Winkelmann, R. (2016). Why the right climate target was agreed in Paris. *Nature Climate Change*, 6(7), 649–653. https://doi.org/10.1038/nclimate3013

Simmons, C. T., & Matthews, H. D. (2016). Assessing the implications of human land-use change for the transient climate response to cumulative carbon emissions. *Environmental Research Letters*, 11(3), 035001. https://doi.org/10.1088/1748-9326/11/3/035001

Skeie, R. B., Fuglestvedt, J., Berntsen, T., Peters, G. P., Andrew, R., Allen, M., & Kallbekken, S. (2017). Perspective has a strong effect on the calculation of historical contributions to global warming. *Environmental Research Letters*, 12(2), 024022. https://doi.org/10.1088/1748-9326/aa5b0a

Strefler, J., Bauer, N., Kriegler, E., Popp, A., Giannousakis, A., & Edenhofer, O. (2018). Between Scylla and Charybdis: Delayed mitigation narrows the passage between large-scale CDR and high costs. *Environmental Research Letters*, 13(4), 044015. https://doi.org/10.1088/1748-9326/aab2ba

Tachiiri, K., Hajima, T., & Kawamiya, M. (2015). Increase of uncertainty in transient climate response to cumulative carbon emissions after stabilization of atmospheric CO_2 concentration. *Environmental Research Letters*, 10(12), 125018. https://doi.org/10.1088/1748-9326/10/12/125018

UNEP. (2016). *The Emissions Gap Report 2016*. Nairobi: The United Nations Environment Programme (UNEP). Retrieved from http://uneplive.unep.org/theme/index/13#egr

UNFCCC. (1997). *Proposed Elements of a Protocol to the United Nations Framework Convention on Climate Change*. In *Conference in Brazil in response to the Berlin Mandate*. Brazil: UNFCCC. Retrieved from http://unfccc.int/cop5/resource/docs/1997/agbm/misc01a3.htm

UNFCCC. (2015). *Paris Agreement (No. FCCC/CP/2015/10/Add.1)* (p. 16). Paris, FRANCE: United Nations Framework Convention on Climate Change. Retrieved from http://unfccc.int/paris_agreement/items/9485.php

UNFCCC. (2016). *Aggregate Effect of the Intended Nationally Determined Contributions: An Update (No. FCCC/CP/2016/2)* (p. 75). Marrakech: United Nations Framework Convention on Climate Change. Retrieved from http://unfccc.int/focus/indc_portal/items/9240.php

van Vuuren, D. P., van Soest, H., Riahi, K., Clarke, L., Krey, V., Kriegler, E., … Tavoni, M. (2016). Carbon budgets and energy transition pathways. *Environmental Research Letters*, 11(7), 075002. https://doi.org/10.1088/1748-9326/11/7/075002

van Vuuren, D. P., Stehfest, E., Gernaat, D. E. H. J., van den Berg, M., Bijl, D. L., de Boer, H. S., … van Sluisveld, M. A. E.. (2018). Alternative pathways to the 1.5°C target reduce the need for negative emission technologies. *Nature Climate Change*, 8(5), 391–397. https://doi.org/10.1038/s41558-018-0119-8

Part II

Myths and metaphors of unreason

6 Why the metaphor of complementary dualism, and metaphor itself, are foundational to achieving sustainability

Edmond Byrne

> Men do not know how that which is drawn in different directions harmonises with itself. The harmonious structure of the world depends upon opposite tension like that of the bow and the lyre.
>
> (Heraclitus, ca 500 BCE, fragment 45, 1907: 150)

The metaphor of complementary dualism, the idea that polar opposites are inherently necessary and mutually obligatory for system progress and endurance and indeed in conceiving reality itself has found resonance across human civilisations throughout history. Eastern traditions such as Taoism and Zen Buddhism espouse the complementary intermingling opposites of yin (black; freedom) and yang (white; structure/constraint) in a way similar to the early Greek philosopher Heraclitus, who considered that the universe could be understood by observing the *relationship* between opposition and unity, such that one can only begin to grasp reality by seeing apparently opposing forces as parts of one whole. Nietzsche considered the ancient Greek tragedy in *The Birth of Tragedy* (Nietzsche, 1872), and decried the use of Socratic rationalism, which he attributed to the demise of original Greek tragedy, since it compromises the necessary balance between the complementary opposites found through opposing Apollonian and Dionysian tendencies. Apollo and Dionysus, as sons of Zeus, are respectively recognised in this context as the god of truth and knowledge (Apollo; broadly representing structure and reason) and the god of the vine (Dionysus; broadly representing unrestraint and disorder). Later, the twentieth century Nobel prize winning atomic physicist, Niels Bohr, incorporated the Taoist yin and yang symbol when designing a coat of arms ahead of his receipt of the Royal Danish Order of the Elephant knighthood in 1947, incorporating the motto: "*contraria sunt complementa*" (Figure 6.1). This was consistent with his philosophy, informed by his paradigm breaking work around his 'principle of complementarity', whereby he articulated the reality that atomic entities could espouse apparently contradictory properties through existing simultaneously in both wavelike and particulate form.

The metaphor of complementary dualism is also, as Heraclitus points out in the quotation which opens this chapter, a metaphor which appears both paradoxical and counter intuitive. A simpler, more rational (though not reasoned) approach would, by contrast, support an attempt to optimise systems, by seeking progress and viewing reality through the "either/or" preferment of one tendency at the expense of the

DOI: 10.4324/9781003143567-8

Figure 6.1 Niels Bohr's coat of arms.

Source: GJo, 2010.

other. In this way, efficiency is invariably prioritised over redundancy, deterministic control over random chance, structure over chaos, and linear progress over contingent process. Indeed, this has been the hallmark of the project of modernity, which has promoted a paradigm of control, certainty, reduction, and separation. This paradigm has been incredibly successful over the past four centuries, and, inspired by neo-Cartesian rationality, it now transcends all of our increasingly globalised society(/ies). Cartesian dualism opposes and critically differentiates from Heraclitian/Taoist dualism in that the former is antagonistic (envisaging mutual exclusivity), while the latter is agonistic (envisaging interconnection and overlap). The neo-Cartesian worldview can also be held responsible however for an increasingly dangerous anthropogenic hubris, in part a consequence of casting away the myth, mystery, mysticism, and enchantment of the pre-modern worldview. The once dominant (pre-modern) worldview, which determined that a singular truth could be revealed from a divine created order, was thus displaced by another that ordained that a singular truth could be revealed through a largely reductionist conception of "science". However, as neuropsychiatrist Ian McGilchrist (more anon) has put it: 'there may be no single truth, but no single truth does not mean no truth'. In addressing a science which would seek reductionist certainty, McGilchrist proposes:

> The origins of science lie in open-mindedness, flexibility in applying models, empirical observation of the world of experience …Don't get me wrong: detailed scientific knowledge is hugely important. We rely on such minute

> information to inform the bigger picture. But it is a necessary, not sufficient, condition, of being a good physician. Without a way of understanding and interpreting it at a deeper level, more detailed knowledge will achieve precisely nothing, and will lead us ultimately to let our patients down. It will close our reality down into what we imagine to be certain, where an appropriate awareness of the limitations of our knowledge would have liberated us and our patients into a world much richer than we can suspect.
>
> (McGilchrist, 2011: 1069)

A complex, contingent and entangled, though ultimately a more fruitful (and authentic) complementary agonistic dualism is thus a more difficult concept to hold and pursue than a simpler antagonistic dualism; "good or bad", "either/or", "for or against", "control or be controlled", conception of incompatible opposites. The latter however, by seeking to stamp out diversity, actually denies the possibility of procuring emergent (radically new) knowledge through transcending both extremes, or of achieving authentic evolutionary progress, or novel creativity. By focussing exclusively on (seeking to increase) system efficiency and constraint, this only serves to reduce system resilience and capacity for authentic growth. Such an approach, as has been shown in the context of ecosystem network flows (Ulanowicz, et al., 2009) and extended in terms of economic and financial sustainability (Goerner, et al., 2009; Kharrazi, et al., 2013; Levin and Lo, 2015; Fath et al., 2019)), is inherently unsustainable.

An intriguing aspect of this metaphor, however, is that it is replicated in the structure of the organ which facilitates our very construction of reality, the brain. Indeed, it is contended that we seek to construct a conception of reality which mirrors the structural composition of the brain, mediated through the relationship between the respective hemispheres (McGilchrist, 2009). The nature and import of such a structure has been described by the aforementioned Ian McGilchrist, in a seminal book on the nature of the brain *The Master and his Emissary* (McGilchrist, 2009). In it, McGilchrist describes two asymmetrically different though complementary hemispheres, extending the work of others such as Ornstein (1997). He explains with great precision how the two hemispheres conceive the world in radically different ways and compete to impose their perspective of the world around them, and then (literally) construct it accordingly. However, he also contends that it is precisely *because* our brains (and those of other animals) are constructed with two anatomically separate asymmetrical hemispheres which are connected via a rather small amount of neural tissue, centred around what is called the corpus callosum, rather than a single spherical lump (which would be, after all, a more efficient use of head-space), that we have been able to evolve and progress as we have done. Indeed, he emphasises the degree of physical separation between the respective hemispheres by pointing out that while most of the nerve cells of the corpus callosum promote communication between the hemispheres, nevertheless a significant proportion of them actually act in an inhibitory fashion (McGilchrist, 2009: 17). Yet, he points out that the resultant separation and complementary duality does not just merely act as a useful means of refining control, it is essentially 'a tool for grappling with the world. It's what brings the world about.' (McGilchrist, 2009: 19).

Moreover, in the case of humans, the frontal lobes (in both hemispheres) have developed significantly (representing as much as 35 per cent of the human brain, as against about 17 per cent for lesser apes, and about 7 per cent for relatively intelligent animals, such as dogs (McGilchrist, 2009: 21)). This has had the effect of enabling us to stand back from the world, from 'the immediacy of experience …to plan, to think flexibly and inventively, and in brief, to take control of the world around us rather than simply respond to it passively.' (McGilchrist, 2009: 21), traits which enable us to exhibit both calculated detachment and empathetic trust; thus turning us 'famously, into the "social animal", and into an animal with a spiritual dimension' (McGilchrist, 2009: 22).

In terms of the hemispherical differences, McGilchrist describes the approaches to reality adopted by each hemisphere. The left hemisphere takes a literalist, rationalist, mechanistic, atomistic, explicit, generalised, functionalist, decontextualised, and "either/or" approach to the world around it and it attempts to shape reality on these terms. The right hemisphere, however, takes an integrative (facilitating "both/and") approach, while seeking both context and (inter)connection, is comfortable with the implicit, and can "get it" beyond the functional and utilitarian world of rational models, deterministic algorithms, and (necessarily) reductionist words (Table 6.1). It thus communicates through (and understands/appreciates the inherent value of) art, images, poetry, music, and indeed metaphor. In this way, the right hemisphere can (exclusively) access implicit (though authentic) feelings and connotations through music, art, poetry, and other creative or spiritual experiences, including in a communal setting, to (metaphorically) *move* a person or group, and/or to attain overall understanding, clarity, or "aha!" moments, in ways which literally 'cannot be put into words'. The communicative discrete building blocks of the left hemisphere by contrast are wholly inadequate to such tasks, since (mere) words and denotative language alone literally cannot reach places that that the right hemisphere can reach via these media and constructs. Metaphor, as with irony and humour, is a facility accessed and understood exclusively by the right. If, for the benefit of the left hemisphere it needs to be explicated (through surrogate words), it is unavoidably reduced and stripped of its full (implicit) meaning.

Of course, both hemispheres are required to operate effectively. McGilchrist elaborates on how both hemispheres function independently but concurrently by instancing the example of a bird, who when it is feeding, employs both its left hemisphere for the necessary precision requiring a "local" strategy of seeking out grain among grit, while at the same time it employs the right hemisphere to capture an equally necessary big picture "global" approach and keep an eye out for potential predators:

> In general terms, then, the left hemisphere yields narrow, focussed attention, mainly for the purpose of getting and feeding. The right hemisphere yields a broad, vigilant attention, the purpose of which appears to be awareness of signals from the surroundings, especially of other creatures who are potential predators or potential mates, foes or friends; and it is involved in bonding in social animals. It might be that the division of the human brain is also the result of the need to bring to bear two incompatible types of attention on the

Table 6.1 Characteristics of respective brain hemispheres.

Left Hemisphere Thinking	*Right Hemisphere Thinking*
Asymmetrical: doesn't recognise right side thinking ("*either-or*" hemisphere)	Asymmetrical: accommodate left side thinking (*integrative* hemisphere)
Analysis, categorisation	Whole picture "*gestalt*"
External "view from nowhere"	Relational "view from somewhere"
Seeks control, certainty	Embraces contingent uncertainty
Symbols, models, musical notes, (denotative) language	Music, spirituality/religion, (connotative & emotive) language: metaphor, irony, humour, poetry
Reduction of whole into separate discretised parts	Interconnectedness, interactions, "betweenness"
Technocratic	Technosceptic
Mechanistic/deterministic; cause and effect	System emergence; propensities
Knowledge as fixed: obtained from facts	Knowledge as contextual: (unique) personal experience required
Individualistic, competitive	Relationships/collective, cooperative
Self-considered "objective" and "value-free" (though (ironically!) actually value laden)	Recognises values as inherent
Utilitarian (e.g. economic) values	Relational moral values and empathy
Ever optimistic (even hubristic/in denial)	Pessimistic/melancholic/realistic
Seeks decontextualisation (as context!)	Embraces context
Stumped by paradox and ambiguity	Embraces and values paradox and ambiguity
Closed system in discretised time (*zeitraffer*/timelapse)	Open systems in flux (process)
Utility and predictability	Novelty and creativity
"Law of excluded middle" (Plato/Aristotle)	"Logic of the included middle" (Nicolescu, 2008)
Build on understanding through adding piece by piece sequentially	Develop understanding through recursive process, considering various aspects of whole at once
Dogmatic	Pragmatic

Source: Based on McGilchrist, 2009.

> world at the same time, one narrow, focussed, and directed by our needs, and the other broad, open, and directed towards whatever else is going on in the world apart from ourselves.
>
> In humans, just as in animals and birds, it turns out that each hemisphere attends to the world in a different way – and the ways are consistent. The right hemisphere underwrites breadth and flexibility of attention, where the left hemisphere brings to bear focussed attention. This has the related consequence that the right hemisphere sees things whole, and in their context, and broken into parts, from which it then reconstructs a 'whole': something very different. And it also turns out that the capacities that help us, as humans, form bonds with others – empathy, emotional understanding, and so on – which involve a quite different kind of attention paid to the world, are largely right hemisphere functions.
>
> (McGilchrist, 2009: 27)

He concludes in his articulation of the differences between the right and left hemispheres respectively as 'at its simplest, a world where there is 'betweenness', and one where there is not. These are not different ways of *thinking about* the world: they are different ways of *being in* the world' (McGilchrist, 2009: 31, emphasis is the original).

It is on this neurophysiological basis that McGilchrist proceeds to develop his thesis which contends that, while the two hemispheres emerged as a means of progressing biological evolution (enabling both decontextualised focus and global breadth), at different points in human history the relative ascendency of either hemisphere towards the other has changed. He suggests that in the modern period (over the past four centuries), and in particular in more recent times, we have privileged the rationalistic decontextualisation of the left hemisphere over the integrative "betweenness" of the right hemisphere. The result is 'the world as machine metaphor' which we've inherited and developed from Descartes and Newton (Montuori, 2017: 147). The ultimate result though, he contends, may be to our existential detriment, since 'alone they [the two hemispheres] are destructive....right now [with left hemisphere dominance] they may be bringing us close to forfeiting the civilisation they helped us create' (McGilchrist, 2009: 93).

The dominance of the left hemisphere is particularly problematic, and since it is the "either/or" hemisphere, it envisions that either *it* has the unique and "correct" view of the world or it doesn't; and *it* must be right:

> Each hemisphere delivers a vital aspect of experience, and nothing good comes from relying over much on one alone. But that itself is the view of only one of the hemispheres. The other thinks that it can go it alone.
>
> (McGilchirst, 2011: 1069)

The right hemisphere, asymmetrically, takes an integrative "both/and" view, and can see the value in a plurality of perspectives. (Hedlund-de Witt's integrative worldview framework is useful in this context, seeing as it compares respective traditional (premodern), modern, postmodern, and integrative worldviews (Hedlund-de Witt, 2013; see also Byrne, 2017a: 56).) Indeed, it finds the singular approach of the left to be an inadequate conception of reality (mirroring by analogy, the differences between an intolerant group and a tolerant group – whereby at the extreme, the latter are faced with a conundrum around whether to tolerate intolerance, while the former group has no such problem).

The left hemisphere bias has manifested itself through the dominant modern worldview, which has underpinned enlightenment modernity, and to an ever-increasing extent has rolled forward in contemporary society. It represents a paradigm of separation and reduction (as described above), and we have constructed our modern contemporary world based on this paradigm; one (at least from a right hemisphere perspective) envisaged largely through a linear "one-eyed" conception of progress, a sort of (metaphorical) machine lavishly oiled by techno-optimistic hubris, as critiqued for example by Barry (2017). "Sustainability", by this worldview is epitomised for example, by utopian ideals such as India's 'One Hundred Smart Cities' project, where cities compete to be included in grand "entrepreneurial urbanisation" projects (Datta, 2015). These would effectively produce privatised technocratic control and monitoring centres, moderated through the likes of smart access cards, with the promise of being "efficient", "clean", "smart", and "sustainable". Contemporaneously, lakes literally burn as a result of extraordinary concentrations of toxic flammable waste in the sprawling IT hub of Bangalore (Bhasthi, 2017; Bhat 2017), a metaphorical cry of anguish from a neglected environment subject to extreme and ongoing anthropogenic degradation. The idea of clean smart cities juxtaposed with market-driven restricted access raises serious concerns around equity, social exclusion, and democracy, quite apart from ecological implications for such "sustainable" constructs (Unnikrishnan and Nagendra, 2015; Mundoli, et al. 2017).

The modern worldview and its outworking has, it must also be noted, brought us great good, not least in terms of material wealth, an unprecedented abundance of goods, energy, food, and information. Yet, it has also brought us to a contemporary crisis of unsustainability and all that that entails; including unprecedented biodiversity loss and unprecedented environmental degradation, increasing societal inequality (coupled with emerging global issues around food, water, and energy equity and security), and "social recession" characterised by increasing health issues around anxiety, depression, suicide, obesity, well-being, etc., on top of the increasing impacts and hardships around climate change. As Wessels rhetorically queries: 'is progress truly possible if its wake continually generates loss – loss of connections to place and community, loss of clean air and water, loss of other species who are truly part of our ancestral family tree?' (Wessels, 2013, xiv). In this context, McGilchrist's characterisation of the brain's hemispheres and the implications for how this impacts on the way society both envisions and structures its lived-in world has resonated with many, prompting them to consider the implications of his model for contemporary (un)sustainability challenges and narratives (Gare, 2013; Kras, 2015; Ehrenfeld, 2017, 2019).

Complementary dualism and sustainability narratives

The pervasive concept of complementary dualism is one which has informed conceptions around sustainability (Byrne, 2017a). It has been applied by Ulanowicz when quantitatively applied to ecosystem networks (Ulanowicz, et al., 2009), and qualitatively for socio-technical systems by Stirling (2014) (see also Byrne, 2017a for a description of each). Essentially, Ulanowicz has shown that flourishing (sustainable) ecosystems operate between the mutual agonistic tendencies of constraint (efficiency/control) and freedom (disorder) (Ulanowicz, 2009a, 2009b), while evolution occurs (and can only occur) in the space between (genetic) causal determinism and directed chance in Kauffman's zone of the "adjacent possible". (Kauffman, 2010: 64). This challenges a reductionist neo-Darwinian conception of biological evolution fundamentally and exclusively predicated on a linear conception of progress driven by competitive 'survival of the fittest'. As Ulanowicz (2017: 14) puts it: 'Natural selection no longer involves only elimination but is driven by a mutualism that is more fundamental than even competition.' This aligns with the neurobiologically informed critique of McGilchrist, who sees hemispherical difference and cooperation writ large on the universe and reality in Heraclitean fashion:

> The relationship between the hemispheres is not straightforward. Difference can be creative: harmony is an example. Here differences cohere to make something greater than either or any of the constituents alone… Before there can be harmony there must be difference.
>
> The most fundamental observation one can make about the observable universe - is that there are at all levels forces that tend to coherence and unification, and forces that tend to incoherence and separation. The tension between them seems to be an inalienable condition of existence, regardless of the level at which one contemplates it. The hemispheres of the human brain, I believe, are an expression of this necessary tension. And the two hemispheres adopt different stances about their differences: the right hemisphere towards cohesion of the two dispositions, the left hemisphere towards competition between them.
>
> (McGilchrist, 2009: 128–129)

On considering complementary dualism as a universal phenomenon, Byrne (2017b) highlights how the universe itself goes about its business of existing (and expanding) amid the mutually opposing though necessary drivers towards dissipation (in the form of second law entropy) and attraction (e.g. gravitational force). To highlight the universality of this phenomenon, Byrne (2017b) also demonstrates how complementary dualism has informed complexity informed approaches across a range of disciplinary spheres, including quantum physics, thermodynamics, electrical power systems, organisational management, and process theology. Indeed, process thinking, which has developed most notably from the works of Whitehead (1929), is based on an abiding complementary dualism recognised right through from early civilisation(s), and presents itself as a (right hemisphere) counterbalance to a reductionist neo-Cartesian representation of reality and progress. In this vein, Ulanowicz identifies:

> an ongoing tension between countervailing tendencies that plays out in living systems. That is, the prevailing dynamic is not one of uniform progression towards some maximal efficiency, but rather a Heraclitean dialectic between order building and decay. This new dynamic also mirrors well the Eastern emphasis on the importance of the nonexistent, and the Chinese dialectic between Yan (constraint) and Yin (freedom). With its roots in both Western and Eastern philosophies, process ecology holds forth the possibility of a more equitable common developmental road towards scientific progress.
>
> (Ulanowicz, 2017: 13)

The narrative of agonistic dualism can be framed too in a transdisciplinary context, in particular with reference to Nicolescu's 'logic of the included middle' (Nicolescu, 2008: 7), which would contend that emergent creative evolutionary knowledge and physical and social constructs (and hence progress) can only materialise through a productive interplay between respective agonistic and interdependent opposites (Mullally, et al., 2017: 32–37). This can be applied too in conceiving education and learning more generally, which can be characterised as a dialectic relationship between general subject excitement provided by the teacher, followed by precision and mastery performed by the student, and finally general re-presentation by student application and 'the creative freedom that issues from the discipline of hard work' (Schindler, 1991: 72).

Such a complexity informed view of reality would envisage such ("both/and") terms among disciplines, whereby each respective discipline legitimately conceives the world and reality at different levels (for example, characterising a person from the perspective of various disciplinary "object world" views such as for example, at the level of, respectively, atomic physics, biochemistry, biology, neuroscience, psychology, the humanities (the human condition), and sociology (persons in relation to others). Each disciplinary perspective offers wholly legitimate knowledge, and each is capable of adding unique insight, but none is capable of unilaterally describing *all* of human reality. Moreover, only together can disciplinary knowledge and insights facilitate an emergent and transcendent conception of (multi-level) reality from the resultant "in-betweenness". This is the sort of transdisciplinary thinking, it is argued, the embodiment of the metaphor of complementary dualism, which behoves authentic sustainability narratives, and which is ultimately required for making a necessary transition from a world and society constructed as it currently exists, based on assumptions of separation and reduction, to one which is authentic fit-for-purpose in conceiving and cohering with complex reality.

The left hemisphere conception of the world by contrast would seek to construct techno-optimistic models and conceptions of sustainability, which would see sustainability as a largely unproblematic and linear one-way journey coherent with conceptions of societal progress. "Sustainability", as thus envisaged, is underpinned by quantifiable metrics and is to be engaged within the context of a growth-based consumerism (thus providing companies who engage with it with a valuable marketing opportunity for growth). By such measure, John Ehrenfeld would contend that 'it has become merely a label for strategies actually driven by standard economic and institutional mechanisms around efficiency' (Ehrenfeld and Hoffman, 2013: 3), and a model of consumption which 'places no limits on demand or want' (Ehrenfeld, 2008: 127). This contrasts starkly with Ehrenfeld's purposely qualitative proposed definition of sustainability, the metaphor of *sustainability-as-flourishing* (Ehrenfeld and Hoffman, 2013: 3). Mirroring the brain hemispherical discourse above, he proposes 'we need to move away from the purely objective, quantitative and "rational" reasoning to consider the spiritual, experiential, and pragmatic.' (Ehrenfeld and Hoffman, 2013: 19). Work by Elphinstone and Critchley (2016) has suggested a positive correlation between left hemisphere mechanistic thinking and the materialistic goals and values which underpin consumerist society, while Dittmar et al. (2014) suggest a link between enhanced materialism and reduced well-being. Each of these outcomes in turn oppose the realisation of sustainability as framed by Ehrenfeld.

Of course, to extend the metaphor with another; a boat with one oar just goes around in circles. Continuing the theme, Petersen counters by conceiving of sustainability as more a contingent, context-dependent *process*, than some linear pre-determinable finite journey (see Byrne, 2017a: 49):

> Like a pot of gold at the end of the rainbow, sustainability is more of a moving target never quite to be reached. Using a navigational metaphor thus captures the concept more comfortably: sustainability discourses help us steer in a sea of future challenges and navigate around the rocky patches of undesirable solutions. In this capacity, as a navigational device, the specific sustainability discourses are also locally defining the legitimacy of new socio-material arrangements, such as technological systems.
>
> (Petersen, 2013: 2)

The foundational value of metaphor in general

> Only the right hemisphere has the capacity to understand metaphor.
>
> (McGilchrist, 2009: 115)

McGilchrist's above proposition has significant import. He elaborates, suggesting that 'this is not a small matter of a quaint literary function having to find a place *somewhere* in the brain.' (McGilchrist, 2009: 4), but actually helps us understand the fundamental role metaphor plays in 'our understanding of the world, because it is the *only* way in which understanding can reach outside the system of signs to life itself. It is what links language to life' (McGilchrist, 2009: 115).

He outlines how the left hemisphere (it being the "explicit" hemisphere) cannot help but recognise metaphor as being nothing more than a mere superficial collection of words, colours, notes, blocks, or fabled stories, devoid of potential for any additional, implicit, connotative, emergent, or transcendent value. McGilchrist argues though that it is this left hemisphere view that gets in the way of us understanding the real value of metaphor; for metaphors are not just merely useful devices to comparatively or approximately describe reality, but in fact it is the other way around: our basic understanding of the world around us is primarily *framed* through metaphor in the right hemisphere. It is only then "translated" into left hemisphere words and language:

> Metaphor (subserved by the right hemisphere) comes *before* denotation (subserved by the left). This is a historical truth, in the sense that denotative language, even philosophical and scientific language, are derived from metaphors founded on immediate experience of the tangible world.
>
> (McGilchrist, 2009: 118, emphasis is the original)

To put this more succinctly, citing Lakoff and Johnson (1999: 123); 'metaphorical language is a reflection of metaphorical thought'. However, metaphor necessarily loses some meaning when converted into (mere reductive) words, since:

> Metaphor *embodies* thought and places it in a living *context*. These three areas of difference between the hemispheres – metaphor, context and the body – are all interpenetrated with one another.
>
> (McGilchrist, 2009: 118, emphasis is the original)

The right-left dialectic is maintained in the interplay between the implicit metaphor and explicit language:

> The point of metaphor is to bring together the whole of one thing with the whole of another, so that each is looked at in different light. And it works both ways, as the coming together of one thing with another always must. You can't pin one down so that it doesn't move, while the other is drawn towards it: they must draw towards each other. As Max Black says: 'If to call a man a wolf is to put him in a special light, we must not forget that the metaphor makes the wolf seem more human than he otherwise would'.
>
> (McGilchrist, 2009: 117)

In this context, it is hardly surprising that a key outcome of an increased dominance of left hemisphere rationality (over right hemisphere (common sense intuitive) "reason") is a loss in the perceived value and power of metaphor (and also indeed, as Evan Boyle eloquently identifies elsewhere in this collection, of myth). And this has hugely significant implications for both narratives and conceptions of sustainability, and the ultimate realisation of same. For a start, it demonstrates why there are competing conceptions of sustainability. Such conceptions range from the right hemisphere "sustainability-as-flourishing" model espoused by John Ehrenfeld, which would see it as at heart, an ethical issue, to a left hemisphere conception which would reduce sustainability to mere quantifiable techno-scientific metrics such as carbon emissions and energy efficiencies. Indeed, Ehrenfeld explicitly examines the implications of McGilchrist's work for sustainability narratives more recently (Ehrenfeld, 2019). The left hemisphere focusses on constructing algorithmic models of the world as it aims to (ever more perfectly) mimic and predict. It strives to do this increasingly through a range of technological silver bullets including geoengineering, nanotechnology, synthetic biology, virtual reality, artificial intelligence, etc., all aimed at assuring unrelenting progress in the face of any problem, even increasingly existential ones concerning (un)sustainability (around climate, food, water, energy, etc.), each euphemistically labelled "grand challenges".

However, a right hemisphere approach would critique the left, by suggesting that this simply proposes more techno-optimism as a proposed "fix", but in fact one which inevitably can only lead to failure through inevitable knock-on "unintended consequences". In other words, like an addict, it merely seeks to satisfy a need with more of the same of that which is implicated in the addiction in the first place (Ehrenfeld, 2008: 35). Thus, the reductionist neo-Cartesian worldview which has caused the problem of unsustainability promises to solve it with even more of the same.

A right hemisphere approach on the other hand would hold that its preferred integrative worldview is necessary in order to construct a sustainable world which actually coheres with the actual "real" world in all its authenticity (rather than a world reduced to algorithmically modelled virtual reality). It would thereby seek to develop such a world through empirical experienced reality and associated metaphor. Any transition to a sustainable societal construct it would suggest can only be accessed through the use of metaphorical thinking, rather than through (restating) rational facts alone.

If this perspective is accepted, it promises to reveal a tantalising insight into perhaps why, despite knowing a myriad of facts about climate change, biodiversity loss, and degradation across several other environmental and social indicators, we are generally and collectively unmoved to take action. It also holds the prospect of metaphor (and its cousins; narrative, story, and myth, as well as other right hemisphere constructs such as art and music) is really the only means of sufficiently moving people, communities, societies, and individuals so as to precipitate the type of transformational change that is required to achieve authentic sustainability (as flourishing).

Indeed, it appears that the penny may be finally dropping, so to speak, in this regard in relation to funding of research into areas such as climate change and energy transitions to identify just two. While up to very recently research funding agencies at national and international levels have generally only considered research in the STEM areas (Science, Technology, Engineering and Mathematics) as being legitimate disciplines for carrying out such research, there appears to be quite a perceptible turn in recent times towards funding research which explicitly demonstrates multi-, inter-, and trans-disciplinary aspects and which clearly and explicitly incorporates and values disciplinary input from the humanities and social sciences. It may well be that this has less to do with a turning away from the narrow focus that a left hemisphere rational approach would imply to a more global right hemisphere approach, and more to do with the fact that an exclusively "scientific" based approach has largely singularly failed to communicate the realities of climate change with public(s) in a way which would precipitate significant transformative change. Nevertheless, research projects in this area are routinely, and to an ever-greater extent, required to allocate portions of the budget allocation to public understanding and communications.

A bolt-on approach here, whereby social scientists are sought out and then charged with going out and communicating or finding the best way to manipulate societal behaviour and views once the physical scientists have done the "real" work, i.e. the heavy lifting of gathering the data, developing suitable algorithms, and designing appropriate technologies, would merely be just another outpouring of a left hemisphere separation (and prioritisation) of disciplinary tasks. Only a genuine (right hemisphere) transdisciplinarity, which would value the real and central input that the social sciences and humanities can and must provide can meet with ultimate success.

Sustainable salvation through myth, metaphor, and the sacred?

The concepts of myth, metaphor, and the sacred are all closely related and tied up in the right hemisphere of the brain while all are equally incomprehensible and superficial to the reductionist left. As Boyle demonstrates in his chapter on myth in this collection, myth is a story that is not just known but "believed" and is regarded as important or sacred. Keohane (2017: 161) highlights 'the central importance of sacred symbols to sustainable future communities' while citing the value of the sacred in developing the origins of respective civilisations in Mesopotamia, Greece, and Rome, for example, as well as in their ultimate destruction:

> By virtue of sacred symbols, a cosmos becomes structured and ordered, synchronically and diachronically. [On the other hand] ...the collapse of civilizations into dark ages, similarly, it is not military defeat or even natural disaster that is the crucial thing, but the dissolution and eclipse of the sacred symbols that organize the cosmos, that make any particular form of life meaningful, worth fighting for and worth defending.
>
> (Keohane, 2017: 161–162)

This is an utterly foreign and ridiculous concept to the rational left hemisphere, which would conceive of the sacred (as with myth and metaphor) as just an emptied, fossilised, antiquated pre-modern conception with little or no use beyond a status as a quaint museum piece among other pre-modern interesting peculiarities. However, the fact remains that the ancient human conception of the sacred has succeeded spectacularly over the ages in preserving and sustaining that which no other means could preserve. Striking examples include that of Bsharri grove, also known as the 'Cedars of God', one of the very last extant historic cedar groves in the Lebanon, and Mount Athos in Greece, a sacred space which managed to maintain its trees and vegetation throughout the millennia, unlike most other Greek mountains which have long been shorn by grazing capricious goats. These remnants of the past have been preserved by virtue of their historic statuses as sacred places, places where it is utterly and eternally taboo to defile, and which therefore have been rendered immune to the economic whims of market forces, whereby 'nothing is sacred' and everything has a market(able) monetary value.

Perhaps therefore, therein lies a key for how we might manage to protect the Amazonian forests or those of Indonesia, where the market forces of globalised agri-food system dictates that indigenous species and communities are less valuable than globally traded marketable commodities. Similarly, for our oceans, which have become dumping grounds for everything from plastics to excess atmospheric carbon while we continually overfish them. If these were to be designated as "sacred" in our collective minds, and if it were deemed to be taboo to desecrate them, then regardless of *any* monetary or market driven incentive, they too could have a far healthier prognosis. Indeed, perhaps this is the only way we can save them from ourselves. The relatively recent legal turn towards indoor smoking bans could only have worked because it quickly became a cultural norm; to such an extent that it is now expressly taboo to smoke indoors in a public place, and is thus simply not done, regardless of any nominal fine or penalty. Such penalties thus remain not called in, because the new cultural norms and constraints dictate that there is no need for legal policing, because it is simply unacceptable to smoke in company in an enclosed public space. The breathable common air in such spaces has been effectively designated as "sacred", and not for wanton contamination by known carcinogens.

We can thus seek to reinvigorate this right brained conception of "sacred", with or without its traditional religious connotations, if we seek to re-enchant our world and environment as something inherently worthy of (beyond materialistic) awe and respect. While the conception of sacred is typically associated with the explicit value set and beliefs of various pre-modern indigenous peoples, it can also find value in contemporary scientific settings. Examples include the work of eminent biologists David Suzuki, author of *The Sacred Balance: Rediscovering our place in nature* (Suzuki and McConnell, 1997) and Stuart Kauffman, who comes at things from a secular scientific biological perspective, in his book, a critique of reductionism entitled *'Reinventing the Sacred: A new view of science, reason, and religion'*. Kauffman implores that we can refashion our attitude to the world around us:

> if we reinvent the sacred, invent a global ethic, come together, and gradually find reverence – meaning, awe, wonder, orientation and responsibility in the world we share. Can I logically force you? No. Can the creativity in the universe invite you? Oh, yes. Listen.
>
> (Kauffman, 2010: 442)

Indeed, Kauffman would consider the self-organised "partially lawless" creativity that is fundamental to the evolution of the universe to be "God enough" for him, and indeed for many others, ahead of the idea of a six day Abrahamic "Creator God":

> I believe the latter is so stunning, so overwhelming, so worthy of awe, gratitude and respect that it is God enough for many of us. God, a fully natural God, is the very creativity in the universe. ...This view of God can be a shared religious and spiritual space for us all.
>
> (Kauffman, 2010: 9)

Others, coming from a religious and theological perspective, such as former Harvard theologian Gordon Kaufman would agree in conceptualising God as Creativity rather than as a hands-on controlling Creator God (Kaufman, 2000). We thus see how secular and religious interpretations can cohere, and even meet, when posited in a right hemisphere context, in ways that are simply impossible from the perspective of the left. This was clearly manifest in an open letter from several eminent scientists (including Nobel Laureates) to the religious community at a 1990 Global Forum of scientists in Moscow entitled *'Preserving and Cherishing the Earth: An Appeal for Joint Commitment in Science and Religion'*:

> As scientists, many of us have had profound experiences of awe and reverence before the universe. We understand that what is regarded as sacred is more likely to be treated with care and respect. Our planetary home should

> be so regarded. Efforts to safeguard and cherish the environment need to be infused with a vision of the sacred.
>
> (Global Forum, 1990: 2)

It is therefore perhaps unsurprising that constructs around the sacred, as well as myth and metaphor cohere so well in both spiritual/religious settings and integrative secular settings (such as described above), since all are accessed via the right hemisphere, while religious language itself is fundamentally mediated through myth and metaphor and the idea of the sacred.

John Ehrenfeld elaborates that his conception of *sustainability as flourishing* requires an ethic of *care* across four domains; *care* for oneself, *care* for other human beings, *care* for the material world (the environment), and *care* for that which is not in the material world (the transcendent and spiritual), thereby recognising various levels of interconnectedness (Ehrenfeld and Hoffman, 2013: 87–88). Again, this only makes any sense from the perspective of the right hemisphere. And again, though he speaks from a secular perspective, this language can also have resonance in religious and theological contexts. Directly echoing Ehrenfeld's four domains of *care*, Pope Francis, in his encyclical on sustainability and inequality, Laudato Si' (Francis, 2015: 45) suggests that 'disregard for the duty to cultivate and maintain a proper relationship with my neighbour, for whose care and custody I am responsible, ruins my relationship with my own self, with others, with God and with the earth.' He emphasises that 'everything is interconnected' and thus 'social love moves us to devise larger strategies to halt environmental degradation and to encourage a "culture of care" which permeates all of society' (Francis, 2015: 140).

Moreover, coming from a similar perspective, Josef Zycinski, one time archbishop of Lublin, wrote of a contemporary 'elimination of any kind of moral taboo', which has manifested itself as 'an alienation in which our ties to God, neighbour and nature are subordinated to illusory visions remote from God's plan of creation'. He suggests that this "alienation" flows from 'man's aspiration for absolute autonomy' and is thus (metaphorically or theologically) akin to 'the contemporary persistence of original sin' (Zycinski, 2006: 242). The value and need for a complementary dualism is evident here again; humans clearly have a particularly well-developed autonomy and freedom to generate relationships with each other and the world around us (rather than, as Laplace would have it, being deterministically controlled (by his "Demon"), as programmed automata). However, it would seem grossly hypocritical of us then if we were to seek to exercise this freedom abusively to try to wholly control (as separate, uncaring agents) the world around us (and thus seek to deny autonomy for others or the natural world around us), and thereby seek, in a fashion of "controlling gods", to wholly dictate and manipulate our self-created world to the detriment of others (people, species, etc.). This approach, suffused with both myopia and hubris (and typically laden with left hemisphere controlling techno-optimism), cannot succeed in producing some sort of utopian (or dystopian) outcome, but can only result in a sort of modern Promethean conclusion (itself the subtitle of the Frankenstein story). We would thus be fated to make the same mistakes over and over, in a futile attempt to escape from our current predicament. Einstein's oft quoted (if perhaps misattributed) definition of insanity springs to mind!

Conclusion

The metaphor of complementary dualism, writ large in the physical manifestation of the hemispheres of the brain, is seen to be both robust and pervasive. Sustainable progress, it is argued, can actually only be achieved through suitable contingent *balance* between control and freedom or between structure and chaos, in an ever-changing and evolutionary interconnected world and universe. Moreover, metaphor itself is key to both recognising and realising sustainability, since it can stimulate people and societies to be *moved* in a way that any amount of cold rational facts cannot. We have therefore come full circle: Metaphor, the language through which both of CP Snow's famous two cultures communicates (albeit implicitly in science), facilitates and indeed demands the healing of the rift between these two cultures.

By this measure, only a humble and open transdisciplinarity, moulded by recursive and complementary creative learning, can help realise a vista of authentic sustainability. If we accept, as McGilchrist suggests, that the relative influences of the respective hemispheres of that malleable complex organ that is the human brain have varied over time, history and context, then there are important consequences. Given that the relationship between respective hemispheres has helped provide a basis upon which we have conceptualised and built the very world around us, there has to be hope that we can recalibrate this complementary relationship as necessary to achieve a proportionate hemispherical balance (as McGilchrist has pointed out). Such a recalibration would facilitate the brain in helping realise its full potential to deal with complex contemporary and emerging existential challenges. Perhaps also, such a recalibration can rebalance hubristic (left hemisphere) tendencies with an appropriate level of (right hemisphere) groundedness and melancholy, thus imbuing the human temperament with an adequate balance to successfully tackle present and imminent challenges, while also, critically for us, facilitating recognition of their deep significance. In this way, the metaphor of complementary dualism as well as the construct of metaphor itself can be regarded as requisite companions in conceiving a successful and sustainable onward journey along the path of human civilisation.

References

Barry, J. 2017. Bio-fuelling the hummer? Transdisciplinary thoughts on techno-optimism and innovation in the transition from unsustainability. In: E. Byrne, G. Mullally and C. Sage, eds. *Transdisciplinary perspectives on transitions to sustainability*. Abingdon: Routledge, pp. 106–123.

Bhasthi, D. 2017. City of burning lakes: Experts fear Bangalore will be uninhabitable by 2025. *The Guardian*, 1 March 2017. [online] Available at: http://www.theguardian.com/cities/2017/mar/01/burning-lakes-experts-fear-bangalore-uninhabitable-2025 [Accessed 2 September 2019].

Bhat, H. 2017. Because the lake burns. In: L. Bremner and G. Trower, eds. *Monsoon [and other] airs*. London: University of Westminster. pp. 77–82.

Byrne, E. 2017a. Sustainability as contingent balance between opposing though interdependent tendencies. In: E. Byrne, G. Mullally and C. Sage, eds. *Transdisciplinary perspectives on transitions to sustainability*. Abingdon: Routledge, pp. 41–62.

Byrne, E. 2017b. Paradigmatic transformation across the disciplines; snapshots of an emerging complexity informed approach to progress, evolution and sustainability. In: E. Byrne, G. Mullally, and C. Sage, eds. *Transdisciplinary perspectives on transitions to sustainability*. Abingdon: Routledge, pp. 65–82.

Datta, A. 2015. New urban utopias of postcolonial India: 'Entrepreneurial urbanization' in Dholera smart city, Gujarat, *Dialogues in Human Geography*, 5(1), pp. 3–22.

Dittmar, H., Bond, R., Kasser, T., and Hurst, M. 2014. The relationship between materialism and personal well-being: A meta-analysis. *Journal of Personality and Social Psychology*, 107(5), pp. 879–924.

Ehrenfeld, J. R., 2008. *Sustainability by design*. New Haven: Yale University Press.

Ehrenfeld, J. R., 2017. Flourishing and the right brain. [online] Available at: http://www.johnehrenfeld.com/flourishing_and_the_right-brai/ [Accessed 2 September 2019].

Ehrenfeld, J. R., 2019. *The right way to flourish: Reconnecting with the real world*. Abingdon: Routledge.

Ehrenfeld, J. R. and Hoffman, A. J., 2013. *Flourishing: A frank conversation about sustainability*. Redwood City: Stanford University Press.

Elphinstone, B. and Critchley, C., 2016. Does the way you think and look at the world contribute to being materialistic? Epistemic style, metaphysics, and their influence on materialism and wellbeing. *Personality and Individual Differences*, 97, pp. 67–75.

Fath, B.D., Fiscus, D. A., Goerner, S. J., Berea A. and Ulanowicz. R. E., 2019. Measuring regenerative economics: 10 principles and measures undergirding systemic economic health. *Global Transitions* 1, pp. 15–27.

Francis, 2015. *Laudato Si', encyclical letter of the Holy Father on care for our common home*. New York: Melville House. [online] Available at: http://w2.vatican.va/content/francesco/en/encyclicals/documents/papa-francesco_20150524_enciclica-laudato-si.html [Accessed 2 September 2019].

Gare, A., 2013. The grand narrative of the age of re-embodiments: Beyond modernism and postmodernism. *Cosmos and History*, 9(1), pp. 327–357.

GJo, 2010. [online] Available at: https://upload.wikimedia.org/wikipedia/commons/d/da/Coat_of_Arms_of_Niels_Bohr.svg [Accessed 2 September 2019]. [CC BY-SA 3.0 (https://creativecommons.org/licenses/by-sa/3.0)]

Goerner, S.J., Lietaer, B., and Ulanowicz, R.E., 2009. Quantifying economic sustainability: Implications for free-enterprise theory, policy, and practice. *Ecological Economics*, 69(1), pp. 76–81.

Global Forum, 1990. *Preserving and cherishing the earth: An appeal for joint commitment in science and religion*. Moscow: Global Forum. [online] Available at: https://fore.yale.edu/sites/default/files/files/Preserving%20and%20Cherishing%20the%20Earth.pdf [Accessed 2 September 2019].

Hedlund-de Witt, A., 2013. *Worldviews and the transformation to sustainable societies*. PhD thesis. [online] Available at: http://dare.ubvu.vu.nl/handle/1871/48104 [Accessed 3 April 2018].

Heraclitus, 1907. Early Greek thinkers, in *The library of original sources: Volume II (the Greek World)*, O.J. Thatcher, ed. New York: University Research Extension.

Kauffman, S.A., 2010. *Reinventing the sacred: A new view of science, reason, and religion*. New York: Basic Books.

Kaufman, G.D., 2000. *In the beginning ... Creativity*, Minneapolis: Augsburg Fortress Publishers.

Keohane, K., 2017. Sustainable future ecological communities: On the absence and continuity of sacred symbols, sublime objects and charismatic heroes. In: E. Byrne, G. Mullally and C. Sage, eds. *Transdisciplinary perspectives on transitions to sustainability*. Abingdon: Routledge, pp. 158–169.

Kharrazi, A., Rovenskaya, E., Fath, B.D., Yarime, M., and Kraines, S., 2013. Methodological and ideological options quantifying the sustainability of economic resource networks: An ecological information-based approach. *Ecological Economics*, 90, pp. 177–186.

Kras, E., 2015. How we think: How it affects sustainable thinking. *Problems of Sustainable Development* 10(2), pp. 63–69.

Lakoff G. and Johnson, M., 1999. *Philosophy in the flesh: The embodied mind and its challenge to western thought*. New York: Basic books.

Levin, S.A. and Lo, A.W., 2015. Opinion: A new approach to financial regulation. *Proceedings of the National Academy of Sciences*. 112, pp. 12543–12544.

McGilchrist, I., 2009. *The master and his emissary*. Yale: Yale University Press.

McGilchrist, I., 2011. The Art of Medicine Paying attention to the bipartite brain. *The Lancet*, 377, pp. 1068–1069.

Mundoli, S., Unnikrishnan, H. and Nagendra, H. 2017. The 'Sustainable' in smart cities: Ignoring the importance of urban ecosystems. *Decision*, 44, pp. 103.

Montuori, A., 2017. The evolution of creativity and the creativity of evolution. *Spanda*, 7, pp. 147–156.

Mullally, G., Byrne, E. and Sage, C. 2017. Disciplines, perspectives and conversations. In: E. Byrne, G. Mullally and C. Sage, eds. *Transdisciplinary perspectives on transitions to sustainability*. Abingdon: Routledge, pp. 21–40.

Nicolescu, B., 2008. *Transdisciplinarity: Theory and practice*. Creskill: Hampton Press.

Nietzsche, F., 1872. *Die Geburt der Tragödie aus dem Geiste der Musik*. Leipzig: Verlag von E.W. Fritzsch.

Ornstein, R. E., 1997. *The right mind: Making sense of the hemispheres*. New York: Harcourt Brace.

Petersen, R.P., 2013. The potential role of design in a sustainable engineering profile. In: *Engineering education for sustainable development (EESD13)*, University of Cambridge, 22–25 September 2013. Cambridge: EESD.

Schindler, S., 1991. The Tao of teaching: Romance and process. *College Teaching*, 39(2), pp. 71–75.

Stirling, A., 2014. From sustainability, through diversity to transformation: towards more reflexive governance of technological vulnerability. In: A. Hommels, J. Mesman and W. Bijker, eds. *Vulnerability in technological cultures: new directions in research and governance*. Cambridge, MA: MIT Press.

Suzuki, D. and McConnell, A., 1997. *The sacred balance: Rediscovering our place in nature*. Vancouver: Greystone Books.

Ulanowicz, R.E., 2009a. *A third window: Natural life beyond Newton and Darwin*. West Conshohocken: Templeton Foundation Press.

Ulanowicz, R.E., 2009b. The dual nature of ecosystem dynamics. *Ecological Modelling*, 220 pp. 1886–1892.

Ulanowicz, R. E., Goerner, S. J., Lietaer, B. and Gomez, R., 2009. Quantifying sustainability: Resilience, efficiency and the return of information theory. *Ecological Complexity*, 6, pp. 27–36.

Ulanowicz, R.E., 2017. Towards a more global understanding of development and evolution, *Progress in Biophysics and Molecular Biology*, 131, pp. 12–14.

Unnikrishnan, H. and Nagendra, H., 2015. Privatizing the commons: Impact on ecosystem services in Bangalore's lakes. *Urban Ecosystems*, 18, pp. 613–632.

Wessels, T., 2013. *The myth of progress: Toward a sustainable future*. Vermont: University of Vermont Press.

Whitehead, A. N., 1929. *Process and Reality*, New York: Macmillan.

Zycinski, J., 2006. *God and evolution: Fundamental questions of Christian evolutionism*. Washington, DC: The Catholic University of America Press.

7 Myth beyond metaphor

Myths in transition

Evan James Boyle

Introduction

Myths, throughout much of human history, have served to explain aspects of human behaviour or nature and give guidance to individuals and societies. Moreover, myths often include a change in shape or form; a transformation. In contemporary society, we have reached a point within climate science whereby consensus has been reached yet inaction, for the most part, prevails. The case will be made here that myths can be described, much like metaphors themselves, as "imaginative rationality" (Lakoff and Johnson, 2008: 193). While the rational discussions on the necessity to transition towards more sustainable practices across society seems to be failing, perhaps more imaginative approaches are needed. Myths and their potential to assist in the societal transition to sustainability will be investigated herein.

> Let us consider abstract man stripped of myth, abstract education, abstract mores, abstract laws, abstract government; the random vagaries of the artistic imagination unchanneled by any native myth; a culture without fixed and consecrated place of origin, condemned to exhaust all possibilities and feed miserably and parasitically on every culture under the sun. Here we have our present age… Man today, stripped of myth, stands famished among all his pasts and must dig frantically for roots.
>
> (Friedrich Nietzsche, 2000)

The Oxford Dictionary of English Etymology traces the root of the contemporary use of the word myth to ca. 1830, where it comes to be defined as a 'fictitious narrative usually involving supernatural things' (Onions, 1966: 601). Coming a century earlier (in 1725), Vico's *New Science* interprets myths to contain both literal and symbolic truths simultaneously; he says 'mythos came to be defined for us as *vera narratio*, or true speech' (Vico, 1999: 127). 'Vico's attempt to retain the 'true meanings', or 'truths', of ancient myth for people in the modern age, people who have become, in his view, critical and utterly incredulous, was destined to expose them to the truths of innocent credulity' (Mali, 2002: 202). Moving from truths and 'innocent credulity' to fictions and the 'utterly incredulous' represents the emergence of a new way to think of myth. By revisiting myth, theoretical

DOI: 10.4324/9781003143567-9

considerations can be made on how to effectively appeal to both reason and emotions on the issue of climate change. The Oxford definition, coming after the emergence of the Enlightenment in the eighteenth century represents a shift in meaning of the term myth, reversing its interpretation from preliterate times and inherent implications of truth, to come to represent fictitious accounts, in keeping with the new science of the time. The Kantian perspective of a dawning of an enlightened age cut ties with tradition, beckoning in the age of modernity. While myths previously established social customs and moral lessons, with modernity they came to be a representation of irrationality, unable to stand up to scientific reasoning. It can be suggested, however, that we have never in fact been "modern" (Latour, 1993). If Modernity is a fallacy, the expulsion of myth to the realms of fantasy may have had detrimental impacts upon society.

Mythopoesis and the Myth Gap

The importance of myths has been interwoven throughout much of the work of the psychologist Carl Jung, who viewed myths as of central importance in the search for self-understanding. The modern world, in which myth has been rejected in favour of rationality, may be affected by a process of disconnection alluded to by Jung; 'The man who thinks he can live without myth or outside it is an exception. He is like one uprooted, having no true link with either the past, or the ancestral life within him, or yet with contemporary society' (Jung, 2015: xcvii). Tolkien, in his poem *Mythopoeia*, as well as his essay *On Fairy Stories,* builds on the theme of the importance of myth through a poetic gaze. The poem is dedicated to C.S. Lewis, who as a friend of Tolkien viewed myths as lies (Weinreich, 2008: 330). *Mythopoeia* is Tolkien's forewarning about the abyss of modernity, the implicit problem with an exclusively materialistic worldview (Weinreich, 2008: 344). Taking *mythopoeia* as the making of myths, and the dedication to Lewis, it is clear that Tolkien saw the importance of reclaiming myths in modernity. His own creative journey towards the formulation of a mythological history for England was an endeavour which took up a large portion of his working life. *Silmarillion, The Lord of the Rings,* and *The Hobbit* formed the basis for Tolkien's mythology: 'Having set myself a task, the arrogance of which I fully recognized and trembled at: being precisely to restore to the English an epic tradition and present them with a mythology of their own' (Tolkien, 1981: 250). Taking inspiration from many different mythological sources, none more influential than Beowulf (Tolkien, 1981: 35), he sought to create myths; 'I was from early days grieved by the poverty of my own beloved country: it had no stories of its own (bound up with its tongue and soil), not of the quality that I sought, and found (as an ingredient) in legends of other lands' (Tolkien, 1981: 167). While much of his career was spent in striving to achieve mythopoesis, Tolkien himself saw it as a failed endeavour:

> Once upon a time (my crest has long since fallen) I had a mind to make a body of more or less connected legend, ranging from the large and cosmogonic, to the level of romantic fairy-story – the larger founded on the lesser in

> contact with the earth, the lesser drawing splendor from the vast backcloths – which I could dedicate simply to: to England; to my country…Absurd.
>
> (Tolkien, 1981: 168)

In coining the term, 'The Myth Gap', Jonah Sachs (2012) indicates that we are living in a liminal transitory time in which, for the first time in history, we as a society no longer share stories and myths in the traditional sense, relying now on marketing, advertising, and commodity fetishism for myth production. We can see Tolkien's mythology reimagined into a multi-billion dollar franchise. The lack of symbolic thinking in modernity is unique to our contemporary rationalist society, as Sachs suggests that 'ours is the first generation not to share myth', while consequently, 'held to the standard of literal truth, traditional myths start to crack' (Evans, 2017: 25). The essence of man, it can be suggested, is symbol creation (alongside dreams and rituals). With these acting as the innate functions rather than interpretations of humans as primarily tool-making animals. If we are the creators of symbols, what are the implications of the "Myth Gap"? A modern rational mind's obsession with black and white, or right and wrong, causes the deeper symbolic importance to be lost at a surface level of interpretation, leading to the negative representation of the mythical, which has gained widespread prominence with the coming of modernity. If we take Voegelin's use of the Platonic symbol of metaxy (Voegelin, 1987: 27) as a metaphor in this instance, whereby man exists within two poles such as the infinite and the finite or the mind and the material, it can be suggested that we must navigate the poles, be they object and subject, man and god, mortality and immortality, or in this case science and myth. Aligning with Byrne's chapter on complementary dualism in this collection, there is a need for a balance between opposing poles, while seeing the journey in directional terms rather than as an ultimate destination is of crucial importance when finding a path in the transitory journey towards sustainability (see also Byrne, 2017: 49).

Rationality and the death of myth

With the coming of the age of reason, Kantian philosophy, following in the wake of the scientific revolutions of Bacon, Descartes, and Newton, set about suggesting man's emergence from his self-imposed "nonage". The move away from nonage, defined as the need for guidance throughout the quest for knowledge and self-actualisation, to the individualised pursuit of truth, represents a breaking point for the connection of myths, and with this oral storytelling traditions, to the modern subject of modernity. In *An Enquiry concerning Human Understanding*, Hume (2000 [1748]) sees miracles as transgressions against the laws of nature which he deems to be immovable. In a literal sense, the incompatibility of miracles with the new scientific and philosophical discoveries and trends of the age of reason led to inevitable division between the two, with the latter usurping the former as the means through which we form meaning in the world. The figurative importance which is contained within such stories was also lost. The similarities between stories of miracles of the Old Testament and shared myths are apparent and the dawning of the Enlightenment of the 17th and eighteenth century has had similar effects on the value which both hold within society. As referenced by David Sloan Wilson (2010), the emergence of religion within societies helped increase the levels of social cohesion within these societies. The role of miracles, or myths (or at a more fundamental level, stories), within these religious orders was of great importance to their cohesive power.

Locke's *An Essay Concerning Human Understanding* (Locke, 1836 [1690]) brought forward the suggestion of the Tabula Rasa, whereby humans are born without in-built mental content. Within this, the idea of ancestry, tradition, and continuation lose prominence in a move towards individualisation. Myths, as a form of communication, are based upon their continued telling over time, down through generations, in a form of tradition in order to communicate a message underneath the story. While the Enlightenment was a period of scientific breakthroughs within the natural sciences and new modes of thought within philosophy, the rise of modernity made a sacrificial lamb of myths, slaughtered for the common good of rationality. The relationship between myth and science was adequately summarised by W. B. Yeats who suggested that 'science is the critique of myth' (May, 1991: 25). Benjamin, however, saw modernity, rather as an intensification of mythic forms, reimagined in the image of Enlightenment values (Benjamin, 1999). As will be discussed further, the irrational rationality of modernity finds a medium through new mythic forms. The story wars (as referenced by Sachs, 2012) is the battle for the attention of this mythic consciousness, often fought today by big brands. The positive and negative within myth must be acknowledged in its modern form, along with the need to enact its ruination in order to 'free its positive potential' (Gilloch, 1996: 176).

Communicating science

Tillich (2001 [1957]) illustrated the danger of searching for literal truths in myths, for in doing so all symbolic meaning is lost, suggesting the literal and the symbolic as incompatible. He went on from this to draw out the benefits which science could bring, serving as a source of answers to a wide array of questions. The issue, however, comes not from a lack of answers, but from the inability of science to act as a source of value. Within the transition to sustainability, the answers have been adequately found within the scientific community; what has failed to emerge is a value system. An informational capacity is evident, but a moral capacity through which information can be acted upon has failed to materialise. The climate change debate has reached the point of post-scientific and pre-social consensus (Hoffman, 2015: 74). The problem of competently communicating science to the public has never been more evident than in the case of climate change.

Epstein (1994) developed cognitive-experiential self-theory (CEST), whereby the brain is identified as two parallel processing systems: analytical processing and experiential processing. Within the climate change debate, the conversation has been placed within the cultural domain of scientific discourse and analytical processing.

> The theories, graphs, projects and data speak almost entirely to the rational brain. That helps us to evaluate the evidence and, for most people, to recognize that there is a major problem. But it does not spur us to action.
>
> (Marshall, 2015: 50)

The need to reconcile analytical and experiential processing with regards to climate change by creating emotionally appealing narratives through which action can be taken is the core issue at the centre of inertia in the transitory process towards sustainability (again see Byrne's chapter in this collection). Research on the communication of climate change has found that narratives and stories around the transitory process required to deal with the problem are often easier to comprehend for audiences and may lead to a higher level of buy-in and engagement than traditional analytically driven logical-scientific communication (Dahlstrom, 2014: 13614).

The role of the media in communicating climate change is something which has been investigated at length for close to two decades (Allan et al., 2000; Anderson, 2009; Depoux et al., 2017). While operating as a communication mechanism linking the national policy level and the scientific world with the general public, the media, on the whole, has failed to effectively engage citizens in the transitory process and raise awareness on the urgency of action required. The lack of continued coverage of climate change issues, with its coverage being event driven (such as around the time of COP21), is an underlying failure within media communication (Mullally & Byrne, 2016; Depoux et al., 2017).

As Boykoff (2011) observes, mass media representations of climate policy and science are not effective in framing public response and engagement. It is, however, a key component in merging everyday life with the science and policy of climate change (Boykoff, 2011: 28–29). The three primary objectives of climate change communication have been established as informing and educating individuals about climate change, encouraging a level of social engagement and action, and bringing about changes in social norms and cultural values (Moser, 2010). While the media has a role within these three processes, the current failings illustrate the need for alternative mediums. 20th Century scholars, in moving away from Enlightenment reasoning on myths, viewed myths as containing inherent truths and analysed them as such (Eliade, 1998). While the power of myths to act as vessels for truth has long since been noted within mythology studies, this has not been synthesized with climate change communication. The development of modern mythologies may be the greatest challenge within communication on climate change (Moser, 2010: 36). Schorer, in championing the necessity of myth suggests a myth to be a 'large controlling image that gives philosophical meaning to facts of ordinary life' and beyond this 'mythology is a more or less articulated body of such images, a pantheon' (Schorer, 1960: 355). A critical re-evaluation of the power of myths is needed in the contemporary case, with no one narrative sufficient within the transition. The formation of a pantheon, as referenced by Schorer, will come from the embracing of many contextually and culturally specific myths which can be combined. Ricoeur adds that 'a philosophy instructed by myths arises at a certain moment in reflection and wishes to answer to a certain situation in modern culture' (Ricoeur, 1991: 484). The liminal situation of planetary crisis currently faced is one in which the importance of the emergence of such a philosophy seems self-evident.

Unreal reality and the role of stories

> Metaphors may create realities for us, especially social realities. A metaphor may thus be a guide for future action. Such actions will, of course, fit the metaphor. This will, in turn, reinforce the power of the metaphor to make experience coherent. In this sense metaphors can be a self-fulfilling prophecy.
>
> (Lakoff & Johnson, 2008: 156)

The potential for metaphors to be self-fulfilling prophecies will be expanded here to include myths and stories. As has been suggested previously, fiction has an important role in the formation of our reality. Novels can "provide a way for analysing the unreality of the real" (Szakolczai, 2014: 169). On from this, and in keeping with Lakoff and Johnson (2008), it can be said that metaphors, or in this case myths, can be guiding forces in the construction of our reality, beyond the level of analysis. In the conversation on the transition to sustainability the potential which this consideration offers must not be overlooked. When detached from facts, fiction can be utilized to explain the unexplainable (Levin, 1959).

Humans desire stories through which meaning can be attained (Byatt, 2001, p.166; White, 1984: 19–20). Cave art from the Paleolithic period can be referenced as the earliest known representations of this desire for stories (Clottes, 2008). Since then, throughout history, the emergence of new and innovative mechanism for the communication of stories has occurred. In the digital age, we have returned to walls for the communication of our stories, this time on social media sites rather than caves (Hurlburt & Voas, 2011). The permanence of human desire for stories is combined with innovative means of communication. Heading towards virtual reality, built on the foundations of Information and Communication Technology (ICT), the potential of stories to unify at a global level is of great importance, particularly in the context of the necessity for a global transition (although in no way a panacea). Narrative and stories play an important role in any meaningful attempt at implementing change (Khasnabish, 2007; Holloway, 2014). These narratives and stories, in order to achieve positive change, must be positive in nature.

In discussing stories, Selbin contends that 'we use them not just to narrate our lives – and narration and story are not identical – but to tell, to share news, information and much more: to guide, to warn, to inspire, to make real and possible that which may well be unreal and impossible' (Selbin, 2010: 3). In looking for trivial examples of fiction bringing reality to the unreal, a fertile breeding ground can be found within science fiction. From Clarke's (1968) depiction of digital media in *2001: A Space Odyssey* to Bellamy's (1888) representation of credit cards, examples of fiction's ability to predict developments are vast in number. A rational perspective suggests that these literary representations of future developments perhaps played a role in their later invention. In the case of the transition to sustainability, such a dynamic may prove useful, as myths or collective stories which represent positively a future built on planetary health could act as a catalyst within the process. From a functional perspective, Malinowski (2014) sees the value of myths which take place in the present or future. The value here is on laying a pathway of adjacent possibilities which can then be taken.

The positive myth

The power of myths, in a negative sense, has been shown throughout history and has brought about great suffering. Opting for the low hanging fruit, the 'Hitler Myth' as discussed by Kershaw (1987) is a suitable example of this power. Nazi propaganda around Hitler's role as a leader unified the population to act in accordance with the ideals of the Third Reich. Turning this on its head, the potential of positive myths must be said to hold the same power. Benjamin (1999), although skeptical with regards to the intensification of mythic forms and the commodity fetishism which follows, was quick to suggest the positive confined within the mythic in modernity. With regards to the transition process, a global unification around planetary issues can be achieved through the emergence of a pantheon of myths, as previously referenced by Schorer (1960), with each holding at its core the values of sustainability traced out through the sustainable development goals. This resonates with complexity biologist Stuart Kauffman's envisioning of an emerging 'global ethic' (Kauffman, 2008). He suggests that we 'lack a global ethic to constitute the transnational mythic value structure that can sustain the emerging global civilization' in a way which 'expands our consciousness and naturally seems to lead to an enhanced potential global ethic of wonder, awe, and responsibility within the bounded limits of our capacity for all of life and its home, Earth' (Kauffman, 2007). As Campbell and Moyers have suggested; 'you get a totally different civilisation and a totally different way of living according to whether your myth presents nature as fallen or whether nature is in itself a manifestation of divinity' (Campbell and Moyers, 2011: 121). Much climate change communication of the future outlook for the world is heavily focused in negative messaging, warning of the dangers of inaction within the transition. As has been shown,

> Dramatic, sensational, fearful, shocking, and other climate change representations of a similar ilk can successfully capture people's attention to the issue of climate change and drive a general sense of the importance of the issue. However, they are also likely to distance or disengage individuals from climate change, tending to render them feeling helpless and overwhelmed when they try to comprehend their own relationship with the issue.
>
> (O'Neill & Nicholson-Cole, 2009: 375)

On from this, it is suggested that communication mechanisms that deal with individuals' attitudes, beliefs, and values are more successful techniques for engagement. The development of narratives around these attributes (attitude, belief, and value) is worth considering at a policy level with regards to the communication of the transition. The "educational myth" can be utilised to realign 'valued actional dispositions, guide conduct and sustain effort' (Murray, 1960: 337). Educational mechanisms for engagement on sustainability have a precedent in Irish policy, illustrated by the 2006 'Power of One' campaign on energy efficiency. The campaign proved successful in raising awareness of the savings associated with implementing energy efficiency measures; however, this did not translate into long-term behavioural change (Diffney et al., 2013). In the context of this advertising campaign, a rational monetary logic was applied to engage citizens. Research suggests, however, that audiences find narratives easier to comprehend than logical communication, thus leading to greater engagement (Dahlstrom, 2014: 13614). Envisioning the transition to sustainability has been considered (Tilbury, 2007); however, the value of positive communication of the transitory future, through shared stories or myths, has yet to be explored at any great depth. We can turn now to an example of mythopoeia in modernity, whereby a mythic pathway was outlined and in turn followed by society.

Goethe's Faust as mythopoesis

Goethe's Faust is the most influential work of German literature, coming to shape a vast range of academic and artistic work. Both Weber and Nietzsche take inspiration from its depiction of the transition to modernity, with the character of Faust acting as an archetype of the modern subject. German tales of Doctor Faust, or Faustus, emerged in the fifteenth century, based on a real magician and alchemist, and acted as inspiration for first Marlowe and then later Goethe. In Goethe's Faust, the protagonist signs a blood oath with Mephistopheles, to gain all knowledge and power during his life in return for his soul in death. The modern subject, hypnotised by technological acceleration and the commodification of social life, ever increasing in the contemporary case, draws parallels with Goethe's Faust. Goethe himself, worked both as a natural scientist and poet in his life. Topics of interest during his scientific career included botany and optics. Alongside these endeavours, he was also critical of the emergent scientific specialisations of the time, seeing the danger of academic silos, long before the inception of this book. He saw the necessity of unifying scientific and poetic modes of enquiry, despite admitting it seemed beyond the horizon (Wellmon, 2010: 153). Goethe's poem is a modern myth, the unification of such work with science being the central concern of this chapter but also a key consideration of Goethe himself, although he was unaware of the mythic life his poetic work would take on.

During the rise of modernity, Faust emerged as a myth in a liminal time. With humanities new-found awareness (and coining) of the Anthropocene perhaps a similar liminal period is in action whereby the potential of myth may be acknowledged. 'Given the practical impossibility of a modern myth, due to the inventions of printing, one could even risk stating that the Faust is *the* modern myth, the only work in the modern world that manages to perform the impossible feat and actually become a myth' (Szakolczai, 2014: 177). This "practical impossibility" of a modern myth, as referenced by Szakolczai, stems from the classical interpretation of myths as being verbally shared through time with no known author or authors (see five key characteristics of myth below). Here, we will turn away from the idea of myth creation to look at mythic revival, with the potential to bypass this “practical impossibility”.

Mythopoesis or Mythoanastaino?

When discussing the potentials of myth, we must be quick to acknowledge the illegitimate from the real. We have discussed myth at the level of myth creation in the contemporary sense; mythopoesis in modernity. The term deriving from the Ancient Greek Poiesis, meaning to create. Narratives and stories which communicate and engage on the values of the transition to sustainability on an emotive level are a necessary consideration, but on from this the idea of these as “new myths” must be considered. The unravelling of myths from modernity requires revisiting the myths of old to seek guidance. Myths are everlastingly elastic (Symonds, 1890). This elasticity must be viewed as an opportunity. Hyman (1955, 472) goes as far as to suggest that no one can invent myths. We must start from this point of departure to look at the use, and revival, of myths throughout history, with a specific focus on works of literature in modernity. Here, we can suggest coining a new term, Mythoanastaino, coming from Anastaino, meaning to revive. ‘Myths can be weakened but hardly annihilated by disbelief; for a successful mythology is one which encourages people to invent new and more reputable reasons for believing in it after the old ones are no longer tenable’ (Ruthven, 1976: 60). Tolkien’s journey towards the creation of a mythology (rather than a revival of what came before) ultimately went unfulfilled, instead become entangled in the “mythic forces” of Hollywood, a bond which will now remain. Breaking free of dormant mythologies is an arduous process. The great works stand upon a platform outlined by what went before. We can think of Blake and Yeats and on from this Mann and Joyce. A reluctance to let go of traditional mythologies may be seen as an impedance to originality (Ruthven, 1976: 70), but the contortion of mythic form to the shape of an ever-intensifying consumer culture and commodity (and technology) fetishism must be considered within analysis of the great works of literature. The value of the mythic may not act as a hindrance but rather offer a direction through an uncertain, liminal situation. The ability of the spirit to create new myths, at a deeper level than shared stories, should be questioned. While the value of these shared stories, with their unifying potential has been outlined, further considerations of what can be achieved through a revival of traditional mythologies are warranted. The classical interpretation of myth which will be addressed here has been outlined along five key characteristics (Winzeler, 2012: 104):

1. Myth is a story rather than a statement of belief or doctrine.
2. It includes personalities that may be human, animal, or supernatural, or in a combination of these, who do things or to whom things happen.
3. It is a story that is not just known but “believed” and is regarded as important or sacred.

4. It has no known author or authors; myth is different in this way from religious revelation or prophecy, to which it may otherwise be similar.
5. It involves events or activities that are in some way extraordinary, "larger than life", if not necessarily supernatural, and cannot be confirmed or disproven.

Starting from this point, we can analyse the value of revival over creation. Poesis, as creation, has connotations of starting anew. Locke's "Tableau Rosa" and a disconnection from ancestry and tradition is a process of individualisation. In the search for mythopoesis, at a more profound scale than the shared stories in the world of the story wars, the idea of individualisation is evident, whereby the creation of new mythologies functions only to isolate contemporary myth from a long-standing lineage of mythology. Ancient mythologies are products of the collective, with no trace to individual authorship (characteristic four above), acting instead as cumulative creations, borne out of many generations. As was noted by Sachs (2012), as the first generation not to share (rather than create) myths, caution must be taken in starting from scratch. Revival could be a more fitting undertaking, looking to the past for inspiration on how to proceed. As both cultural and social conditions change with time, myth too transforms in an act of revival, transcending, and discarding earlier considerations without total destruction (Eliade, 1998). Here we can muse on the potential to revive myths in a contemporary case to analyse the transition to sustainability. The everlasting elasticity of the myth of Oedipus can be used to muse over the potential of traditional myths. Foucault, when looking at Oedipus the King, saw the play as transitional in nature (Foucault et al., 2010). From this starting point, and moving beyond the play to the wider level of the myth, Oedipus can be used as a metaphor for the ways in which "mythoanastanio" can occur.

Oedipal theme

In consulting the Oracle at Delphi, Oedipus was informed that he would kill his father and marry his mother. Unaware of his adoption, and in order to avoid such circumstances, Oedipus did not return to his home in Corinth. Oedipal myths can be found across cultures, without having a striking resemblance to the Greek form, but rather being different manifestations under the broad themes of fate and wisdom and knowledge (Kluckhohn, 1959). In keeping with this we may investigate the potential of the broad mythic form to the contemporary case, to act as an illustration of mythic potential. Taking the maternal to represent the earth, or "mother nature", and paternal connotations of the "breadwinner", we are faced with a similar quandary to the one faced by Oedipus. A situation has been reached where the current economic model must be restructured in order to transition to a sustainable society. A "marrying" of society with the natural world must be ensured in order to move forward. By killing the father, we must look for alternatives. If the focus should move away from seeing nature as a resource to be turned to profit towards a more sustainable from of economics, thought must now be given to what options are available. To make a transition towards "sustainability" alongside continued endeavours for economic growth at all costs is oxymoronic (Robinson, 2004). A discussion on alternative economics such as, for example, the circular economy, degrowth, sustainability entrepreneurs, the slow money movement, resource-based economy, the shared economy, must occur alongside any discussion on a transition towards sustainability. The mythic trope of murdering the father can be used here metaphorically. Just as the economy is of central importance we must look to nature, a marrying of society with Mother Nature. Heidegger's idea of *Herausfordern,* meaning to challenge, in this case the challenges which modern society places upon nature, can be said to have reached a breaking point. In the Anthropocene, we are for the first time fully aware of the negative impact which society places upon nature. While acting alongside global economics and growth at all cost logics, further from this we must address the disconnection from nature. The modern myth of the social contract gives form to the imagining of the super city, but beyond this nature is lost (Midgley, 2003). If the reduction of the essence of the natural world to mere resource cannot be dissolved, any hope of a transition beyond the "myth" of technological "progress" is absurd. Marrying the mother, just as murdering the father, are fundamental aspects within any discussion of a transition towards sustainability. To present it again as a tragedy will only lead to following a self-constructed mythic path towards climate chaos. The elasticity of mythic forms must be utilised (Symonds, 1890).

Conclusion: The potential for consolidating Myth and Science

Across a number of areas such as policy, advertising, and education, a great importance must be placed on framing the future through the use of myths and collective stories, utilising numerous mediums, disciplines, and technologies. Myth can be seen from both perspectives. The ability of shared stories and narratives to unify has been addressed. The task now is to develop shared stories which lead to engagement on climate action, enabling the transition towards sustainability. At the level of mythos, it is crucial to move beyond esoteric interpretations. While the answers are not clear, revival of traditional mythologies which can be applied to the contemporary may be the only way out of the inert social response to climate change. The myth of Oedipus has been used here only as a metaphor; to bring thought towards how traditional myths can be applied to current circumstances. Here the role of the Arts becomes self-evident. A transdisciplinary approach, by definition, should move beyond the walls of the academy. The artistic community may prove most fruitful. If the essence of mythic events is recurrence (Gilloch, 1996: 10), then moving away from the mythic is only an act of ignorance. The freedom of myths, and their study, from any strictly defined academic discipline gives great potential to their usage. While championing transdisciplinary approaches to transition studies, mythology fits comfortably, unrestricted in form or method of investigation.

While myth and science are not compatible from a rational perspective, this does not suggest that alternative forms of compatibility are not possible. Cassirer sought to interpret myth from two perspectives. The sociological essence of myth seeks to achieve unity between members of a society, while the psychological function can provide a sense of unity with life itself, or nature as a whole, moving beyond the social contract. In aiming to find harmony between science and myth, we must look for the mythic within science. It can even be suggested that science in and of itself is a mythic force. Aligning with much of Popper's (1957) work, science can be disproven but never proven, remaining hypothetical; in a sense mythological. If we invert this conclusion, it can then be suggested that myth was to primitive Man what science is in modernity. 'In the life of the human race the mythical is an early and primitive stage, in the life of the individual it is a late and mature one' (Mann, 1947: 422). In thinking of the human race and primitive societies, mythology was ubiquitous in order to make sense of the physical world due to a distinct lack of control over it. Modern society has gained this control through science. While we can speculatively catastrophise about the dissolution of control moving forward, it proves more useful to think at the individual level as alluded to by Mann (1947). The mythical in the individual comes with maturity. In a time of individual disillusionment at corrupted mythic forms, with a lack of control, the revival of myth can be explored.

Bibliography

Allan, S., Adam, B. and Carter, C., 2000. *Environmental risks and the media.* Hove: Psychology Press.

Anderson, A., 2009. Media, politics and climate change: Towards a new research agenda. *Sociology compass*, 3(2), pp. 166–182.

Bellamy, E., 1888. *Looking backward, 2000–1887.* Toronto: W. Bryce.

Benjamin, W., 1999. *The arcades project.* Cambridge: Harvard University Press.

Boykoff, M.T., 2011. *Who speaks for the climate?: Making sense of media reporting on climate change*. Cambridge: Cambridge University Press.

Byatt, A.S., 2001. *On histories and stories: Selected essays.* Cambridge: Harvard University Press.

Byrne, E.P. 2017. Sustainability as contingent balance between opposing though interdependent tendencies; A process approach to progress and evolution. In: E. Byrne, G. Mullally and C. Sage, eds. *Transdisciplinary perspectives on transitions to sustainability.* Abingdon: Routledge, pp. 41–62.

Campbell, J. and Moyers, B., 2011. *The power of myth.* New York: Anchor.

Clarke, A. 1968. *2001: A space Odyssey.* New York: New American Library.

Clottes, J., 2008. *Cave art.* London: Phaidon.

Dahlstrom, M.F., 2014. Using narratives and storytelling to communicate science with nonexpert audiences. *Proceedings of the National Academy of Sciences*, 111(4), pp. 13614–13620.

Depoux, A., Hémono, M., Puig-Malet, S., Pédron, R. and Flahault, A., 2017. Communicating climate change and health in the media. *Public Health Reviews*, 38(1).

Diffney, S., Lyons, S. and Valeri, L.M., 2013. Evaluation of the effect of the Power of One campaign on natural gas consumption. *Energy policy*, 62, pp. 978–988.

Eliade, M., 1998. *Myth and reality.* Illinois: Waveland Press.

Epstein, S., 1994. Integration of the cognitive and the psychodynamic unconscious. *American Psychologist*, 49, pp. 709–724.

Evans, A., 2017. *The myth gap: What happens when evidence and arguments aren't enough.* New York: Random House.

Foucault, M., Davidson, A.I. and Burchell, G., 2010. *The government of self and others: Lectures at the Collège de France 1982–1983.* Springer.

Gilloch, G., 1996. *Myth & metropolis–Walter Benjamin and the City*. Oxford: Polity Press.
Hoffman, A.J., 2015. *How culture shapes the climate change debate*. Palo Alto: Stanford University Press.
Holloway, J., 2014. *Change the world without taking power: the meaning of revolution today*. North Carolina: Lulu Press.
Hume, D., 2000 [1748]. *An enquiry concerning human understanding: A critical edition, Volume 3*. Oxford: Oxford University Press.
Hyman, S.E., 1955. The ritual view of myth and the mythic. *The Journal of American Folklore*, 68(270), pp. 462–472.
Hurlburt, G.F. and Voas, J., 2011. Storytelling: From cave art to digital media. *IT Professional*, 13(5), pp. 4–7.
Jung, C.G., 2015. *Collected works of CG Jung: The first complete English edition of the works of CG Jung*. Abingdon: Routledge.
Kauffman, S.A., 2007. *Beyond reductionism: Reinventing the sacred. Zygon*, 42(4), 903–914.
Kauffman, S.A., 2008. *Reinventing the sacred: A new view of science, reason, and religion*. New York: Basic Books.
Kershaw, I., 1987. *The Hitler Myth: Image and reality in the third Reich*. Oxford: Clarendon Press.
Khasnabish, A., 2007. Insurgent imaginations. *Ephemera: Theory and Politics in Organization*, 7(4), pp. 505–526.
Kluckhohn, C., 1959. Recurrent themes in myths and mythmaking. *Daedalus*, 88(2), pp. 268–279.
Lakoff, G. and Johnson, M., 2008. *Metaphors we live by*. Chicago: University of Chicago Press.
Latour, B., 1993. *We have never been modern*. New York: Harvester Wheatsheaf.
Levin, H., 1959. *Some meanings of myth. Daedalus*, 88(2), pp. 223–231.
Locke, J., 1836. *An essay concerning human understanding*. London: T. Tegg and Son.
Mali, J., 2002. *The rehabilitation of myth: Vico's 'New science'*. Cambridge: Cambridge University Press.
Malinowski, B., 2014. *Myth in primitive psychology*. Worcestershire: Read Books.
Mann, T., 1947. *Essays of three decades*. New York: AA Knopf.
Marshall, G., 2015. *Don't even think about it: Why our brains are wired to ignore climate change*. New York: Bloomsbury Publishing USA.
May, R., 1991. *The cry for myth*. New York: WW Norton & Company.
Midgley, M., 2003. *The myths we live by*. London: Routledge.
Moser, S.C., 2010. Communicating climate change: history, challenges, process and future directions. *Wiley Interdisciplinary Reviews: Climate Change*, 1(1), pp. 31–53.
Mullally, G. and Byrne, E., 2016. A tale of three transitions: A year in the life of electricity system transformation narratives in the Irish media. *Energy, Sustainability and Society*, 6(3).
Murray H., 1960. *Myth and mythmaking*. New York: George Braziller, pp. 324–337.
Nietzsche, F.W., 2000. *The Birth of Tragedy*. Oxford: Oxford University Press.
O'Neill, S. and Nicholson-Cole, S., 2009. "Fear won't do it" promoting positive engagement with climate change through visual and iconic representations. *Science Communication*, 30(3), pp. 355–379.
Onions, C.T., 1966. *The Oxford dictionary of English etymology*. Oxford: Oxford University Press.
Popper, K., 1957. *Philosophy of science: British Philosophy in the Mid-Century*, London: George Allen and Unwin.

Ricoeur, P., 1991. *A Ricoeur reader: Reflection and imagination*. Toronto: University of Toronto Press.

Robinson, J., 2004. Squaring the circle? Some thoughts on the idea of sustainable development. *Ecological economics*, 48(4), pp. 369–384.

Ruthven, K.K., 1976. *Myth: The critical idiom*. London: Methuen & Co.

Sachs, J., 2012. *Winning the story wars: Why those who tell - and live - the best stories will rule the future*. Massachusetts: Harvard Business Press.

Schorer, M., 1960. The necessity of myth. In: H. Murray, ed. *Myth and mythmaking*. New York: George Braziller, pp. 355–360.

Selbin, E., 2010. *Revolution, rebellion, resistance: The power of story*. London: Zed Books.

Symonds, J. A. 1890. *Essays Speculative and Suggestive. Volume 2: Nature myths and allegories*. London: Chapman and Hall. pp. 126–149.

Szakolczai, A., 2014. Theatricalized reality and novels of truth: Respecting tradition and promoting imagination in social research, In: K. Keohane, ed. *Imaginative methodologies in the social sciences: Creativity, poetics and rhetoric in social research*. Surrey: Ashgate.

Tilbury, D., 2007. Learning based change for sustainability: Perspectives and pathways. *Social learning towards a sustainable world*, pp. 117–132.

Tillich, P., 2001 [1957]. *Dynamics of faith (Volume 42)*. Michigan: Zondervan.

Tolkien, J.R.R., 1981. *The letters of JRR tolkien*, H. Carpenter, ed. Boston: Houghton Mifflin.

Vico, G., 1999. *New science*. London: Penguin.

Voegelin, E., 1987. *The new science of politics: An introduction*. Chicago: University of Chicago Press.

Weinreich, F., 2008. *Metaphysics of Myth: The Platonic Ontology of 'Mythopoeia'. Tolkien's Shorter Works*, Zollikofen: Walking Tree Publishers. pp. 325–347.

Wellmon, C., 2010. Goethe's morphology of knowledge, or the overgrowth of nomenclature. *Goethe Yearbook*, 17(1), pp. 153–177.

White, H., 1984. The question of narrative in contemporary historical theory. *History and Theory*, 23(1), pp. 1–33.

Wilson, D.S., 2010. *Darwin's cathedral: Evolution, religion, and the nature of society*. Chicago: University of Chicago Press.

Winzeler, R.L., 2012. *Anthropology and religion: What we know, think, and question*. Lanham: Rowman & Littlefield.

8 The hare and the tortoise

Metaphorical lessons around sustainability

Connor McGookin, Brian Ó Gallachóir, and Edmond Byrne

Fables have been used for centuries to relay important messages in a playful manner, through fantasy tales of objects, plants, or animals. Their origin is reputed to Aesop, a slave, and storyteller believed to have lived in ancient Greece somewhere between 620 and 560 BCE. The Perry Index (Perry, 1952) provides a definitive account of "Aesop's Fables" or "Aesopica", listing 725 fables along with known sources and Greek/Latin testimonies about Aesop (Perry, 1933).

As metaphoric tools, the underlying message is often quite explicitly highlighted. This makes the message very accessible, even "child proof", as noted by Martin Luther who dedicated a surprising amount of time translating some of Aesop's fables into German, praising them as a useful tool for all ages. "In short, after the Bible, the writings of Cato and Aesop are in my judgment the best, better than the harmful opinions of all the philosophers and jurists". (Springer, 2011, p. vii). Fables combine both merriness and wisdom (Zsolnai, 2015). They are a fiction whereby the animated characters act as human beings and thus their behaviour and circumstances are experienced by the audience as a true reflection of the real world. According to Luther, this ability to tell the truth about real life in the world is the hidden strength of the fable "Die Warheit sagen von eusserlichem Leben in der Welt" (Steinberg, 1961: 86).

In the fable of the hare and the tortoise, the tortoise declares, having been mocked by the boastful hare: 'I will beat you in a race!'. There are a number of variations of how the race plays out; two of the most common are that either the overconfident hare naps at the start of the race (Gibbs, 2002: 117), or does so having established a considerable lead (Zsolnai, 2015); in both cases he wakes from his nap just in time to see the tortoise crossing the finish line.

'Slow and steady wins the race' is the moral of the short story. In terms of pressing contemporary issues around climate change, climate action, and sustainability, this message appears paradoxical in the context of the often-panicked sense of urgency being reiterated by those intimately involved in the area, including climate scientists, environmentalists, etc. However, as shall be explored through this chapter, there is an inherent value to the tortoise's approach; while there is urgent need for transformative and dramatic change to the course of our civilisation, we must be careful to ensure that such transition is not hastily misdirected (with potentially even worse consequences). In this context, the warning of the

DOI: 10.4324/9781003143567-10

Roman proverb attributed to Octavius Caesar Augustus; "festina lente" ("hasten slowly" or "make haste slowly") seems suitably apt (Suetonius Tranquillus, 1957).

This chapter will explore a number of narratives around issues of sustainability, in particular concerning climate change, casting the hare and tortoise in a variety of roles. This will be done not simply understanding them as fast and slow but using playful characterisations like the hare's pace rendering him over-eager and anxious when dealing with the tortoise's stubborn disinterest. Although similar in some cases, each narrative's characterisations are independent of the others and play with the metaphor quite differently. Despite this, an interesting number of thematic commonalties emerge, demonstrating the seemingly universal truth of the fable's lesson. It is also important to note that while the discussions may seem to cast the hare in quite a negative light, the interdependent nature of the actor's relationship places value in both throughout the narratives.

Narrative one: Environmentalism as a social movement

In this narrative, the hare represents environmentalists and actors concerned about radical societal change, in particular looking at the emergence of conservation environmentalism as a social movement, which is often characterised by an air of anxiety and apprehension. While the tortoise represents wider society, maintaining the "status quo" of consumptive-based growth practices, dismissive of the hares' concerns.

The emergence of conservation environmentalism as a social movement is accredited by many to Rachel Carson's *Silent Spring*, published 1962. (Carson, 1962) Combining a number of examples whereby various pesticides had been linked to ecological destruction, she drew the eerie picture of a spring without birds. Despite attempts to ban the book and heavy media criticism, it precipitated such public concern that it prompted President John F. Kennedy to order an investigation. This led to a complete ban of the DDT insecticide (which had featured in Carson's book) in the US, and the creation of a number of environment institutions and regulations (Bouwman et al., 2012: 240). However, the real legacy of *Silent Spring* was a new public awareness that our actions were damaging the environment. A decade later, in 1972, the Club of Rome's 'Limits to Growth' represented a pioneering scientific endeavour which sought to capture the ecological boundaries that our global society was stretching and would ultimately exceed under our current patterns of consumption and growth (Meadows et al., 1972). Its primary message was that, despite its apparent vastness, planet Earth is a finite resource. The 'prophets of doom' involved in the study have also been referred to as 'Malthus with a computer' (Cole et al., 1973). They affirmed Malthus's beliefs that without the ironically named "positive checks" to our population provided by nature through natural disasters and disease, the human species would inevitably exceed the ecological limits of our planet (Malthus, 1798).

The conservation environmentalist, which emerged in the 1960s, was concerned with highlighting the need for a dramatic societal shift away from the well-established neo–classical economic model of continuous growth in the pursuit of so-called progress. This apprehension in general was met by varying degrees of incomprehension, ignorance, disregard, and/or hostility from wider society. When challenged by this global existential (though seemingly far off and hypothetical) threat and excited calls for a rapid and radical change of direction, the dominant public response was largely one which amounted to indifference. Slowly but surely, the stubborn tortoise continued unperturbed. A failing of those early environmentalists was their inability to change dominant attitudes that embraced consumption-based economic growth. An emerging middle class who had only just begun to taste the luxuries of economic affluence were reluctant to part ways with the system that had seemingly provided it, and which promised to continue doing so.

The movement experienced initial success in growing public concern for environmental issues, which resulted in a number of regulation victories in the US during the 1970s, such as the 1972 Clean Water Act, the 1974 Safe Drinking Water Act, and the 1976 Toxic Substances Control Act. However, in response to the growing threat, firms set up front groups promoting the corporate agenda while posing as public-interest groups, beginning the counter movement of "Corporate Activism" (Beder, 1997). Through the 1980s, as corporations began to complain about the effect of regulation on profitability, the public's attitude switched and became increasingly sympathetic towards business interests. The narrative that emerged was that environmentalism was going too far, threatening the status quo of growth-based consumerism and hence potentially endangering public interests like jobs, welfare, and health that relied heavily on continuous economic growth. As retailing analyst Victor Lebow explains,

> Our enormously productive economy demands that we make consumption our way of life, that we convert the buying and use of goods into rituals, that we seek spiritual satisfaction, our ego satisfaction, in consumption. We need things, consumed, burned up, worn out, replaced, and discarded at an ever-increasing rate.
>
> (Lebow, 1955: 3)

Metaphorically, one might say consumption became the world's most popular religion, and a universal one at that, cutting across all cultures and continents amid an increasingly globalised hegemonic consumerist system. In what is often referred to as the greening of industry, environmentalism's slogan of sustainability was taken and reworded as sustainable growth. This meant economies could continue to expand consumption as before so long as industry was slightly more conscious of environmental concerns like direct pollutants and process efficiencies. Ultimately, the regulation of the 1970s only moved pollutants overseas to countries where environment laws were either lax, unenforced, or non-existent.

This left the passionate environmentalists dejected. The hare's eagerness leaves him quite prone to disappointment, unlike the tortoise who demonstrates an air of calm by virtue of her perseverance and grit. A well-documented problem within environmental activism is burnout. With access to often quite provocative data, the burden of knowledge and frustration experienced when trying to relay this message to the public can leave individuals vulnerable to bitterness, depression, exhaustion, and illness (Macy and Brown, 2010). Rather than due to the high level of work commitment (voluntarily), this is caused by the feeling of resignation and disillusionment from stress factors, such as the feeling of responsibility for global problems, and most importantly the repeated failure of efforts (Thomashow, 1996).

Today, this struggle can still be seen with climate change nongovernmental organisations (NGOs). Some within these groups may have been fighting for several decades, but to seemingly little avail (as many environmental indicators continue to deteriorate). Change isn't easy, the course of a civilisation changes significantly slower than these actors would like. In a call to arms, the message increasingly relayed by some groups is characterised as being shrouded in fear, the panicked message of impending doom. As frustration and panic increase, such groups have the potential to slip into an ever more isolated existence. These actors may well be right, and as John Foster indicates in his book, *The Sustainability Mirage* (Foster, 2008), if we are serious about addressing this issue then we need a collective effort, to get onto a "wartime footing", analogous to that of the British during the second world war. However, a crucial leadership feature of Winston Churchill, Foster points out, is that he didn't peddle fear; despite Britain being dragged into a war that it didn't want to be in, but entered out of a sense of duty, in 1939, against a looming risk of failure and loss. Foster suggests that Churchill's great trait was the ability to rally Britons by turning the narrative around; this was an existential war of freedom that we daren't loose; otherwise, life would not be worth living under a proposed thousand-year Reich of Nazi totalitarianism. 'You ask, what is our aim? I can answer in one word: Victory.' Britons were thus mobilised to want to do whatever it takes to win the war, and glad to endure short-term rationing for the promise of greater long-term good.

The shrill tone from some environmentalists can sound very negative and off-putting for "swing voters", and even for some who would generally support greater or speedier action. This limits the effectiveness of concerned groups in having an impact as the message reaches an ever-narrowing audience. Downhearted, such groups may find it increasingly difficult to engage with the actors in opposing camps and instead favour 'preaching to the converted', comfortable with recognition from their own. This of course is exacerbated by contemporary social media echo chambers. As seen in Williams, et al. (2015), groups in social media outlets discussing issues around climate change are often segregated into the polarised "sceptic" or "activist" groups. This means that most users only interact with like-minded others, reinforcing their own beliefs or values.

The slower the tortoise moves, the more agitated the hare becomes. The pleas of encouragement turn to a frustrated whine and the tortoise grows ever less fond of the hare, and then may go even slower out of spite. In the public eye, action on climate change becomes almost pointless, as many perceive by association with these groups, that the lifestyles they preach are "too radical" or "extreme".

Narrative two: Climate change – An expert's opinion

Following similar characterisations to the previous narrative, here the hare (climate scientists) has run far enough ahead to see the approaching danger; however, the tortoise (neo-classical economists, wider society, etc.) is resistant to a change of direction (and/or speed) as it is content with its current surroundings. This narrative will also look at the limited effectiveness of the hare hastily following a single disciplinary approach, as opposed to taking the time to give greater consideration to alternative and potentially conflicting perspectives through multi- and trans-disciplinary approaches.

Over half a century since the impact of the greenhouse effect on climate (the connection between atmospheric CO_2 and global mean temperature) was firmly established (Manabe and Wetherald, 1967), climate scientists' anxiety and fear has developed into a state of panic and agitation. While there is now general agreement on the need for societal change, the urgency is still overlooked. Across the various disciplines, there are still conflicting views on the severity of the situation. The hare's panic and urgency is met by seemingly unstoppable technological/societal momentum towards ever-increasing complexification. As with any other meta-structure or system, changing the course for our civilisation requires overcoming significant inertial forces.

Many infuriated climate scientists now argue that it is already too late, that it is likely even if we stopped producing CO_2 almost immediately, we will still reach disastrous global temperature rises (depicted below in the background of Figure 8.1). In this light, the necessity of rapid societal transformation is quite apparent. However, rather than inspiring change when people are warned of the severity of our ecological crisis, the extent and sense of inevitability of it can become disempowering. As seen in the first narrative, fear fails to have the desired empowering effect (O'Neill and Nicholson-Cole, 2009). In fact, the effect is quite the opposite; when faced with apocalyptic or doomsday scenarios, people can often feel a sense of hopelessness. It is quite difficult to comprehend how any personal changes or advocacy may help such a complex global issue and thus it may seem easier to dismissively plunder on with one's well-established habits.

The public tends to favour arguments for maintaining the "status quo", preached by actors like those who would adhere to neoclassical economics, arguing for example, that there is a cost-optimal level for climate change mitigation. This argument would have it that there is a potential under a cost-benefit analysis that we spend too much too soon and any damage avoided may be less than that spent or that we don't have the right balance between mitigation and adaptation spending. There are also arguments that in the short to medium run, climate change may well bring gains as well as losses for the economy and human welfare through the twenty-first century (Tol, 2018). In addition, some have attempted to demonstrate how the value added from a tonne of carbon emitted outweighs the negative social costs; with the message that fossil fuels add more value than they destroy (Tol, 2017). In the eyes of the concerned, this is quite shocking. Many climate scientists are somewhat baffled by these perspectives, highlighting that committing to embarking on a vigorous campaign to mitigate climate change is fundamentally a moral issue, not a long-run economic issue (Rosen and Guenther, 2015). An alternative view would hold that not only is this adding unnecessary risk to an already unstable system but it completely overlooks ongoing significant ecological destruction, as well as associated negative societal impacts such as potential for loss of life, social disruption, population migration,

Figure 8.1 'Steady as she goes' (by Nina Wohlfahrt).

species extinction, and so on, as a result of climate change. For the neo-classical economist however, artefacts and natural capital that have no defined monetary value fall into the category of economic "externalities".

As is rather humorously depicted in the cartoon in Figure 8.1, action in the political sphere is achingly slow. While there has been a near universal shift in opinion leading to agreement around the need for action, actual change has been limited. The 2015 Paris Agreement is the most recent milestone in global action against climate change. While it is clearly significant in being the first widely accepted global target for climate change mitigation, it fails to provide any clear means by which the rather ambitious targets may be achieved. It seems that the tortoise has agreed with the hare that it should move faster but it is constrained by its own physical limits. Truly meaningful action is unlikely to occur until the issue makes its way on to the ballot paper, which requires fundamental changes to the public's attitude/beliefs. Yet due to the seemingly far away threat of climate change, people are unlikely to support serious action on the issue until such time that the damages become more apparent and it is likely already too late, sometimes referred to as the "Giddens paradox" (Giddens, 2009). As depicted above in Figure 8.1, the consequences of missing the fast-approaching left turn will only become clear once we have passed it.

Inaction or delayed action is further supported by techno-optimists claims that we are safe to continue with business as usual (or almost business as usual), since over time new technologies like renewable energy along with improvements in efficiencies supported by carbon sequestration or capture and storage technologies will lead to a nearly zero carbon future. However, quite apart from the fact that this seeks to sidestep the issue of tackling consumption-based growth (and ultimately the prospect of long-term global heating (Chaisson, 2008) as the earth produces more energy than it can dissipate), this overlooks reliance on finite resources and offers nothing but a stalling tactic as opposed to a solution. The idea that any technological "silver-bullet" solution (depicted on the right track in Figure 8.1) is available is fanciful. It has been claimed to be shown that with respect to the energy transition, that 100% renewable energy is simply not sufficient to achieve carbon neutrality; we must also transition our lifestyles (Millot et al., 2018). A study on the French energy system projecting two scenarios of business as usual consumption ("digital society") and a reduction in consumption ("collective society") found that attaining carbon neutrality by 2050 was unrealistic for the business as usual scenario. Thus, by this measure, technology alone is incapable of resolving the problem, as the issue of climate change is much more deeply engrained in our socio-economic system. It not only challenges us to rethink our energy sources but also our consumption practices and wider environmental concerns.

Experts in the field of psychology and in the social sciences recognise that greater effort is required which goes beyond education of the public through factual information before significant change can occur. Changing attitudes is a prerequisite to the behaviour shifts that are required to successfully transition to a low carbon future. This is a slow and tedious process, as it requires overcoming a number of psychological factors that influence behaviour, including contextual support, social norms, action difficulty, and habits (Arbuthnott, 2009). This can come as quite a frustrating realisation to climate scientists or other actors concerned about climate change, who see the need for immediate action. However, it is clear, as seen in the previous narrative, that changing attitudes and beliefs is critical to broader societal change.

In seeking to make (r)evolutionary and transformational change therefore, experts might do well to learn to think outside their own disciplinary "silos" and recognise that there is a clear need for greater communication across the disciplines. While to the hare this may appear too time consuming a process, its necessity is strikingly apparent (Byrne, et al. 2017). A successful transition to a low carbon future will require a shift in focus to multi- and trans-disciplinary thinking and approaches, giving greater consideration to the interactions between people, technology, the economy, and environmental limits, all in the context of broader ethical considerations (Barry, 2017; Ó Gallachóir, et al., 2017).

In this rather unique race, the hare's existence is dependent on the tortoise, presenting an interesting challenge to the hare; it cannot simply run ahead.

This demands more robust solutions or arguments be developed to overcome the tortoise's resistance to changing direction. Without application in broader socio-economic contexts, solutions or arguments will do little to tackle issues around sustainability like climate change, and only exasperate an already convoluted challenge.

Narrative three: Niche versus mainstreaming

The hare represents early adapters, grassroots innovators, movers, and shakers on the path of reduced unsustainability. These are the proportion of society who strive to reduce their individual footprint, often finding themselves marginalised and stereotyped in media and public perception. Again, the tortoise represents wider society, which determinedly plods along familiar societal structures. One of the most significant barriers to any new technology is consumers' scepticism around unproven or "alien" alternatives to well-established socio-technical regimes.

Growing environmental concerns coupled with increased consciousness of diet and health has brought ever-growing changes in opinion to food practices. One such cause for concern is large-scale intensive industrial farming. Concerns around this have precipitated emergent alternatives such as an embrace of organically produced food, vegetarianism, and veganism. Vegetarianism and its more radical partner veganism are often simply characterised as a boycott of animal products, but may also be considered as a form of political stance against the principle of limitless consumption (Sage, 2014). While the practice of avoiding the consumption of meat is not a modern ideal, the social context and broader identity of veganism as an anti-consumerism or degrowth movement is particular to the last half century. To a certain degree, veganism has established a perception as one of the most effective means of reducing one's own personal environmental impact. However, its criticism of normative consumption practices means it is often quite aggressively met with scepticism and dismissal, thus potentially marginalising the movement.

Another issue the movement faces, shared with its less radical dietary choice companion in the organic food movement, is the perception of such a lifestyle as a luxury choice. Niches often enter the market as more expensive alternatives, with reasons other than financial being motivators. This can bring with it an air of aloof (or elite) smugness that creates tension between early adopters and wider society. The hare has raced too far ahead, and in the eyes of the tortoise would appear to be taunting it. The impulsive reaction is to reject these taunts and defend oneself (and more importantly, one's threatened values) by dismissively criticising the other.

Another separatist movement which has emerged out of the degrowth movement is the ambition of moving off-grid or becoming self-sufficient. The primary motivation is a strong desire to remove oneself from an established society that is over-reliant on unsustainable practices such as the global food network, fossil fuel energy consumption, etc. In his book, *Confessions of a Recovering Environmentalist*, Paul Kingsnorth describes how his disappointment with what he perceived as the death of environmentalism drove him to seek solitude and personal fulfilment through the move to a life of self-sufficiency in the Irish countryside (Kingsnorth, 2017). However, innovators who have chosen this route experience a life of significant hardship and as a lifestyle it often fails to pass the generation test. As the next generation of off-grid children mature, they often start to desire the luxury items and comforts that wider mainstream society promises, questioning the beliefs of their parents.

As with the above discussion on dietary choice, actors within this group are again marginalised and the lifestyle is stereotyped (in this case for good reason) as being very harsh and impractical. Radical innovators pose a direct challenge to our beliefs, values, and lifestyles. In response to this challenge, they are quite often targeted by efforts to dismiss them through marginalisation. Niches of this nature thus have the potential to be a cul-de-sac. But even here the hare fulfils an important role in scouting ahead, and as with natural evolution, there is a period of trial and error as a variety of paths are considered before the best route becomes apparent. Of course, it would also be unfair not to point out that early adaptors, pioneers, and visionaries so to speak do achieve a degree of gratification through obtaining personal sustainability goals. As Frost wrote in the conclusion of his poem *The road not taken* (Frost, 1916), 'Two roads diverged in a wood, and I - I took the one less travelled by, and that has made all the difference'. While it was against his instinct to choose the path less trodden, Frost was ultimately grateful that he experienced a journey less shaped by others.

However, it is quite clear that without a connection to wider society, the efforts of such actors can prove to have limited impact, and this can be exhausting for them. It is important not to "polarise" or "radicalise" the ideals behind these lifestyle changes. Tackling climate change does not require that we all move to the woods and live on plants and leaves. In fact, if we did all do so, we may actually make things worse. Instead, we must find ways of precipitating widespread transformative change around our consumptive behaviour. This may be accomplished by arriving at a tipping point of mass consciousness, whereby the link between eating meat and carbon emissions is firmly established in the public mind, similar perhaps to the quite rapid bolt in people's consciousness around the problems of single use disposable plastics that occurred following the airing of BBC's Blue Planet II in 2017 (BBC, 2017). The final episode of the series dramatically captured the threat plastics are posing to life in our oceans, demonstrating the power popular media has to mainstream environmental concerns.

Niches are more likely to influence change when they show certain degrees of compatibility with current practices. While this requirement may blunt the scope for niches to be radically innovative, it may be essential in encouraging wider adaptation. Electric vehicles fall under this umbrella, offering an alternative to traditional internal combustion engines. While they still suffer from a more limited range and a perception as being more expensive, growing environmental concerns (fuelled by concern over diesel particulate emissions and political leadership that have imposed imminent bans on such vehicles) have nevertheless sparked ever-increasing interest. The function or service provided by this technology is exactly the same as its predecessor, allowing for a continuation of the regime of personal travel but potentially reducing its CO_2 footprint by changing the fuel source. This will in time mean a significant reduction in emissions as the carbon intensity of the electrical grid decreases. However, as it does little to challenge the practice of personal mobility, it will not change attitudes or behaviour. While it has the potential to encourage greater consideration for how the electricity is provided, which may support the development of renewable micro-generation, a transition to electric vehicles in itself is unlikely to reduce the demand for transport energy. More transformative options may lie in deep investment in shared public transport options, a move away from using over a tonne of materials (by car manufacturers) to transport individuals around short distances, and most usefully (for reasons of public health and well-being as well as social and environmental) local and regional planning and national incentives which would prioritise both bicycle use and pedestrianisation.

Remaining in the energy sphere, and the transition to a low carbon energy future, a movement growing traction is the idea of community energy or energy regions, whereby small groups take the initiative to transition their town or region to a low carbon economy. This has been seen to be quite effective in bridging the gap between innovators and other actors. As well as providing a testbed for new technologies, pilot projects are important in demonstrating the success of new ideas and thus overcoming the stigmas surrounding unproven or unfamiliar technologies (Van Bueren and Broekhans, 2013). Once there is a verified proof of concept, adaption by the swing group becomes easier to encourage. This may demonstrate why the rate of change requires a lot of groundwork before reaching a tipping point and a sudden acceleration.

An interesting pattern that has been seen in this movement is that change of this fashion is non-linear; a great deal of effort must be expended with little evidence of progress before some sort of tipping point or pivotal moment is reached. This was reported for the Austrian region of Güssing's transition to energy independence (Hecher et al., 2016) where the rate of change was almost non-existent for the first ten years of the project but then a sudden rapid acceleration over the next five years. We see this representation of the Pareto principle, often referred to as the 80/20 rule. Initially, 80% of the effort must be expended for 20% of the gain and then the remaining 80% requires only 20% of the effort.

This makes the value of the hare as drivers of change quite apparent, without these actors there would be no race. Though it may seem to the actors that their efforts have little effect, reaching this societal tipping point requires a great deal of action before change can occur. From this, the actors within the hare category may take a degree of reassurance, they must avoid falling victim to burnout and press-on, while ideally making sure not to leave the tortoise too far behind. Once the hare has established a clear line of sight for the tortoise, it can easily make its way. Thus, as a guiding voice, it is vital the hare continues to pull the tortoise forward.

Narrative four: A maturing civilisation

Here the hare represents our current dominant market-driven neo-classical economic model that focusses on the individual, whereby gratification is achieved through the acquisition and ownership of material stuff. The current era that has become known as "the Anthropocene" (Crutzen, 2002), in which our impacts on the environment are of such scales that we as a species are now considered a geological force. The tortoise represents the transition to a more collective, relational, community-based society. There is an imperative need for our civilisation to mature from the insatiable hare to the conservative tortoise whose ambition is shaped around qualitative growth (of mind/spirit) as opposed to further quantitative (physical) growth. In this light, the hare's identity is shaped by his consumption while the tortoise's is defined by her relationship to others. The young hare, taking a short-term perspective seeks instant gratification, while the more mature tortoise seeks a (more sustainable) slow burner for the long road ahead.

During the 1960s, a famous study on self-control and willpower at Stanford University's Big Nursery School developed what became dubbed 'the Marshmallow Test' (Mischel, 2014). As part of the study, pre-schoolers were placed in a room with nothing apart from the table and chair where they sat. On the table was a bell and a dinner plate with treats (of the pre-schoolers choice), in one corner there was a single treat and the other two. The dilemma the children faced was quite simple, at any time they may ring the bell thus ending the experiment and allowing them to eat one treat. However, if they could endure long enough for the researcher to return, they would be rewarded with two treats. Years later, the participants of the study were surveyed to see how they had progressed. From this, a clear relationship between an ability to delay gratification and success was established. Those who had the self-control to earn two treats also had the willpower to focus more effectively in school and later in life achieving higher-earning positions.

Our society faces a similar dilemma to these pre-schoolers. Consumption of fossil fuels and other natural resources can provide us with the short-term gains of monumental social and technological acceleration. However, this has ultimately come at the cost of a global ecological crisis. Avoiding the often-apocalyptic narrative surrounding our current ecological crisis will require that we learn sufficient self-control to sustainably manage the (finite) resources of our planet. Through delaying the gratification, we may achieve much greater satisfaction from the growth in our collective consciousness.

As Charles Eisenstein suggests in his book *Sacred Economics*; 'Just as life does not end with adolescence, neither does civilization's evolution stop with the end of growth. We are in the midst of a transition parallel to an adolescent's transition into adulthood.' (Eisenstein, 2011: 109). The rapid technological and societal growth of the twentieth century is depicted as our civilisation's early adolescent growth spurt. Fuelling this rapid physical grow requires the consumption of a vast quantity of resources. As with the hare (or any confident teenager) who is irritatingly self-assertive of their capabilities, the pursuit of progress and growth/consumption casts itself as the undisputed purpose of civilisation. Moreover, as with a baby or young child, they are necessarily the centre of their universe as they increasingly consume and grow in an anti-entropic fashion. Their parents meanwhile are pre-disposed to self-sacrifice for their children and for the greater good of intergenerational success and progress, prioritising inter- and intra-generational qualitative bonds and connections as they no longer seek (nor require) personal quantitative growth.

Metaphorically, we now face somewhat of a "quarter-life crisis" as we enter the early adulthood of our civilisation; there is a stubborn refusal to tackle the challenge of altering our deeply ingrained habits of consumption (or indeed, to even face up to or admit the problem). As with someone undergoing the transition from teenager to adult, there is a period of denial, nominally in the early twenties (though it can extend further) whereby we cling to the hope our youth will never end. We as a society appear to converge on a rather similar dismissal that while it is agreed there is a need for change, it cannot happen immediately and will thus be planned for an undetermined future date. St. Augustine like, the insincere acknowledgement is made that we are not yet quite ready.

However, this presents a great deal of risk. If we can't kick our bad habits before maturing, then we risk a lock-in. The old saying of 'you can't teach an old dog new tricks' holds quite true for societal transitions. As can be seen in Figure 8.2, while a successful transition can lead to stabilisation, failing to achieve this can lead to lock-in and backlash (Vandevyvere and Nevens, 2015).

In the marshmallow test, a very interesting observation was that if the reward was not visible to the participants then they were less likely to succeed in their attempts at self-control. Our society seems to demonstrate a very similar struggle when faced with the moral challenge of future disasters resulting from current actions. Just as one may struggle to be morally challenged by something that is physically distanced by space, time also presents an ethical barrier. As seen in the previous narratives, changes to attitudes and beliefs play an important role in shaping our ability to value environmental goals. We appear to have accepted the fallacy that we will never have to deal with the consequences of our actions, as the true severity won't be apparent until an undeterminable future date. Mum or Dad will just bail us out! However, as maturing adults, Mum or Dad may no longer be around for us.

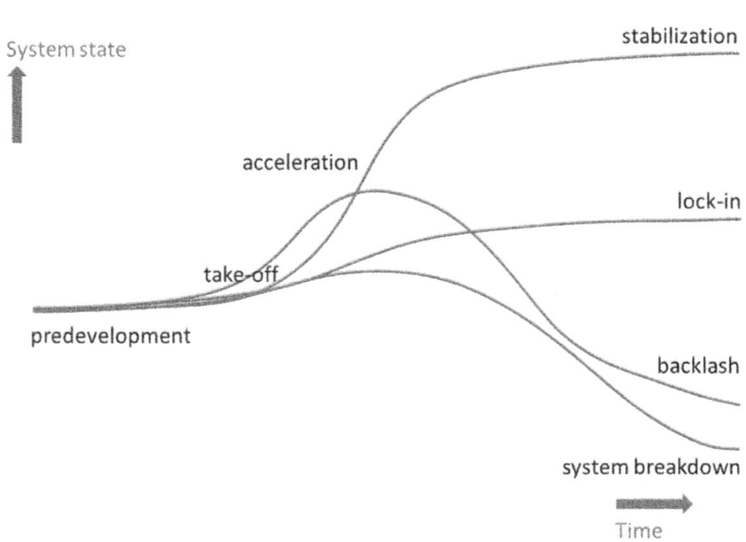

Figure 8.2 The multi-phase perspective on systemic transitions.

Source: Vandevyvere, H. and Nevens, F. (2015).

The hare's value set is extrinsic (self-centred), while the tortoise has greater consideration for intrinsic (greater than self) values. As John Foster presents in *The Sustainability Mirage* (Foster, 2008), our civilisation appears to have made the Faustian bargain. Through the use of fossil fuels, we have been granted these magical powers, however, ultimately that must be paid for with an eternity in hell. We have made an unthinkable deal with the devil, that we will enjoy lives of excessive luxury, risking the potential suffering of future generations. The hare whose principle concern is personal attainment of material possessions accepts this price, whereas the tortoise challenges the agreement. The tortoise seeks fulfilment through interaction with others and the legacy of its existence. It can see a greater value in restraining its personal desire for consumption and instead to flourish through interpersonal connections. Achieving harmony with itself and its environment in order to better the lives of itself, its peers, and its world, as well as that of future generations.

As is summarised in the accompanying cartoon (Figure 8.3), in the hypothetical race to/of sustainability, the techno-optimistic, materialistic hare will ultimately lose to the patient, family-orientated, techno-critical tortoise. While there is value in the hare as a disruptive and innovative force (in this case purchasing

Figure 8.3 'Slow and steady wins the race' (by Nina Wohlfahrt).

a high-end electric car), so long as the ambition is still narrowly shaped by technological solutions, it will have limited effect in tackling the broader challenge of consumption. Truly meaningful change requires a wider societal shift in values towards a more collective society.

Closing discussion

The lesson apparent throughout this chapter would appear to be that those who are environmentally conscious must be careful not to allow their passion to exasperate them. The sustainability race is not a 100m sprint or even a 400m track race, it is a marathon comparable perhaps to something like the Tour de France. The hare's pace will inevitably tire, in contrast to the tortoise's slow and steady endurance.

Clearly there is great value in the passion (and eagerness) of the hare, which plays an essential role in starting the race and can be viewed as a strong guiding force. While change can never happen at the rate that the hare desires, quite clearly it may never happen at all if the hare were not pulling the tortoise along. As with the natural flow of evolution, the niche innovators provide a valuable lesson as they scurry ahead down a variety of paths, which allows the tortoise to make a favourable choice. Societal transitions don't occur linearly, there is quite a lag between the response of the wider public and action of the hares until some sort of pivotal moment is reached bringing sudden acceleration, before finishing in a consolidated state. However, this urgency and pace leaves actors susceptible to frustration and burnout. The passion must be restrained, racing ahead means the hare cannot prevent the tortoise from taking a wrong turn and generates tension between the various actors.

Ultimately, the relationship is one of dialectical complementary opposites, as with many sustainability narratives (Byrne, 2017). If the hare is to see the desired change, it is imperative that it exhibits a degree of patience, recognising that transformative societal change can only occur whenever the tortoise makes its particular path. It's wise to remember that the race is ever evolving and indefinitely long!

Acknowledgement

The authors would like to thank Nina Wohlfahrt for her creative flair that provided the caricatures for this chapter.

References

Arbuthnott, K. D. 2009. Education for sustainable development beyond attitude change, *International Journal of Sustainability in Higher Education*, 10(2), pp. 152–163.

Barry, J. 2017. Bio-fuelling the hummer? Transdisciplinary thoughts on techno-optimism and innovation in the transition from unsustainability. In: Byrne, E., Mullally, G., Sage, C. (eds) *Transdisciplinary Perspectives on Transitions to Sustainability*. Abingdon, Oxon, UK: Routledge 106–123.

BBC. 2017. Blue Planet II, Episode 7. In: O. Doherty and W. Ridgeon, producers. *Blue Planet II*, London: BBC.

Beder, S. 1997. *Global Spin: The Corporate Assault on Environmentalism*. Hartford: Chelsea Green Publishing.

Bouwman H., Bornman R., van den Berg H. and Kylin H. 2012. *Lessons from Health Hazards|DDT: Fifty Years Since Silent Spring, Late Lessons from Early Warnings: Science, Precaution, Innovation*. Copenhagen: European Environment Agency.

Byrne, E.P. 2017. Sustainability as contingent balance between opposing though interdependent tendencies: A process approach to progress, evolution and sustainability. In: Byrne, E., Mullally, G., Sage, C. (eds) *Transdisciplinary Perspectives on Transitions to Sustainability*. Abingdon, Oxon, UK: Routledge 41–62.

Byrne, E., Mullally G. and Sage, C. 2017. *Transdisciplinary Perspectives on Transitions to Sustainability*. Abingdon: Routledge.

Carson, R. 1962. *Silent Spring*. Greenwich: Fawcett Publications.

Chaisson, E. J. 2008. Long-term global heating from energy usage, *Eos, Transactions American Geophysical Union*, 89(28), pp. 253–254.

Crutzen, P. J. 2002. The "anthropocene". *Journal de Physique IV (Proceedings)*, 12(10), pp. 1–5.

Eisenstein, C. 2011. *Sacred Economics: Money, Gift, and Society in the Age of Transition*. Berkeley, CA, USA: North Atlantic Books.

Foster, J. 2008. *The Sustainability Mirage: Illusion and Reality in the Coming War on Climate Change*. London: Earthscan.

Cole, H. S. D., Freeman, C., Jahoda, M. and Pavitt, K. L. R. 1973. *Models of Doom: A Critique of the Limits to Growth*. New York: Universe Books.

Frost, R. 1916. The road not taken. In: R. Frost, ed. *The Mountain Interval*. New York: Henry Holt.

Gibbs, L. 2002. *Aesop's Fables*. New York: Oxford University Press.

Giddens, A., 2009. *Politics of Climate Change*. Cambridge: Polity.

Hecher, M., Vilsmaier, U., Akhavan, R. and Binder, C. R. 2016. An integrative analysis of energy transitions in energy regions: A case study of ökoEnergieland in Austria, *Ecological Economics*, 121, pp. 40–53.

Kingsnorth, P. 2017. *Confessions of a Recovering Environmentalist*. London: Faber & Faber.

Lebow, V. 1955. Price competition in 1955, *Journal of Retailing*, 31(1), pp. 5–10.

Macy, Y. and Brown, M. Y. 2010. The greatest danger: Apatheia, the deadening of mind and heart. In: R.S. Gottlieb, ed. *Religion and the Environment*. New York: Routledge. pp. 365–378.

Malthus, T. R. 1798. *An Essay on the Principle of Population, as It Affects the Future Improvement of Society: With Remarks on the Speculations of Mr. Godwin, M. Condorcet and Other Writers*. London: J. Johnson.

Manabe, S. and Wetherald, R. T. 1967. Thermal equilibrium of the atmosphere with a given distribution of relative humidity, *Journal of the Atmospheric Sciences*, 24(3), pp. 241–259.

Meadows, D. H., Meadows, D. L., Randers, J. and Behrens, W. W. 1972. *The Limits to Growth*, New York: Universe Books.

Millot, A., Doudard, R., Le Gallic, T., Briens, F., Assoumou, E. and Maïzi, N. 2018. France 2072: Lifestyles at the core of carbon neutrality challenges. In: G. Giannakidis, K. Karlsson, M. Labriet and B. Ó Gallachóir, eds. *Limiting Global Warming to Well Below 2°C: Energy System Modelling and Policy Development*. Cham: Springer International Publishing. pp. 173–190.

Mischel, W. 2014. *The Marshmallow Test: Understanding Self-Control and How to Master It*. New York: Random House.

Ó Gallachóir, B., Deane, P. and Chiodi, A. 2017. Using energy systems modelling to inform Ireland's low carbon future. In: Byrne, E., Mullally, G., Sage, C. (eds) *Transdisciplinary Perspectives on Transitions to Sustainability*. Abingdon, Oxon, UK: Routledge: 170–185.
O'Neill, S. and Nicholson-Cole, S., 2009. "Fear Won't Do It" Promoting positive engagement with climate change through visual and iconic representations. *Science Communication*, 30(3), pp. 355–379.
Perry, B. E. 1933. The text tradition of the Greek life of Aesop. *Transactions and Proceedings of the American Philological Association*, 64, pp. 198–244.
Perry, B. E. 1952. *Aesopica: A Series of Texts Relating to Aesop or Ascribed to Him or Closely Connected with the Literal Tradition that Bears His Name*. Urbana: University of Illinois Press.
Rosen, R. A. and Guenther, E. 2015. The economics of mitigating climate change: What can we know?, *Technological Forecasting and Social Change*, 91, pp. 93–106.
Sage, C. 2014. Making and unmaking meat: Cultural boundaries, environmental thresholds & dietary transgressions. In *Food Transgressions: Making sense of contemporary food politics* (M. Goodman & C. Sage, eds). Abingdon, Oxon, UK: Routledge: 181–203.
Springer, C. P. E. 2011. *Luther's Aesop*. Kirksville: Truman State University Press.
Steinberg, W. 1961. *Martin Luther's Fabeln*, Halle: Niemeyer.
Suetonius Tranquillus, G. 1957. *De Vita Caesarum, Liber I, Divus Iulius*. Baltimore, MD: Penguin Books.
Thomashow, M. 1996. *Ecological Identity: Becoming a Reflective Environmentalist*. Cambridge: MIT Press.
Tol, R. 2017. The private benefit of carbon and its social cost. *Department of Economics Working Paper Series 0717*. Brighton: University of Sussex Business School.
Tol, R. S. J. 2018. The economic impacts of climate change, *Review of Environmental Economics and Policy*, 12(1), pp. 4–25.
van Bueren, E. and Broekhans, B. 2013. Individual projects as portals for mainstreaming Niche innovations. In: J. Hoffman, ed. *Constructing Green: The Social Structures of Sustainability*. Cambridge: The MIT Press.
Vandevyvere, H. and Nevens, F. 2015. Lost in transition or geared for the S-curve? An analysis of Flemish transition trajectories with a focus on energy use and buildings. *Sustainability*, 7(3), pp. 2415.
Williams, H. T. P., McMurray, J. R., Kurz, T. and Hugo Lambert, F. 2015. Network analysis reveals open forums and echo chambers in social media discussions of climate change, *Global Environmental Change*, 32, pp. 126–138.
Zsolnai, L. 2015. *The Spiritual Dimension of Business Ethics and Sustainability Management*. New York: Springer.

Part III

Metaphor, myth, and mind

9 Myth, metaphor, and parable in the psychoanalytic concept of development

Ian Hughes

Environmental sustainability, sustainable development, and psychology

Sustainability lies at the heart of two critical contemporary international agreements which together serve to define the major challenges currently facing the global community. The Paris Agreement, which is based on the concept of environmental sustainability, aims to strengthen the global response to the threat of climate change by keeping global temperature rise this century to within acceptable limits (Paris Agreement, 2016). The 2030 Agenda for Sustainable Development, which encompasses the Strategic Development Goals (SDGs), is based on the broader concept of sustainable development and aims to end all forms of poverty, reduce inequalities, tackle climate change, and improve the quality of life and well-being of every human being (United Nations, 2015). Achieving the aims of both the Paris Agreement and the 2030 Agenda will require transformative change in the technological, political, social, and ideological systems that currently underpin global society.

The models of change used to describe transformation towards environmental sustainability tend to highlight the role of new technologies and innovation to help move societies away from unsustainable practices in energy, transport, heating, and agriculture to more sustainable sociotechnical systems (Markard, et al., 2012). Models of change used to describe transformation in sustainable development, while including technology, encompass a broader range of change processes, including power relations, political change, cultural evolution, changing economic paradigms, and transitions from conflict to peace.

Psychology makes important contributions to the theory and practice of both environmental sustainability and sustainable development. With regard to the former, behavioural economics is increasingly being used to design behavioural change strategies to "nudge" citizens towards more environmentally sustainable behaviours (Lehner et al., 2016). Psychology also informs the fields of risk communication, the ABC (attitude, behaviour, and choice) model of social change (Shove, 2010), perceptions of global warming, psychological barriers that limit individual and collective action (Swim et al., 2009), as well as the understanding of impacts on human well-being and adaptation responses (Clayton et al., 2015).

DOI: 10.4324/9781003143567-12

With regard to sustainable development, psychology's diverse contributions include understanding the psychological costs of unhealthy and unsustainable aspects of the natural and social environment (Jaipal, 2017), psychology and inclusive development (Gupta & Vegelin, 2016), conflict reduction and peace building, and understanding the dynamics and constraints of social change processes at all scales, from the individual to the global (Norström et al., 2014).

While psychologists are therefore playing an active role in furthering the goals of both environmental sustainability and sustainable development, there is a conviction that psychology can and must do more. As Schmuck and Schultz point out, a fundamental change in relationships, between individuals, groups, generations, and species, lies at the root of the transformations required: "To achieve sustainability large scale changes are needed aimed at intergenerational equity, intragenerational equity, and interspecies equity" (Schmuck and Schultz, 2002). From this perspective, psychological transformation is indispensable for achieving both environmental sustainability and sustainable development.

This chapter aims to provide an input into the discussion of psychology's role in furthering transitions to sustainability by exploring the psychoanalytic concept of development and the myths and metaphors that are used therein. Its rationale is that the myths and metaphors identified may provide a basis, as part of the wider transdisciplinary project on Metaphors of Transformative Change, for greater discussion of, and understanding of, the multi-disciplinary nature of the fundamental transitions to sustainability that we are currently facing and the critical role that psychology must play in these transitions.

The story of development in psychoanalysis

Just as we go through stages of physical growth and development, so too we go through stages of psychic growth and development. Physical development is marked by changes in the size and maturity of our bodies; psychic development by changes in the inner structures and capacities of our minds. According to psychoanalysis, however, the two processes do not necessarily follow the same timeline, nor do they proceed in the same linear manner. While physical growth is a progressive process, psychic development is a less deterministic process which proceeds unevenly and may not necessarily advance beyond a given stage (Waddell, 2002).

Life stages, internal objects, and states of mind

The story of development that psychoanalysis tells places fundamental importance on the earliest years of childhood. It is during these years that the foundations of the psyche are laid down. In the Freudian view, the primary psychic structures of the mind are already established, if tentatively, by around age five. The rest of life, in this view, builds on and returns to the structures established by this time, albeit under very different internal and external conditions. It is a fundamental tenet of psychoanalysis that these foundations can affect our capacities for relationship with self and others for the rest of our lives (Freud, 1965).

During the first years of life, every child faces two enormous psychic challenges. The first is to overcome what is known as primary narcissism; the second is to come to terms with the fact that (s)he lives in a world of non-exclusive relationships. In psychoanalysis, two powerful myths are used to encapsulate the dramatic nature of these formative experiences: the myth of Narcissus and the myth of Oedipus.

The myth of Narcissus – Early omnipotence to separation

The ancient myth of Narcissus tells us how the handsome Narcissus was doted on by the nymph Echo, whom he rejected. In retaliation, the gods decided to punish Narcissus by making him fall in love with his own image in a mountain pool. Intoxicated by his own reflection, Narcissus was unable to love anyone but himself.

According to psychoanalysis, in the earliest stage of life, new-born infants are unable to distinguish between themselves and the world around them. There is a magical, omnipotent quality to their experiences. They cry and are automatically enveloped in a warm soothing embrace. They are hungry, and warm milk is quickly conjured up to satisfy their needs. Physical discomfort from a soiled nappy is magically dispelled whenever it is required. In this initial state of primary narcissism, as Freud called it, the infant is the world and the world responds to their every need.

From this experience of themselves as the sum-total of existence, the growing baby is faced with coming to terms, through painful experience, that they are not everything that exists. The infant's life as the centre of its mother's and father's attention does not continue in the magical vein of the earliest days for long. The infant soon experiences the inevitable frustrations that occur as the mother and father slowly withdraw from the intensity of care that was initially necessary. Now the child's every wish is no longer immediately and magically satisfied, and the existence of an outside reality begins to break in. Psychoanalysts refer to this crucial period of development as the beginning of "object relations" and it marks the beginning of the infant's psychic development.

According to psychoanalyst Melanie Klein, in these earliest days, the infant only has the capacity for a single state of mind, which is called the paranoid schizoid state of mind (Klein, 1975). When the infant is being fed and cared for, the emotional state is one of total bliss and satisfaction with the world. However, when the infant is suffering from hunger pains, or its existential needs are not being met, they become overwhelmed by fear of dying, and feelings of persecution and hatred. Unable yet to distinguish clearly between itself and the outside world, the infant experiences these intense emotions of either bliss or terror as pervading all of reality. The term paranoid schizoid for this state of mind reflects the twin realities that the infant's dominant anxiety is for self-preservation in the face of perceived persecution (paranoid), and that the infant is capable only of experiencing split emotions, either over-idealised good or catastrophically bad (schizoid).

The crucial first step in the infant's psychic development is the establishment of capacities within the infant's mind to ameliorate these intense emotions. In Kleinian terms, the infant develops the capacity to experience a second state of mind, known as the depressive position, in which both good and bad feelings can be held simultaneously (Klein, 1935).

In this process, the infant's carers play a vital role. Of crucial importance is what psychoanalyst Wilfred Bion termed the "containing function" of the parents. According to Bion, the mother and father bring not only their loving and caring functions to their infant, but also crucially their thinking functions as well (Bion, 1962). Of primary importance is the parent's capacity to hold the baby's anxiety, and their own, and to remain calm and go on thinking in the face of the child's intense distress. This parental capacity gradually dispels the intensity of the infant's pain and fear. Over time, through repeated experiences of having the chaos of their psyche "contained", the baby begins to internalise this capacity within its own mind. A "containing" function, which represents a rudimentary capacity to think, becomes an internal object in the infant's mind. For Bion, thinking arises primarily as a function for processing emotional experience by containing the chaos of intense emotion, rather than expelling it. With it, impulse life becomes bounded by thought rather than by the infant merely acting to shut out or expel intense emotions. This capacity marks the beginnings of the child's ego, the seat of rational thought and the inner structure that we most readily identify as the "self".

The infant's internalised capacity for containment gradually allows the "good"-idealised-loved aspects of experience and the "bad"-catastrophised-hated aspects of experience to be recognised as different aspects of the one entity. The psychic structures and capacities which enable the depressive position allows these two opposite emotions to be held at the same time and a more nuanced and realistic emotional and cognitive position to be reached. This state of mind is termed the depressive position because it is accompanied by the sadness of mixed emotions.

The infant's earlier paranoid-schizoid state of mind is characterised by an exclusive concern for self-preservation, a sense of paranoia and persecution, an idealisation of the good and a catastrophising of the bad, and a focus on their own interests. The depressive position, by contrast, is marked by a more realistic assessment of the balance between good and bad, a reduction in persecutory anxiety, and an ability to feel concern for others.

The development of an internalised capacity for containment of emotion makes a fundamental contribution to the development of the psyche and is a prerequisite for continued psychic development. By around the stage of weaning, it should have enabled what Melanie Klein characterises as the most important developmental shift in psychic functioning, namely the shift from the paranoid schizoid state of mind, characteristic of early infancy, to the depressive position. Over time, and provided the caring environment supports it, the depressive position becomes established as the child's predominant state of mind (Spillius, 1994).

The myth of Oedipus – From omnipotent control to relationship

Once the baby is conscious that the mother is a separate person, an awareness soon follows –once again forced on the child by experience – that the mother also loves and cares for others, including the father. With this awareness, the child enters the Oedipal phase of development. Freud used the myth of Oedipus to illustrate the intense emotions experienced by the baby at this second formative stage of early psychic development (Freud, 1924).

In Greek mythology, Oedipus was the King of Thebes. Abandoned at birth, he did not know the identity of his parents. On a pilgrimage to Delphi as a child, he was told in prophecy that he was fated to kill his father and marry his mother. As an adult, this prophecy came true when Oedipus slay his father during a roadside quarrel between strangers. As a result, he received the throne of Thebes and the hand of the widowed queen Jocasta, his mother, in marriage.

For Freud, this myth encapsulates the child's primitive wishes to deny the dawning reality that he is not the sole focus of his mother's love, and the powerful feelings of envy and rage that accompany the infant's recognition of this painful fact. By wishing his father dead, the child fantasises that he can continue his exclusive relationship with his mother. While Freud has been justly criticised for the gendered nature of the Oedipus myth, the underlying psychic challenge of coming to terms with living in a three-way relationship remains true for both genders.

The powerful emotions of the Oedipal phase test the primitive structures of the infant's psyche, in particular the child's ability to maintain the depressive position in the face of powerful regressive feelings. As before, the containment of the infant's distress by his or her parents is crucial in rendering their envy, hatred, and sadness manageable. The parental containing function again plays a crucial role in strengthening the child's nascent capacities for self-containment of emotion and in instilling a belief that they are loved, even if that love is shared with others. The challenge of this phase, which is never fully resolved, is the acceptance that the child cannot exclusively possess and control their parent, and more broadly to accept others as independent individuals with their own needs and desires.

The diverging paths of early development: Healthy and unhealthy narcissism

The psychoanalytic picture of development places internal conflicts as the basis of psychic functioning. The human mind, in this view, is a complex of loosely interacting parts with varying degrees of cohesion. Just as the infant's first relationships are not only to people outside him/herself, but also to the fragmented and developing parts of his own mind, so too in adulthood our relationships with others and the relationships within our own minds are inextricably interwoven.

Two of the internal parts of the mind, the "ego" and "superego", are familiar to us in popular discourse. The ego is the part of the psyche that we most readily relate to as the "self". Freud described the ego as the part of the personality that enables the individual to delay immediate gratification. A mature ego acts as the seat of judgement, rationality, and control. Its foundations, as we have seen, are the internalisation of the capacity to contain intense emotion and the rudimentary ability to think.

A second part of the mind is the superego. According to psychoanalysis, as the intensity of the mother and father's care is slowly reduced, the infant's earliest experiences of the parent's absence are of pain, persecution, and fear. In order to cope with these threatening emotions, the infant internalises an image of the parent in their mind. This internal image forms the beginnings of the superego, which serves to comfort the child. It plays the role of an ever-present carer, guarding over the thoughts and behaviour of the child, and eventually comes to act as the source of conscience and guilt. Over time, with repeated experiences of the mother's loss and return, the infant begins to recognise that although the mother may not be physically present, she has not died or disappeared forever, and consequently he himself is not in danger of dying. This capacity for accepting loss and separation and for mourning is connected to his/her future capacity to risk change, to widen experience, to enter new relationships – in short, the capacity to grow and develop.

In the infant's mind, two other psychic parts are also initially present that are less well known in popular discussion – the "ego ideal" and the "narcissistic self". The ego ideal is that part of the mind which holds onto the belief in the child's omnipotence despite all evidence to the contrary. Refusing to adapt to the limitations placed on it by the external world, the ego ideal continues to exert relentless demands for grandiosity and perfection.

The fourth part of the infant psyche – the narcissistic self – contains the child's natural drive for love and admiration and their desire to be looked at and admired. In early infancy, the narcissistic self has a heightened intensity that reflects the infant's existential need for attention. During the course of normal development, the narcissistic self eventually loses its original all-consuming quality and becomes integrated into the ego as a source of imagination, vitality, and healthy self-esteem.

Under normal circumstances, as the child matures, the developing ego manages to moderate the extreme demands for perfection and omnipotence of the ego ideal, and to contain the childish exhibitionism and desperate need for acclaim of the narcissistic self. As a result, as psychoanalyst Heinz Kohut writes, the mature personality becomes dominated by the ego – which exercises a measure of rational control – under the guidance of the superego, which sets realistic ideals and moderates behaviour through a healthy modicum of guilt (Kohut, 1966). During the course of normal psychic development, a person acquires a measure of humility, the recognition of external reality, and the acceptance that others are not here simply to serve their own needs.

If the child's early environment fails to provide the necessary care, the infant may fail to develop the capacity to establish the depressive position. As a result, (s)he may not develop a mature ego capable of utilising thought as a check on impulsive action, or a superego that can act as an ethical guide. Under adverse conditions, the child's psyche may become dominated not by a mature rational ego and an ethical superego, but by the immature parts of the infant psyche, namely the narcissistic self and the ego ideal. The outcome is an adult with a pathologically narcissistic character. The features of narcissistic disorder are well known: a grandiose sense of self-importance or uniqueness; an exhibitionistic need for constant attention and admiration; a lack of empathy and an inability to recognise how others feel; disregard for the personal integrity and rights of other people; and relationships marked out by a sense of entitlement and the exploitation of others.

Latency

After experiencing the intense emotional dramas of Narcissus and Oedipus, the child enters what is called the latency period, which stretches from around five years of age until the beginning of adolescence. This is a period of relative calm, during which the child begins to not only make a home in the external world, but also to take solace in their own internal world. How well the foundational structures of the child's mind settle during this period depends crucially on whether their internal objects and states of mind are predominantly benign rather than malign. If this is the case, the latency years should be a time of both inward withdrawal into the imagination and increasingly confident outward exploration of the world. Make believe and fantasy play a major role in helping the child's mind expand during these years. Latency is also typically a time of obsessive learning of facts and collecting of objects, such as sports cards, books, and dolls and their accessories.

The child's parents once again play a crucial role in supporting the child's psychic development by both allowing the child the space to withdraw into her/his own mind, and providing the encouragement and support needed to venture out into the world. In the story "Peter Pan", J.M Barrie provides a striking image of the parent's role during this stage of development, that of continuing to strengthen the child's capacity to deal with their own thoughts and feelings in the run-of-the mill day to day:

> Mrs. Darling first heard of Peter Pan when she was tidying up her children's minds. It is the nightly custom of every good mother after her children are asleep to rummage in their minds and put things straight for the next morning… When you awake in the morning, the naughtiness and evil passions with which you went to bed have been folded up small and placed at the bottom of your mind, and on the top, beautifully aired, are spread out your prettier thoughts, ready for you to put on.
>
> (Barrie, 1911)

Adolescence

With the onset of adolescence, the child enters the tumultuous transition from child to adult. While the physical changes of puberty take place rapidly within the span of the teenage years, the psychic development demanded by adolescence may take many more years, or even decades, to occur. Freud characterised the challenges of adolescence as the crystallisation of sexual identity, the finding of a sexual partner, and the bringing together of the two main streams of sexuality – physical sexuality and love. As with earlier stages of development, the emotional challenges of adolescence both test the psychic structures already in place and demand the development of new psychic capacities.

The dominance of the depressive position and the relinquishment of the desire for exclusivity and narcissistic control in relationships, achieved more or less imperfectly by the beginning of latency, are both revisited in adolescence with a vengeance. The onset of intense sexual impulses, along with physical and sexual maturity, bring with them a barrage of intense adolescent emotions and anxieties. The child is developing, of course, within the context of social norms and values with which he or she is expected to comply. The socialisation process in adolescence gives rise to feelings of restriction, rebellion, and anxiety. The teenager's anxieties are heightened by the fact that (s)he is capable of sexual and physical acts that the Oedipal age child could only fantasise about. The containment of emotion now takes on a new level of importance, given the real-world consequences that can result from his/her behaviour.

The psychological defences typical of adolescence attest to the difficulties that arise in containing the powerful emotions of this period. The most prominent defence is acting out. Unable to contain his emotions, the teenager will act in order to avoid thinking about threatening and confusing states of mind. The defence of splitting, typical of early infancy, can also play a prominent role in defending against threatening parts of the self. This is recognisable in the teenager's splitting of both self and others into all good and all bad which are passionately felt to be irreconcilable. For many adolescents, intelligence and retreat into academic pursuits is their best defence against real thinking. For others, drugs, alcohol, theft, and other forms of delinquency are the forms that acting out takes (Waddell, 2002).

When the young adult's internal structures are under such strain, the external environment, particularly family, school, and peer group, acquires enormous significance. If psychic development goes well, there will be a gradual move to a state of mind less prone to projection, acting out and denial, towards a state of mind more able to think about, accept, and identify with the feared aspects of the self. If the environment is conducive, the adolescent mind acquires a strengthened capacity for empathy and concern for self and others. Ultimately, the struggle of adolescence is a struggle towards an internal capacity for intimacy with self and others. As psychoanalyst Earnest Jones wrote, over the course of adolescence, the capacity to love grows stronger at the expense of the desire to be loved (Jones, 1922).

Adulthood

As stated already, in the psychoanalytic story psychic development is not a deterministic process that unfolds according to a predetermined plan. Instead it occurs unevenly, depending on internal capacities and environmental circumstances. In adulthood, earlier states of mind, whether infantile, latency, adolescent, or adult, are ever present and can shift in and out of dominance. Adult character and mental health both depend on having developed the internal structures and capacities required for mature adult functioning and maintaining the appropriate mental flexibility for negotiating the internal conflicts that are the very basis of mental life.

By mid-life, however, the various defences a person has adopted to avoid pain in the past may have acquired the appearance of fixed character. In contemporary society, it is tempting in adulthood to mistake status for identity. The routine of family and work life too may provide a holding structure that can serve to freeze psychic development and obviate the need for further creative engagement with others and with the world around us.

But development, at whatever stage, is grounded in the capacity to go on engaging with experience. To engage in this lifelong task, Heinz Kohut stresses the importance of retaining some element of the infantile narcissistic self as a source of vitality and creativity. Donald Winnicott highlights the importance for an adult of retaining the capacity to imagine and fantasise and to actively generate experiences that feel deeply real, personal, and meaningful.

T.S Eliot wrote, "For a man who is capable of experience finds himself in a different world in every decade of his life; as he sees it with different eyes, the material of his art is continually renewed. But, in fact, very few poets have shown the capacity for adaptation to the years. It requires, indeed, an exceptional honesty and courage to face the change" (Eliot, 1957). What is true of poets is true in this case for us all. "The difficulties of developing a mind and a life of one's own, of becoming one's self and finding one's place in the world", Margot Waddell writes, "requires constant mental and emotional work, at any age" (Waddell, 2002).

A parable of psychic development

From the first moment there is consciousness. I start as everything. An emotional being from the very beginning, but not yet a thinking being. There is no inside and outside – no boundaries. I am essentially a state of mind – the paranoid schizoid state of mind in which "I-The-Universe" am either all good -blissful-and-benign, or all bad-threatening-and-terrifying.

Slowly I become a separate being with an inside and an outside. The inside and outside of my physical being are clearly separated by my skin. I can take in things from the outside world – food, air, water – and make them part of me. Psychically, although my inside and outside are not so clearly separate, I can also take in psychic elements from "outside" to help me grow. Gradually four internal persons come to occupy my mind – ego, superego, narcissistic self, and ego-ideal. These internal persons are made up of aspects of people from the outside world. The nature of these internal persons depends on the nature of the persons around me, on whether they nourish or poison.

Gradually I become aware of a second mood that takes over from my all good/all bad state, the depressive mood. Before I was either a bright cloudless day or a dark stormy night. Now I am a mix of storm and sunshine. My internal persons respond differently to my different moods. If the depressive mood prevails, ego and superego grow stronger while narcissistic self and ego-ideal wither way. If the paranoid-schizoid mood dominates, ego and superego find it difficult to grow and narcissistic self and ego-ideal remain the dominant characters controlling my mind.

When ego and ego-ideal are the principal characters I can think rationally and care for others; I find purpose in loving relations with others. When narcissistic self and ego-ideal dominate, I am emotionally reactive, paranoid of others, and care only for myself; I find purpose in amassing wealth and power and in dominating others. A shifting balance between these states of mind remains a constant battle for me – a battle for life.

Whether I am to fully experience the joys and sorrows of life will be decided by the characters in my mind. If they can overcome the anxieties that guard against new experiences, I will experience time and again the emergence of my "as-yet-unknown-self". If anxieties prevail, they will maintain their fearful grip against my mental development.

Although I will come to understand multitudes during my lifetime, much of the story of how I came to be will forever remain a mystery to me but will guide the way I live my life.

Reimagining sustainability

If a crucial part of the challenge of sustainability is the psychological challenge of developing greater intergenerational, intragenerational, and interspecies equity, can the story of development in psychoanalysis help us imagine a way forward? Three aspects of the psychoanalytic view of development outlined here can offer psychologically informed ways of thinking about the transitions to environmental sustainability and sustainable development. These are: framing sustainability transitions as ongoing conflicted and creative processes that are never complete; overcoming individual and collective pathological narcissism; and recognising the need for the sacred as a psychological necessity in times of fundamental change.

Development as an ongoing creative process that is never complete

Psychoanalysis sees development as an ongoing process that is never complete. It is a dynamic process that is subject to premature arrest, blockages, and reversals. It is not a map with pre-laid roads and its destination, far from being prescribed, has to be created and discovered. The individual's freedom for creation in this task of development is bounded by both the inheritance they are born with and the circumstances with which they are faced. Mind, in the psychoanalytic view, is the intermediary function that navigates the twin impositions of nature and nurture and plays a vital role in how an individual fares in life (Craib, 1994).

Humanity's progress towards a sustainable future is similarly not a prescribed journey but one that needs to be collectively created. Our ability to navigate towards a more just and sustainable global society depends critically on the capacities of our collective mind, and on the prevailing moods that can either empower or disempower the capacities that enable us to progress. Like the individual mind, contemporary global society is fragmented and conflicted, and prevailing norms and values – particularly narcissistic values – are playing a critical role in blocking the transition to sustainability.

The necessity of overcoming infantile narcissism

The outcome of individual psychic development is fundamentally determined by the presence or absence of moral relationships. Progression from infantile narcissism to the capacity to care for others, and for the environment, is only possible when the child's predominant experiences are ones of love and concern. When the predominant experiences are of denigration, exploitation, and abuse, psychic development is frozen in the primitive narcissistic state that prioritises greed, self-concern, and hatred or indifference.

As Schmuck et al. (2002) emphasise, the transition to sustainability demands a collective psychological transition towards greater intergenerational, intragenerational, and interspecies equity. This psychological transition in turn demands the containment of the forces of infantile narcissism that influence contemporary economic, political, social, and religious systems. The infant's psychological transition away from primary narcissism is, as we have seen, a painful and protracted one and involves, crucially, the capacity to overcome denial that the world is not as the infant would wish. In the transition to sustainability, the collective capacity to overcome denial of the reality and potential catastrophe of climate change is a similarly painful necessity (Postel, 1992).

As Donald Winnicott, Heinz Kohut and other psychoanalysts have stressed, the retention of some elements of childish narcissism – such as boldness of ambition, belief in success, and refusal to give in to failure – are necessary though if the transition to sustainability is to be successfully navigated (Winnicott, 1982). Other features of the child's way of experiencing the world, including spontaneity, imagination, and directness of awareness are also vital aspects of adult mental life that are necessary for the navigating sustainability transitions. And as the psychoanalytic story of development reminds us, even though maintaining the vitality of childhood becomes more difficult with age, every new human generation has the potential – but only as potential – to change the character of humanity for the better.

Recognising the sacred

The psychoanalytic story of development identifies four key psychic functions that are performed, initially at least, by forces either outside the infant or beyond the infant's control. The first is the life force or vitality that animates us. Just as every human being is born with a life force that drives physical development, so too we are born with a force that drives psychic growth. Freud termed this force libido. In early psychanalytic theory, libido was primarily characterised as an essentially pleasure-seeking energy. Contemporary psychoanalytic theory, by contrast, views psychic development as motivated not primarily by pleasure seeking, but by relationship-seeking. Connection with others, rather than the satisfaction of our own material needs, is what contemporary psychoanalysis believes drives human psychological development.

A second function that is performed initially outside the infant and beyond its control is the containment of unbearable anxiety. As we have seen, the containment of the infant's incomprehensible and uncontainable emotions is one of the essential psychic functions which the parents bring to the parent-infant relationship. The infant's experience of care and concern eventually leads to the establishment of this containing function within the child's own psyche, a function which marks the beginnings of the capacity to think and is essential if psychological development is to proceed.

A third psychic function that is performed initially outside the infant by the parent, and later beyond the infant's control, in unconscious knowledge. For the growing child, much of their experience of the world is beyond their comprehension, and it is for the parent to make the incomprehensible bearable. Later in life, even when education renders experience more amenable to understanding, the unconscious remains as a constant part of the self that knows things that are not consciously known. The unconscious, rather than being a noun and a place where the repressed is hidden and stored, has come to be seen as a verb – as a term for the collective of defences and identifications and inner objects that constitute the dynamics of the psyche and which operate moment to moment beyond conscious awareness.

The final psychic function is hope grounded in love. Initially the responsibility for the infant's growth and development, and the love and encouragement needed for such, is assumed by the parents. As the child grows, aspects of this too are internalised and become an inner force willing engagement with the world in the service of a better future alongside an inner confidence that a better self is possible.

Freud famously explained religious belief in an omnipotent being as rooted in the universal human experience in infancy of what is presented here as vitality, containment, unconscious knowledge, and hope grounded in love, all of which emanate from beyond the self. He dismissed religion as a neurotic refusal to develop beyond an infantile understanding of the world and to relinquish what he viewed as the comforting but childish worldview that religion can provide (Freud, 1985).

As this chapter illustrates, however, the psychoanalytic story of development that has emerged since Freud's time indicates that some of the core psychic functions that religions perform are indispensable for psychic growth and for attaining a truer knowledge of the world. If a fuller engagement with the process of sustainable development entails an overcoming of the infantile narcissism that characterises much of the modern world, Freud's rejection of the sacred may itself be unsustainable. On the contrary, we may need to collectively construct a psycho-social environment that acknowledges our deepest vulnerabilities and dependencies on each other and on nature. We need to construct a psychologically nurturing environment in which care and concern, not greed and indifference, dominate.

To accomplish this, in ways secular and religious, we may need to collectively recognise the sacred in ourselves and in nature, not primarily as an objective reality, but as an individual and collective subjective need, and as an urgent necessity without which we may not survive. In times past, humankind has developed religious belief systems in part to deal with the need for human survival. In our own time, the threat of environmental collapse necessitates the evolution of a new spirituality and an accompanying moral revolution as essential components of our transition towards sustainability.

Acknowledgements

Roger Duck, Mapsar Limited.
Margot Waddell, on whose book 'Inside Lives' this chapter draws extensively.

References

Barrie, J.M., 1911. *Peter Pan*, London: Everyman.

Bion, W.R., 1962. A theory of thinking, *International Journal of Psychoanalysis*, 43. pp. 306–310.

Clayton, S., Devine-Wright, P., Stern, P. C., Whitmarsh, L., Carrico, A., Steg, L., et al. (2015). Psychological research and global climate change. *Nature Climate Change*, 5(7). pp. 640–646.

Craib, I., 1994. *The Importance of Disappointment*. Abingdon, UK: Routledge.

Eliot, T.S., 1957. *On Poetry and Poets*. London: Faber and Faber.

Freud, S., 1924. The dissolution of the oedipus complex. In: *The Standard Edition of the Complete Psychological Works of Sigmund Freud, Volume XIX (1923–1925): The Ego and the Id and Other Works*. London: Hogarth Press. pp. 171–180.

Freud, S., 1965. *New Introductory Lectures on Psychoanalysis*. New York: Norton.

Freud, S., 1985. Civilisation and its discontents. In: *Collected Works*, Vol 12. London: Penguin.

Gupta, J. and Vegelin, C., 2016. Sustainable development goals and inclusive development, *International Environmental Agreements*, 16. pp. 433–448.

Jaipal, R., 2017. Psychology at the crossroads: Sustainable development or status quo?, *Psychology and Developing Societies*, 29: 2. pp. 125–159.

Jones, E., 1922. Some problems of adolescence, *British Journal of Psychology*, 13. pp. 31–47.

Klein, M., 1935. A contribution to the psycho-genesis of manic-depressive states, *International Journal of Psychoanalysis*, 16. pp. 145–174.

Klein, M., 1975. Notes on some schizoid mechanisms. in M. Klein, ed. *Envy, Gratitude and Other Works, 1946–1963*, London: Hogarth.

Kohut, H., 1966. Forms and transformations of narcissism. *Journal of the American Psychoanalytic Association* 14. pp. 243–272.

Lehner, M., Mont, O. and Heiskanen, E., 2016. Nudging: A promising tool for sustainable consumption behavior? *Journal of Cleaner Production*, 134, Part A. pp. 166–177.

Markard, J., Raven, R., Truffer, B., 2012. Sustainability transitions: An emerging field of research and its prospects. *Research Policy*, 41 (6). pp. 955–967.

Norström, A.V., Dannenberg, A., McCarney, G., Milkoreit, M., Diekert, F., Engström, G., Fishman, R., Gars, J., Kyriakopoolou, E. and Manoussi, V., 2014. Three necessary conditions for establishing effective Sustainable Development Goals in the Anthropocene. *Ecology and Society*, 19(3). pp. 8–16.

Paris Agreement on Climate Change, 2016. Available at: http://unfccc.int/paris_agreement/items/9485.php.

Postel, S., 1992. Denial in the decisive age. In: L.R. Brown, ed. *State of the World*. New York/London: Norton. pp. 3–8.

Schmuck, P. and Schultz, P.W., 2002. Sustainable development as a challenge for psychology. In *Psychology of Sustainable Development*. Boston, MA: Springer. pp. 3–17.

Schmuck, P., Wesley, P. and Schultz, P., 2002. *Sustainable Development as a Challenge for Society*. In: P. Schmuck, P. Schultz, and P. Wesley, eds. *Psychology of Sustainable Development*. Boston, MA: Springer.

Shove, E., 2010. Beyond the ABC: Climate change policy and theories of social change. *Environment and Planning A*, 42. pp. 1273–1285.

Spillius, E.B., 1994. Developments in Kleinian thought: Overview and personal view. *Psychoanalytic Inquiry*, 14(3). pp. 324–364.

Swim, J., Clayton, S., Doherty, T., Gifford, R., Howard, G., Reser, J., Stern, P. and Weber, E., 2009. *Psychology and global climate change: Addressing a multi-faceted phenomenon and set of challenges. Report of the American Psychological Association Task Force on the Interface Between Psychology and Global Climate Change*, http://www.apa.org/releases/climate-change.pdf.

United Nations, 2015. *Sustainable Development Goals*. Available at: http://www.un.org/sustainabledevelopment/sustainable-development-goals/.

Waddell, M., 2002. *Inside Lives: Psychoanalysis and the Growth of the Personality (The Tavistock Clinic Series)*, 2nd edition. London: Karnac Books Ltd.

Winnicott, D.W., 1982. *Playing and Reality*, Abingdon, UK: Routledge.

10 The elusive target

Towards an understanding of the metaphors about dementia and sustainability

Cormac Sheehan

The dementia conundrum

The terms Alzheimer's disease (AD) and dementia are often used interchangeably. Dementia is not a specific disease but a category of over 400 diseases, of which AD is the most common. Observed symptoms of dementia include changes in cognition, memory, and physical activity and abilities. The interchangeability of AD and dementia is understandable, given that AD accounts for eighty per cent of dementia cases, and points to the complex and changing nature of categorisation. For the purposes of this chapter, dementia will be used as the preferred term.

According to the World Health Organization's International Statistical Classification of Diseases and Related Health Problems (ICD-10), symptoms of dementia can be mild, moderate, and severe. The most common symptom is a loss or decline of memory, both non-verbal and verbal, especially with regard to the learning of new information. Thinking, calculation, language, and performance of daily activities also become impaired. Changes in emotional well-being, social behaviour, and motivation, for example increased apathy, are also evident. Diagnosis requires the presence of such symptoms for six months and should be objective and supported by cognitive testing. Mild cognitive impairment (MCI) is often the threshold for the development of AD and other dementias, but this is not always the case. The single biggest risk for the development of dementia is age. This is reflected in the common usage of the term senility to refer to dementia, from the Latin senilis, "of old age", or the Proto-Indo European "sen", meaning old.

Dementia has been evident throughout history. Alois Alzheimer's "discovery" in 1906, and "rediscovery" by Katzman in the 1970s, are predated significantly by ancient Egyptian texts, which according to Boller and Forbes record that age can be accompanied by a major memory disorder. Throughout history, what we now call dementia has had many synonyms: 'Alienation, amentia, anoea, dotage/second childhood, silliness, foolishness, idiocy, imbecility, insanity, lethargy, morosis, simplicity, stupidity, and senility' (Boller and Forbes, 1998, p. 126). Even today, the use of the term dementia is not universally accepted, with French-speaking countries still using the terms senility and AD (Sheehan et al., 2016). Dementia is, however, currently the predominantly used term, although this could change again as our understanding of the neurobiological changes and aetiology of dementia increases.

DOI: 10.4324/9781003143567-13

Globally, the story of dementia is also diverse, with variations in care modalities, diagnosis, prevalence, and rates. In 2015, the global rate of dementia was estimated to be 45 million; by 2050 this is set to rise to one hundred and thirty-one million. The largest increase is predicted in low- to middle-income countries, with aging populations and improvement in diagnosis and awareness as the main reasons for this increase. Finland, for example, currently reports 60 dementia deaths per 100,000 deaths, whereas Cambodia reports less than 1 per 100,000. While it is uncertain what the true rate of increase of dementia will be, dementia should be considered a global endemic. The burden of dementia falls predominantly on family carers, with over ninety-four per cent of care in low- to middle-income countries provided by family members. In higher income countries, an estimated sixty to seventy per cent of care is provided by family members (Sheehan et al., 2016).

Writing on the complexity of AD, Locke explains how interest in, and research on, dementia was 'seriously neglected for many years, in part because they fell into a no-man's land, neither fish nor fowl, neither neurological nor psychiatric disorder, and geriatrics did not exist as a specialty' (Locke, 2013: 37). In 1906, Alzheimer observed potential evidence of changes in the brains of patients exhibiting dementia symptoms. However, the situation described by Locke persisted until the 1970s, when Robert Katzman, a pioneer in research into dementia, called for the abandonment of the common wisdom that ageing leads to senility. Instead, he argued, dementia should be considered a pathology, distinct from the normal ageing process. Yet dementia still resists easy diagnosis and categorisation. As Locke explains, the pathology of dementia is difficult due to inconsistent diagnostic practices in settings other than highly specialised units. At present, cases of AD can be definitively diagnosed only retroactively, post-mortem, and numerous cases do not undergo autopsy. To complicate matters further, one can harbour AD pathology in the brain but exhibit no outward signs of dementia (2013: 5).

Metaphors of dementia and sustainability

As George Lakoff explains, in classical theory of language, metaphor was seen as part of language, not of thought, much in the same way as dementia was assumed to be part of ageing process, not a pathology. Modern theories of metaphor, however, have moved away from viewing metaphor as merely poetic and linguistic expression.

For Lakoff, 'the generalizations governing poetic metaphorical language expressions are not in language, but in thought. They are general mappings across conceptual domains….and apply not just to novel poetic expressions, but to much of ordinary everyday language' (1993: 202). Lakoff contends the true locus of metaphor is in the domain of conceptualisation, specifically how 'we conceptualise one mental domain with another' (Ibid: 202). The act of interpreting, writing and speaking, is an act of cross-domain mapping. Everyday concepts, which are abstract, can be interpreted as metaphors, like 'time, states, change, causation and purpose' (Ibid: 203). From this contemporary theory of metaphors, it is assumed that metaphors and interpretation of metaphors are cross-culturally diverse, and open to reimagining, reuse, criticism, and retirement.

While cross-cultural diversity is to be anticipated, there is evidence to suggest that metaphors have universal functionalities. Kimmel (2004) argues that these functionalities relate to the general job that metaphor accomplishes in connecting culture and cognition. Anthropologists have reflected on the role of metaphor in cultural thought systems as creators, sustainers, delimiters, conceptual 'glue', or contesters of the status-quo. Metaphor enables a movement from an abstract concept to a concrete image; it triggers affect and/or experience; it bridges logical gaps, it relates parts to a larger whole, and it maps out nonverbal phenomena or behaviour (2004: 276). In her classic short essay *Shakespeare in the Bush*, Bohannan postulates that 'human nature is pretty much the same the whole world over; at least the general plot and motivation of the greater tragedies would always be clear—everywhere—although some details of custom might have to be explained and difficulties of translation might produce other slight changes' (1966). Bohannan found that while there are functional universals underpinning metaphors, the cross-mapping domain leading to interpretation of experience and understanding are diverse. While living among the Tiv of Nigeria and Cameroon, and failing miserably at her attempts to understand the emic kinship structure, Bohannan found herself re-imagining Shakespeare's Hamlet with a group of elder Tiv:

> That night Hamlet kept watch with the three who had seen his dead father. The dead chief again appeared, and although the others were afraid, Hamlet followed his dead father off to one side. When they were alone, Hamlet's dead father spoke.
>
> "Omens can't talk!" The old man was emphatic.
>
> "Hamlet's dead father wasn't an omen. Seeing him might have been an omen, but he was not."
>
> My audience looked as confused as I sounded. "It was Hamlet's dead father. It was a thing we call a 'ghost.'"...."What is a 'ghost?' An omen?" "No, a 'ghost' is someone who is dead but who walks around and can talk, and people can hear him and see him but not touch him." They objected. "One can touch zombies." "No, no! It was not a dead body the witches had animated to sacrifice and eat. No one else made Hamlet's dead father walk. He did it himself." "Dead men can't walk," protested my audience as one man. I was quite willing to compromise. "A 'ghost' is the dead man's shadow." But again, they objected. "Dead men cast no shadows." "They do in my country," I snapped.
>
> (1966)

Metaphors are used, as in the case of Hamlet, as a "conceptual glue" (Kimmel, 2004: 276), but in this case it was impossible to 'bridge logical gaps'. If a dead person can't talk, or a dead person can't be understood as a rhetorical device, then interpretation of the metaphor will either be rejected, or as in this case of the Tiv elders, adapted to their familiar cultural domains - a ghost is an omen sent by a witch. Like the functional universals presented by Kimmel, others have postulated that storytelling, which by its nature depends on metaphor, is also universal.

Kurt Ranke (1967) coined the term Homo Narrans; for German folklorist Ranke, storytelling is a precondition. Ranke argued that humans have no choice in telling stories, and that it is a basic human need 'to perceive the world in the fullness of its contents and functional dimensions in order to be able to tell about it.' The importance of storytelling is undeniable in the structure and maintenance of cultural values, codes, and behaviours.

In her seminal work 'Illness as Metaphor and AIDS and Its Metaphors', Susan Sontag (originally published in 1978) states that 'illness is not a metaphor'. Although terms associated with specific illnesses can be used metaphorically, Sontag is interested in the sometimes malicious and lurid metaphors that place the person with an illness as potentially responsible for their own illness:

> My point is that illness is not a metaphor, and that the most truthful way to regarding illness - and the healthiest way of being ill - is one most purified of, most resistant to, metaphoric thinking. Yet it is hardly possible to take up one's residence in the kingdom of the ill unprejudiced by the lurid metaphors with which it has been landscaped.
>
> (2009: 3)

Sontag argues that we must understand illness metaphors to be 'liberated from them' (1991: 3). She uses metaphors throughout 'Illness as Metaphor and AIDS and Its Metaphors' in a 'mock exorcism of the seductiveness of metaphorical thinking'(Ibid: 91). However, Sontag also argues, 'one cannot think without metaphors' (Ibid: 91). This view is supported by Lakoff's assertion that the locus of metaphors is in thought. Sontag's brilliant analysis of metaphors concerning TB and AIDS is not for discussion here. However, her assertion that metaphors should be retired or abstained from use is particularly valuable in a discussion of metaphors about dementia and sustainability:

> But that does not mean there aren't some metaphors we might well abstain from or try to retire. As, of course, all thinking is interpretation. But that does not mean it isn't sometimes correct to be 'against' interpretation.
>
> (Ibid: 91)

Sontag wrote (metaphorically) that when we are born, we 'hold dual citizenship in the kingdom of the well and in the kingdom of the sick' (Ibid: 3). Prior to writing 'Illness as Metaphor and AIDS and Its Metaphors', she took up citizenship in that "other place". Sontag had been diagnosed with cancer and found the metaphors about cancer to be malicious and lurid. Since the metaphors we use about any illness are an interpretation of an illness, and by extension a conceptualisation of people who have that illness, it is important to ask whether the metaphors we use cause further suffering and pain to those who inhabit the kingdom of the sick?

Tom Kitwood, in his seminal work 'The Person Comes First: Dementia Reconsidered', rejects the common metaphors of dementia. Kitwood argues that personhood is a status or standing bestowed on one human being by others in the context of social relationship and social being, and implies recognition, respect, and trust (1997:8). He argues that the self is derived and bestowed on the person with dementia through social action (Sheehan et al., 2016). Through negative actions, like the use of metaphors about dementia which are inherently about decline and decay (Innes 2009: 1), Kitwood argues that a person with dementia is socially and emotionally affected by such metaphors. He termed this negative social interaction as malignant social psychology. He asserts that even without intent, a caregiver for example, can be malicious, through lack of knowledge. The use of metaphors about dementia without understanding or interpretation can lead to the abstract becoming the objective image of that person, regardless of absence of intent to cause social malignancy.

A counterpoint to the harmful view of metaphors is presented by Golden et al. (2012), where they encourage nursing practitioners to be aware of the metaphors used by carers of those with dementia. Here metaphors are understood as a positive conceptual means to elucidate inner states, and as an economical means of expression that allows for vivid expressions of feelings and sentiments.

It is true that there has been a shift in how dementia is considered within societies, but to what extent that reconsideration is equal in different societies is difficult to tell. We know little of people with dementia in the developing world. The discussion in the developing world is focused on minimising risk factors, predicting future trends, and highlighting the global burden of the dementia endemic. This is a global crisis, whereby an ageing population and chronicity presents new and long-term public health concerns. Metaphors about dementia must be understood not just in terms of the age chronicity, but as interwoven with metaphors of war, flood, and epidemic and metaphors of loss. The interconnectedness of metaphors about dementia suggests that singular conceptual domains are insufficient in understanding such metaphors. Lakoff's term of "cross-mapping domains" is particularly relevant to understanding the domains associated with dementia.

The medicalisation of dementia is, according to Beard et al. (2009), woven into the metaphor of loss. Loss is not restricted to capacities, somatisation or cognition, but rather the primary loss is of the self. Cohen and Eisdorfer (1986) and Fontana and Smith (1989) use terms like "dissolution" and 'unbecoming of the self'. Aronson and Lipkowitz (1981) describe dementia as an extinction of personhood, and Kastenbaum (1987) argues that dementia causes a death-in-life or living death. Johnstone calls the loss metaphors associated with dementia the "Alzheimer's metaphor".

> The very notion of Alzheimer's disease was found to have been conceptualized and used as a clichéd, yet potent and unsettling metaphor for 'losing one's mind' and ultimately 'personal control', which is intricately associated with culturally constructed notions of personhood and eligibility for moral membership of the human social environment.
>
> (Johnstone 2011: 382)

Linking metaphors of loss, Johnstone suggests there is implicit moral judgement on a person with dementia. Simply, if a person loses personhood, they cannot be conceived to be a complete individual, and therefore, their presence and latent immoral behaviours are threatening. Alzheimer's disease is feared and felt to be contagious, literally and morally. The fear of contagion aroused by Alzheimer's disease is evoked and reinforced, according to Johnstone, by the persistent and mutually re-enforcing use of war/military, epidemic/flood, and thief/lurking metaphors about dementia.

In their analysis of social death, Sweeting and Gilhooly (1997) argue that components that form the basis of social death coalesce in the case of dementia – namely people with a lengthy fatal illness, the very old, and those inflicted with loss of personhood. Others 'may view or treat the dementia sufferer as a liminal or non-person, who is demonstrably making the transition from life to death' (1997). However, in their study, they find that the presence of social connections challenges the metaphors of loss and social death. While society may deem a person to be within a "living death" or experiencing a social death, those close, perhaps part of a carer dyad, actively challenge the social death of the person with dementia. As Sweeting and Gilhooly explain, carers – especially relatives and friends – do not allow the person to become socially dead for a number of reasons – namely, the person is physically present, there is hope of a connection to the person, and such an acceptance would impact the quality of care. They found that people caring for those with dementia do not allow their physical appearance – the hair, clothes, and cleanliness of the person with dementia – to become unkempt. This care for the body is the embodiment of social conformity and the antithesis of social death. According to Pearlin et al., '… giving care to a person is an extension of caring about that person' (1990: 583). If we accept both Sontag's assertions of lurid and malicious qualities of illness metaphors, and Kitwood's malignant social psychology, we must also factor in the importance of social connectedness. It is possible that the further away we move from the kingdom of the sick the more malignant metaphors of dementia become. Are metaphors used by those who are in daily contact with people living with dementia different to the dominant societal metaphors on dementia? I will return to this question later in the chapter.

Loss metaphors are interconnected with the conceptualisation of war/military and wave/flood/epidemic metaphors. There must be a reason to wage war against dementia and fight against the tide of dementia – namely to avoid the loss of personhood and alleviate the burden of care and the latent threat to public health posed by dementia. War/military metaphors are well established in metaphors of disease and illness. If we take a simple definition of disease, namely abnormalities in the structure and function of organs and the body systems, then the idea that plagues, blight, microbes, viruses, and bacteria can "invade" the body, and "attack" homeostasis is unsurprising. With regard to AD, Johnstone explains:

> Use of the military metaphor enemy frame in Alzheimer's discourse characteristically included the following signature terms: 'invades', 'attacks', 'strikes', 'hits', 'explodes', 'fights', 'destroys', 'kills', 'wins'. Use of the military metaphor alien invader frame in Alzheimer's discourse, in turn, characteristically included the following signature terms: 'alien', 'invades', 'takes over', 'consumes', 'sucks', 'wastes', 'destroys', 'leaves'.
>
> (Johnstone: 2011: 384)

In recent years, the use of military/war metaphors has increased in academic publications, governmental white papers, and in the media. In their commentary piece, 'The war against dementia: are we battle weary yet?', Lane et al. (2013) offer a comprehensive list of the pervasiveness of military/war metaphors about dementia.

> Military metaphors are widely used when describing health status and illness, and the mainstream media are frequently accused of perpetuating them. Military metaphors are certainly prevalent in newspaper headlines… these metaphors are not exclusively the domain of the popular and commercial press. In the medical literature, military metaphors are used in reference to an extensive range of medical conditions.
>
> (Ibid: 281)

Former UK Prime Minister David Cameron and former U.S. President Barack Obama, in separate speeches, framed dementia in the context of war/military metaphors. Cameron declared that we need 'an all-out fight-back against this disease', while Obama "declared war" on Alzheimer's disease. Zellig contends that these war/military metaphors also reflect contemporary fear of terrorism, since referencing dementia as a ticking time bomb is a pervasive metaphor (Zeilig, 2015).

Another category of pervasive metaphors about dementia is that of epidemics and impending natural disasters. Johnstone explains that dementia metaphors include "epidemic", "plague", "immune/immunity", "afflicted", "you can catch it", "it can affect anyone" (Ibid: 382). These metaphors have been effective, in that dementia is often treated "as if" it is a contagious disease. Zeilig outlines similar evocative metaphors of natural disasters evident in the public domain, for example, the "rising tide", the "wave of dementia", and "silent tsunami" of dementia.

One could argue that such metaphors are part of the discourse of disease and are an effective tool to describe the social construction of disease. The problem, as Sontag noted almost 40 years ago, is that such metaphors have become so pervasive, and as Parsi (2016) contends, are used with such frequency, 'that patients may not even think about how to interpret their illnesses in a different way… What's troubling is when one metaphor (in this case, the military metaphor) becomes the only or dominant way we interpret various illnesses. Thus, we need more and better metaphors'. In the literature concerning the typologies of illness narratives, the most common types of narratives are restitution and quest narratives. In both cases, the emphasis is on the patient to recover from their illness, or to take it upon themselves to overcome public attitude towards their illness.

Let us return now to the question posed earlier in this chapter. As we move closer to the experience of the illness, do the dominant metaphors change, or do people who live with an illness, whether patient or carer, reinforce the dominant metaphors?

Golden et al. (2012) analyse the metaphors used within a carer dyad and argue that metaphors used within this context can aid in the care for people with dementia. The most significant metaphor they found in their qualitative study was that of a journey. Of over 200 metaphors they identified, there were 62 references to the metaphor of dementia as a journey, compared with only four references to loss.

In their systematic review of non-literal language in dementia, Rapp and Wild (2011) point to several studies of non-literal language, including metaphor, which show that patients with right-hemisphere lesions have difficulties in matching orally presented metaphors with an appropriate picture (Ibid: 212). In contrast, left-hemisphere lesioned patients are able to give definite responses to non-literal language. Rapp and Wild also identify several studies in which patients with dementia were given new metaphors and old metaphors. These studies suggest that patients with dementia 'were only significantly impaired for new metaphors, but not for conventional metaphors and idioms' (Ibid: 214).

This presents a challenge to those who seek new metaphors or aim to retire old metaphors about dementia. We know from dementia studies that people with dementia recall earlier and established memories and behaviours and have difficulty with new tasks and information. If the established metaphors about dementia are culturally pervasive, what options are there to change such metaphors? Parsi (2016) argues that some professionals, institutions, and patients just 'want to wage ever-lasting war' on disease. However, war, be it metaphorical or actual, causes damage and can undermine public health goals.

Can there be new metaphors about dementia? Will those with dementia be able to interpret new metaphors about dementia? Will there be a transitional period when old metaphors are retired, and new ones put in place? Or are cultural conceptualisations of dementia so pervasive that it is impossible to create new metaphors?

There is little doubt that over recent years the attitude towards people living with dementia has changed. Even the subtle but pronounced shift of moving from Alzheimer's to dementia as the preferred descriptor is, in a sense, metaphorical. Alzheimer's is a disease; dementia is not a disease. Crucial within this reframing of dementia has been the appearance of people living with dementia speaking publicly and advocating for themselves. Across Europe, several organisations have funded long-term projects for the creation and promotion of dementia awareness. Advocacy and dementia narratives are key to people with dementia achieving agency and autonomy. Zeilig argues that the landscape of dementia has changed rapidly and radically over the last decade:

> The cultural and social landscape has radically altered since then and has most rapidly changed in the last decade. Most people now have a sense about the work of old age psychiatrists – and we are inundated by a rash of films, plays, novels and even operas that represent "dementia". At the same time, we have dementia friends, dementia friendly communities, dementia champions and even a dementia "Challenge". This is not to mention the 2013 G8 summit that focused on galvanising global action on dementia'.
>
> (2015)

According to Sontag, 'Illness is the night-side of life, a more onerous citizenship. Everyone who is born holds dual citizenship, in the kingdom of the well and in the kingdom of the sick. Although we all prefer to use only the good passport, sooner or later each of us is obliged, at least for a spell, to identify ourselves as citizens of that other place' (Ibid: 3). Sontag is speaking here about the fragility of good health. Everyone is susceptible to illness, including dementia, and for Sontag the retirement of metaphors that further alienate or increase social suffering or stigma is needed. Within Sontag's assertion, perhaps unwittingly, is a reference to what has emerged since Kitwood's seminal work and gained rapid pace over the last decade – the concept of living well with dementia. It is possible to analyse this rapid change through the lens of sustainability.

The statistical significance of a health intervention is central to the question of sustainability of care for people with dementia. For example, Zieschang (2013) examined the effectiveness of a three-month period of specialised standardised motor training, where patients with dementia were revaluated nine months later. Overall, 'effective sustainably improved functional performance of patients' was observed (Ibid: 191). This is a type of sustainability, whereby assessment is made and quantified and then revaluated after a period of time. The type of sustainability of living well with dementia, however, is less objective and more abstract, but still observable and perhaps qualitatively measurable. As with all basic ideas of sustainability, the key issue is resources – are there enough for everyone at present and in the future. Sustainability can be judged as a success or failure, based on time, trend, and future predicted outcomes. Within studies that assess sustainability of chronic disease management/interventions, sustainability encompasses, but is not restricted to maintenance of health benefits and core activities, and capacity of community to deliver health programmes (Francis et al., 2016).

Dementia and metaphors about dementia coalesce in matters of memory, previous abilities, and personhood. The conceptualisation that a person living with dementia is the sum of their memories or lack of recall has been challenged. Living well with dementia has little to do with past abilities, it is about the 'aim of improving quality of life and quality of care for people with dementia' (Downs, 2013: 368). At the heart of this challenge is the conceptualisation of what constitutes personal memory, and the location and typologies of memory. Central to the discourse of the reframing of memory is embodiment (Sheehan et al., 2016). Downs explains that embodiment 'provides a direct challenge to the dominant discourse....by critically examining the myriad of ways in which selfhood is manifest in the body' (Downs, 2013: 368). Kontos (2006), who wrote an ethnographic piece on an orthodox care home for Jewish women, argued that the self resists the disease progression because of the authenticity of corporeality. Drawing heavily from Merleau-Ponty, Kontos suggest that those with dementia will often act in the world at a primordial level through actions of the body. For Kontos, selfhood resides within and manifests through the ways of the body. While recalling the actions of one inpatient who was non-verbal, Kontos observed that they would spend their day stitching and knitting clothes. Kontos's observations of selves within the ethnographic framework saw non-representational intentionality – the embodied self. The embodied primordial self is observable, present, and not lost to dementia. It is no less than a socio-cultural and theoretical evolution that argues for the authentic self or the embodied self in the conceptualisation of the self in dementia. It is a counterpoint to metaphors of social death and medicalisation of dementia. Kitwood (1997) postulated the view that everyday life for people without dementia was deeply pathological (perhaps because of the banality and mundaneness of everyday life), whereas those with dementia become an exemplary model of interpersonal life which accordingly is the epitome of how to be human. People with dementia live in the moment or, more accurately, live in the moment of becoming, and Kitwood paradoxically suggests that people with dementia should be considered the more authentic self, true to the nature of humanity. The authentic self, accordingly, is derived from interaction. Freed from the pathology of everyday life, a person with dementia is more authentic. The embodiment paradigm may lead to future problematic metaphors about dementia and may be resisted due to a lack of interaction and experience of people living with dementia. But what image of a person with dementia does society want to see? (Post, 2000). With people with dementia coming into the foreground, the elusive target of dementia may come into focus.

However, the issue of sustainability of people living well with dementia is still unresolved. The global burden of dementia is set to increase demonstrably. Can communities and families cope with such a burden? The emphasis in international and national white papers has been for people to remain in their own homes and communities. The concept of sustainability has an implied timeframe, a future tense, based on an evolving present. Sheehan et al. (2016) suggest that sustainability of care cannot be achieved through medical resources or pharmacological interventions alone, but through validation. If we conceptualise dementia in an Aristotelian frame, dementia is flaw in nature that can be corrected by using 'suggestions from nature itself' (Boal, 2000: 11). Art imitates nature or, more precisely, art is 're-creating the creative principle' (Ibid: 10). Dementia is a fault in nature, whereby realities are uncertain and memories change. In the case of dementia, understanding changes, faculties change, reality changes, memory changes, and by taking those faults of nature as suggestions of what to do, it is possible (for, perhaps, short periods of time for a carer) to help a person to live well with dementia.

Validation therapy has gained increasing recognition in recent years. Although piloted in the early 1980s, the idea of recreating a different reality for inpatients, either through dress or environmental change, reflects the growing distance between pathological treatments of dementia and socio-psychological modalities. The validation therapy (VT) premise is not to filter the reality of the person with dementia, but rather to engage within the present reality. Another way of imagining VT is in allowing the human drama to unfold within shifting realities, regardless of whether those realities are spurious in the eyes of the beholder. VT also places a greater emphasis on non-verbal communication, for example, eye contact, touch, and tone of voice. VT has not been scrutinised by systematic reviews, but there is positive anecdotal information suggesting that VT is commonly practiced within informal care settings. The care for people with dementia needs to be sustainable. The informal care setting is where the majority of care takes place, VT is sustainable, but with the necessary support structures in place. Informal care needs to remain an option, as formal care for all is not economically feasible and therefore unsustainable.

Conclusion

The conceptualisation of dementia is still unclear, and dementia is still a field of competing modalities and philosophies of care. In many respects, the bringing together of different metaphors about dementia in this chapter has raised more questions than answers. The question of sustainability of dementia interventions, formal and informal, is important considering the global endemic of dementia. Predicted future trends and persistent divisions between the developed and developing world create a level of uncertainty of the global impact of dementia. The emphasis in the developing world has been to identify future trends and causal links between dementia and other illness. Less is known about social perceptions of dementia in the developing world. A large qualitative-based study carried out in Central Africa found that while the symptoms of dementia were widely known, 'the biomedical concept of dementia was unknown other than among health professionals' (Mbelesso et al., 2016: 3). The lessons learned from carers, and the reframing of dementia in terms of the authentic self and embodiment, should be to the fore in future work in the developing world, which should not be limited to strictly medical interventions. Mbelesso et al. (2016), for example, have identified that care and spirituality are key aspects to the social construction of dementia in the developing world. Sustainability of dementia care is only achievable if dementia is viewed in such a multidimensional way. Future metaphors are certain to develop within the context of rapidly changing social attitudes and the global endemic of dementia.

References

Aronson, M.K. and Lipkowitz, R., 1981. Senile Dementia, Alzheimer's type: The Family and the Health Care Delivery System, *Journal of the American Geriatrics Society*, 29 (12). pp. 568–571.

Beard, R.L., Knass, J. and Moyer, D., 2009. Managing Disability and Enjoying Life: How we Reframe Dementia through Personal Narratives, *Journal of Aging Studies*, 23. pp. 227–235

Boal, A., 2000. *The Theatre of the Oppressed*. London: Pluto Press.

Boal, A., 1998. *Legislative Theatre*, New York: Routledge.

Bohannan, L., 1966. Shakespeare in the Bush: An American Anthropologist Set Out to Study the Tiv of West Africa and was Taught the True Meaning of Hamlet, *Natural History*, August-September.

Boller, F. and Forbes, M.M., 1998. History of Dementia and Dementia in History: An Overview, *Journal of Neurological Science*, 30;158(2). pp. 125–133.

Bourgoeois, M.S. and Hickey, E.M., 2009. *Dementia From Diagnosis to Management- A Functional Approach*. Hove: Psychology Press.

Cohen, D. and Eisdorfer, C., 1986. *The Loss of Self: A Family Resource for Care of Alzheimer's Disease and Related Disorders*. New York: W.W. Norton.

Downs, M., 2013. Embodiment: The Implications for Living Well with Dementia. *Dementia* 12(3). pp. 368–374.

Golden, M.A., Whaley, B.B. and Stone, A.M., 2012. The System is Beginning to Shut Down: Utilizing Caregivers' Metaphors for Dementia, Persons with Dementia, and Caregiving, *Applied Nursing Research* 25. pp.146–151.

Gubrium, J.F., 1991. *The Mosaic of Care: Frail Elderly and their Families in the Read World.* New York: Springer.

Fontana, A., and Smith, A., 1989. Alzheimer's Disease Victims: The 'Unbecoming' of Self and the Normalisation of Competence, *Sociological Perspectives*, 32. pp. 35–46.

Francis, L., Dunt, D. and Cadilhac D.A., 2016. How is the Sustainability of Chronic Disease Health Programmes Empirically Measured in Hospital and Related Healthcare Services?: A Scoping Review. *BMJ Open* 6:e010944.

Harding, N. and Palfrey, C., 1997. *The Social Construction of Dementia: Confused Professionals?* London: Jessica Kingsley.

Innes, A., 2009. *Dementia Studies*. London: Sage.

Johnstone, M.J., 2011. Metaphors, Stigma and the 'Alzheimerization' of the Euthanasia Debate. *Dementia*, 12(4). pp. 377–393.

Kastenbaum, R.J., 1987. 'Safe Death' in the Postmodern World. In: A. M. Gilmore and S. Gilmore, eds. *A Safer Death: Multidisciplinary Aspects of Terminal Care*. London: Plenum, pp.3–15.

Kontos, P., 2006. Embodied Selfhood: An Ethnographic Exploration of Alzheimer's Disease. In: A. Liebing and L. Cohen, eds. *Thinking About Dementia: Culture, Loss and the Anthropology of Senility*. New Brunswick: Rutgers University Press. pp.195–217.

Kitwood, T., 1997. *Dementia Reconsidered: The Person Comes First.* Berkshire: Open University Press.

Lakoff, G., 1993. The contemporary theory of metaphor. In A. Ortony (Ed.), *Metaphor and Thought* (pp. 202-251). Cambridge: Cambridge University Press.

Lane, H.P., McLachlan, S., Philip, J., 2013. The War Against Dementia; Are we Battle Weary Yet?, *Age and Ageing*, 42(3). pp. 281–283.

Locke, M., 2013. *The Alzheimer Conundrum: Entanglements of Dementia and Aging.* Princeton, NJ: Princeton University Press

Lynam, K.A., 1989. Bringing the Social Back In: A Critique of the Bio-Medicalisation of Dementia, *Gerontologist*, 29 (5). pp.597-604.

Mbelesso, P., Faure-Delage, A., Guerchet, M., Bandzouzi-Ndamba, B., Mouanga, A.M., Tabo, A., Cartz-Piver, L., Preux, P.M., Clement, J.P. and Nubukpo, P., 2016. Social Aspects of Dementia in Central Africa (EDAC Survey), *African Journal of Neurological Sciences*, 35, (1). pp. 1–12.

Kimmel, M., 2004. Metaphor Variation in Cultural Context: Perspectives from Anthropology, *European Journal of English Studies*, 8:3. pp.275–294.

Parsi, K., 2016. War Metaphors in Health Care: What Are They Good For?, *The American Journal of Bioethics*, 16 (10). pp.1–2.

Pearlin, L., Mullan, J., Semple, S. and Skaff, M., 1990. Caregiving and the Stress Process: An Overview of Concepts and their Measures, *Gerontologist*, 30(5). pp.583–594.

Post, S., 2000. *The Moral Challenge of Alzheimer Disease*. Baltimore: Johns Hopkins University Press.

Ranke, K,. 1967. Simple Forms, *Journal of the Folklore Institute*, 4 (1). pp. 17–31.

Rapp, A.M., and Wild, B., 2011. Nonliteral Language in Alzheimer Dementia: A Review, *Journal of the International Neuropsychological Society*, 17. pp. 207–218.

Sabat, S.R., 2005. Capacity for Decision- Making in Alzheimer's Disease: Selfhood, *Positioning and Semiotic People, Australian and New Zealand Journal of Psychiatry*, 39 (11–12). pp. 1030–1035.

Sheehan, C., Blackhorse, C., Ramseyer, D., Murphy, A., Howson, V., 2016. Pressure Play: Reflections on Forum Theater and the Art of Caring for People with Dementia, *Irish Journal of Anthropology*, 19 (1). pp. 19–29.

Sontag, S., 2009. *Illness as Metaphor and AIDS and Its Metaphors*. Penguin Books: U.K.

Sweeting, H. and Gilhooly, M., 1997. Dementia and the Phenomenon of Social Death, *Sociology of Health and Illness*, 19 (1). pp. 93–117.

Zeilig, H., 2015. What Do We Mean When We Talk About Dementia? Exploring Cultural Representations of "Dementia", *Working with Older People*, 19 (1). pp. 12–20.

Zieschang, T., 2013. Sustainability of Motor Training Effects in Older People with Dementia, *Journal of Alzheimer's Disease*, 34 (1). pp. 191–202.

11 The shamanic dream as a metaphor of transformative change

Lidia Guzy

Dreams, visions, revelations, oracles, and prophecies have shaped societal, cultural, and religious changes throughout the history of mankind, with Pythia, the ancient Greek oracle of Delphi, and Martin Luther King's "*I have a dream*" speech as best-known examples. Dreams are imaginative, relying on other worldly visualisations and images of the world. Dreams question, represent, or sustain the world and they have the power to transform.

In this chapter, I would like to discuss the shamanic dream in indigenous cultures as a powerful cultural metaphor of ecological sustainability and as constant transformative social and individual communication and change. I suggest that from these shamanic dreams, contemporary Western cultures can learn and incite inspiration towards imagination and practices of sustainability. The shamanic dream is viewed in this chapter as access to understanding another eco-cosmological worldview where ritual visual imaginary has not been diminished by technological developments, modernity and the commodification of values, symbols and metaphors, but where it plays a crucial role in social analysis, cultural and spiritual advice towards human, biological, ecological, and cosmological coexistence.

Dreams, indigenous peoples, and indigenous knowledge

Ethno-psychoanalytical studies have shown that peoples in different cultures dream, and value dreams, differently (Devereux, 1978; Kakar, 2008; Crapanzano, 1973). Dreams reflect cultural-specific knowledge systems and worldviews. Knowledge systems represent specific ontologies and epistemologies – leading to a culturally particular perception of the surrounding world.

The father of ethnographic fieldwork Bronisław Malinowski noted that "Every human culture gives its members a definite vision of the world, a definite zest for life" (Malinowski, 2014: 534) and this has been approached in anthropology as "culturally organised systems of knowledge" that include ideas about values, cosmology, and conceptions of the natural and social universe (Kearney, 1975: 248). Dreams as cultural visions play an important role in indigenous cultures as they indicate creativity and the production of knowledge. Creativity and the production of indigenous knowledge is the result of distinct indigenous epistemologies (Brabec de Mori, 2016: 80). Such epistemologies are grounded and sustainable

DOI: 10.4324/9781003143567-14

through the material, environmental, and ecological worlds inhabited by indigenous communities and people.

The United Nations Working Group on Indigenous Populations defines indigenous peoples as:

> those which, having a historical continuity with pre-invasion and pre-colonial societies that developed on their territories, consider themselves distinct from other sectors of the societies now prevailing in those territories, or parts of them. They form at present non-dominant sectors of society and are determined to preserve, develop and transmit to future generations their ancestral territories and their ethnic identity, as the basis of their continued existence as peoples, in accordance with their own cultural patterns, social institutions and legal systems.
>
> (Cobo 1987)

Darcy Ribeiro's (1972) anthropological account of the history of the Americas designates indigenous and Native Americans as "testimonial people", those who – while becoming cultural minorities during the European colonisation project and conquest – witnessed with the arrival of European "transplanted people" the loss of their own local majoritarian societal status, facing new hegemonic power structures, unchanged until today. Given the spread of "an increasingly homogeneous industrial civilisation" (Ribeiro, 1957: 299) and the cultural vulnerability, diversity, and smallness of indigenous peoples, the cultural survival of indigenous peoples worldwide is critically endangered. Given the ethnocidal[1] and ecocidal[2] logic of homogenous industrial civilisation, indigenous peoples often cannot "simply continue as parallel systems" (Ribeiro, 1957: 299) if their bio-ecological, cultural, and spiritual environment is irreversibly destroyed by aggressive and often deadly incursions of farmers, legal or illegal loggers, or industrial monopolies[3].

In the context of India, indigenous peoples are generally called Adivasi, a term which is widely used as self-designation for groups of people in India who are also known as "indigenous", "tribals", "Scheduled Tribe's", or "forest dwellers" (vanvasis). Adivasi satisfy the criteria for indigenous status established by the United Nations, making India home to roughly one hundred million indigenous people, more than any other country in the world. In the South Asian and Indian context in particular, the term "analogous" is also used to stress the culturally, economically, and ecologically alternative but contemporary situation of indigenous or tribal people living in South Asia (Kulirani et al., 2015).

Analogous/indigenous/testimonial/tribal societies represent stateless forms of social organisation and community formation. Indigenous worldviews are based on radically different values to the global, homogenous capitalistic worldview and lifestyle. They thus represent radical – and therefore often devalued – alternative models of humanity.

Throughout global history, indigenous/testimonial and analogous peoples have been marked by marginalisation, degrading colonisation processes – on the one hand by European settler societies, on the other hand by local agrarian settler politics (see Lee 2006: 134) and by the general devaluation of their knowledge systems. "Indigenous explanations" appear "meaningless for the modern world" (Brabec de Mori, 2016: 80–81) as there is little recognition of their unique epistemological frameworks that gave birth to such knowledge. The products of indigenous knowledge however have crucial value for the dominant wider society and this chapter presents and discusses the value of indigenous knowledge as important guidelines to new thinking on sustainability.

To understand indigenous knowledge, the ontological turn in French anthropology (Descola, 1992; Descola, 2005; Descola, 2013; Viveiros de Castro, 1998) as well as the discourse on neo-animism (Harvey, 2005) have been directive. In this vein, indigenous knowledge appears not to be anthropocentric and dualistic, escaping a clear separation between nature and culture. An emerging "new kind of ecological anthropology" (Descola, Pálsson, 1996: 2) has been blurring the supposed clear demarcation line between nature and culture opening up new approaches to understanding alternative taxonomies, epistemologies, and ecologies. Based on deep ethnographic studies, anthropological research has shown that contemporary indigenous knowledge systems relate to eco-cosmologies or eco-cosmological worldviews as I call these diverse ontologies, ecologies, epistemologies, and theories (see also: Århem, 1996; Kopenawa and Albert, 2013; Sarma and Barpujari, 2011). Eco-cosmologies relate the human, the animal, ecology, territory, geography, and the cosmos within a non-dualistic world and cosmos full of human, non-human, and other-than-human agencies. Ontological pluralities include the diverse perspectives of humans, non-humans, and other-than-human elements regarding each other, interconnecting with each other as different personalities in a mutually interconnected world and cosmos (see Viveiros de Castro, 2015). This new ecological anthropological thinking offers important perspectives for our reflections on sustainability.

Eco-cosmologies and shamanic dreams

Eco-cosmologies are worldviews and life-worlds intrinsically relating the human with the non-human, the cosmos, and the other-than-human sphere such as trees, animals, rivers, mountains, and spirits. An important element in eco-cosmologies is the absence of the dualistic separation between the human and the surrounding geography and landscape, such as mountain, river, or a tree. Thus, "identity" is not ego, body, or gender centred but spread and integrated with the surrounding ecology, geography, and territory. In this sense, trees represent deceased family members and ancestors, mountains, and/or rivers manifest clan deities directly intertwined with local cosmologies and kinship structures, and the shamanic dream of the ritual specialist openly interconnects the world of the non-humans and other humans with the ritual, cosmological, and ecological calendar of the local community. The shaman in this eco-cosmological setting is the local intellectual and spiritual leader of the community with the capacity to transcend different dimensions of existence through the shamanic dream and its ritual communication.

Shamanic dreams are crucial characteristics of shamanic worldviews, life worlds, and eco-cosmologies. Shamanism is a common term pre-eminently designating a religious phenomenon of Siberia and Inner Asia with the term shaman originating from the siberian Tunguz śaman, a term used in the Tungusic language of the indigenous Evenki people in eastern Russian Siberia to relate to a religious trance specialist. Throughout the immense area comprising the central and northern regions of Asia, the religious/ritual life of societies centres around the ritual specialist – the shaman – who "*remains the dominating figure, for throughout the vast area of Asia in which the ecstatic experience is considered the religious experience par excellence, the shaman, and he alone, is the great master of ecstasy. A first definition of the complex phenomenon of shamanism—and perhaps the least hazardous—is that it is a technique of ecstasy.*" (Eliade, 2005: 269)

Despite its "ism", shamanism should not be regarded as an institutionalised religion, rather it is a complex of different rites and beliefs surrounding the activities of the ritual specialist – the shaman. Piers Vitebsky speaks of "Shamanism" as a common shamanic spiritual worldview (2001) and defines "The Shaman" (1995) as a spiritual trance healer mediating between the world of the living and the world of the spirits, a common phenomenon occurring from indigenous Siberia to the Amazon.

The trance state as the altered state of consciousness of the shaman is crucial in understanding indigenous hermeneutics:

> While in a state of trance, the shaman is regarded as capable of direct communication with representatives of the otherworld, either by journeying to the supranormal world or by calling the spirits to the séance. He is thus able to help his fellow men in crises believed to be caused by the spirits and to act as a concrete mediator between this world and the otherworld in accompanying a soul to the otherworld, or fetching it from the domain of the spirits. The shaman acts as a healer and as a patron of hunting and fertility, but also as a diviner, the guardian of livelihood.
>
> (Siikala, 2004: 8280)

Today shamanism is regarded as a variety of similar phenomena in indigenous North American, South American, and South Asian cultures. Contemporary shamanisms are embedded in analogous ecological knowledge systems relating always to indigenous cosmologies, ontologies (Kopenawa and Albert, 2013), and indigenous hermeneutics. Shamanic dreams have not disappeared from European cultures. Dreams that transmit both personally or communally transformative knowledge from other realms continue to manifest within the Balkan's Christian and Islamic cultures. In the Balkans and Eastern Europe, careers of healers are commonly initiated through shamanic dreams (Kapalo, 2014).

Local shamanisms are thus commonly understood as value systems based on the one hand on cultures of orality, on the other hand on the non-dualistic perspective on human and non-human agencies in a mutually shared world and cosmos (Descola, 1992, 2005, 2013; Viveiros de Castro, 1998). This is exemplified by the importance of dreaming and ecstatic capacities of shamans who interconnect with the other world through therianthropy, the shape-shifting between humans, non-human beings, and agencies (Beggiora, 2013) and by communicating with them through their visionary dreams. "*The knowledge that comes through dreaming is absolute because it comes from a level of symbolic association that is deeper than consciousness*" (Ridington, 1971: 123). This knowledge is acquired by the lonesome experience with the "bush", "the jungle" where encounters with animals transmit transformative knowledge between humans and non-humans such as medicine mythic animals or other than empirical beings. The North American Beaver Indians for example consider animals as educators and as symbols for the varieties of human nature as "*a man can learn his combination of qualities through getting close to the qualities of animals. The experience with a medicine animal in the bush is the culmination of childhood and the beginning of adulthood*" (Ridington, 1971: 122). The same applies for example for the Siberian Nenets who as indigenous reindeer herders send their children "*to learn from the reindeer*". The reindeer for the Nenets is a mythic and central animal that transmits spirituality to the child. This spirituality is regarded as a particular knowledge of life based on human-animal communication and understanding (Toulouze, 2017).

In shamanic worldviews and life worlds (Guzy and Kapalo, 2017), the transformative experiences of animal-human and ecological encounters and theriantropic transformations are transmitted in a rich culture of orality expressed in rituals, songs, performances, and dances. In this knowledge transmission, visions and dreams are the most important expressions of shamanic imaginaries (Noll et al., 1985), realities, epistemologies, and ontologies revealing imagined, dreamt, and lived experiences of local shamanic societies. In this way, the visual mental imagery experiences construe the inner and outer knowledge of life worlds and worldviews.

The fact that shamanisms and with them the shamanic dreams are so widespread in cultures around the world from Asia to Europe to South America indicate a lot about human culture and psychology. As the shamanic dream is transmitted to society in the form of ritual, song, and dance – all crucial elements of human verbal, non-verbal and bodily communication – the shamanic dream is used on the one hand as a metaphor for the non-human/other-than human communication from the otherworld. On the other hand, the shamanic dream is a metaphor for social, cultural, and individual communication, transformation, and healing. Middle Indian shamans (Vitebsky, 1993) for example communicate via spiritual visions and dialogues with the spirits of the deceased and in this way transform personal and collective grief into individual and communal healing. In shamanic cultures, healing is not achieved through concepts of salvation, redemption, or sin but through the act of communication and transformation. It is as well a verbal as a non-verbal, performative, visual communication with humans, non-humans, and other than worldly agents in a trance state of a shamanic dream and its theriantropic transformation.

The shamanic dream indicates thus the existence of a different form of knowledge system that does not anthromorphise everything by explaining all existence in terms that are relevant to humans. Shamanic cultures, ontologies, and epistemologies render many human concepts in explicitly non-anthropomorphic terms.

The human in shamanic worldviews is not dualistically separated from nature but is intrinsically interconnected with the surrounding ecology, geography, and territory. The shamanic worldview is thus rooted in the eco-cosmologies that indigenous peoples believe in, whereby humans deeply interrelate and blur not only with the non-human sphere (including other animals) but also with a spirit world. With its theriantropic characteristics, the shamanic dream is a powerful cultural metaphor of transformation.

A good example for the intersection between eco-cosmology, tribal, indigenous identity, cosmology, territory, geography, ecology, and the shamanic dream is the case of the Donghria Khond from Odisha, India. The Khond became internationally prominent due to their successful resistance against bauxite mining of their sacred mountain of the Niamgiri Hills (Padel et al., 2010) considered by the Khonds and their shamans as the geographic centre of their cosmogony (Beggiora, 2015). Supported by shamanic dreams of desacralising dangers of mining inside the mountain deity of the Niamgiri Hills and international agencies such as Survival and Amnesty International joining the fight against unlawful environmental destruction, exploitation, and impoverishment of indigenous communities, the eco-cosmological fight of the Khond was accompanied by enduring international bad publicity and international ethical outcry. It eventually led the Indian government to halt Vedanta Limited Aluminium company's permits to exploit bauxite from the Khond sacred Niamgiri Hills in 2010. India's Supreme Court further ordered in 2013 that affected communities must be consulted about the project before it could go ahead. The political and judicial landmark decisions opened a new dimension in indigenous agency, eco-cosmological resistance, resilience, and a major victory for human rights on tribal lands in India.

Global modernity and education have a lot to learn from the metaphor of the shamanic dream as transformative cultural instrument. In the shamanic worldview exemplified in its eco-cosmology, the non-anthropomorphic character of communication through visions of the shaman connects the human with ecology, territory, geography, cosmology, and the world of the spirits. It can transform and heal via shamanic visions and dreams that transgress anthropocentric communication patterns thus extending to non-human, other-than human, spiritual, and eco-cosmological dimensions – a broadening of perspective crucial for reflections on sustainability.

Amazonia

The Amazon rainforest – Amazonia – represents one of the world's greatest natural resources where about 20% of earth's oxygen is produced as thus is labelled as the '*Lungs of our Planet*'. The Amazon rainforest also contains the world's highest level of biodiversity with half of the world's species: with over 500 mammals, 175 lizards; over 300 other reptiles species, about 30 million insect types and one third of the world's birds population. The Amazon River gives Amazonia the life streams of the rainforest with the Amazon delta covering 2,722,000 million square miles, and extending to Brazil, Columbia, Peru, Venezuela, Ecuador, Bolivia, and the three Guyanas. Amazonia covers more than half of Brazil.

Anthropogenic forests of Amazonia

Since the publication of *the Handbook of South American Indians* by Julian Steward (1946) and the beginnings of research on Cultural Ecology (Sutton & Anderson 2004), it is known that a large part of Amazonia's bio-diversity is the result of the skilful agricultural, ecological, and botanical knowledge of Amerindian indigenous communities. This leads to the scientific recognition of the dramatic ethno-ecological impact of Amazonian indigenous hunter-gatherers societies for the development of the unique ecological diversity of the Amazonas since millennia (Rival 1999).

Botanical and ethno-botanical research by Balée and Posey (Balée, 1993; Posey, 1984; Posey and Balee, 1989, Balee, 1989) has empirically proven that indigenous people of Amazonia have created "anthropogenic forests" by means of a complex knowledge of "agroforestry" since prehistoric times to this day. Also, Rival's research on historical ecologies in Amazonia among the Huaroani of Ecuador proves the socio-cultural and ecological existence of "anthropogenic" forests as resources that are not "wild" but "bio-cultural", ethno-historical, and manmade according to elaborated ethno-agricultural, ethno-biological knowledge of Amerindian hunter gatherer societies, and their local ancestral histories (Rival, 2002, 2006).

> The Huaorani are very conscious of past human activity, and are perfectly aware of the fact that every aspect of their forested territory has been transformed in equal measure by their ancestors, other indigenous groups, and the forces of nature and the supernatural. Taking the forest to be a legacy from the past, they have developed an understanding of the forest as owing its existence to past human activities. The forest exists to the extent that humans in the past lived and worked in it, and by so doing produced it as it is today for the benefit and use of the living. Their relation with the forest is lived as a social relation with themselves across generations, hence its eminently historical character.
>
> (Rival, 2006: 82)

In the same vein, research by Josep A. Garí on indigenous agro-ecology refutes the global perception of "wild Amazonia" and proves the sustainable indigenous impact on Amazonian ecosystems and biodiversity.

> Indigenous communities conserve, use, cultivate, manage and exchange biodiversity as a fundamental component of their rural lifestyle. The indigenous agro-ecology comprises the whole set of knowledge systems, agro-ecological practices and socio-cultural dynamics that shape indigenous agriculture in the context of biodiversity. The indigenous agroecology provides food security, health care, and ecosystem resilience through a local regime of biodiversity conservation and use.
>
> (Garí, 2001: 21)

Deep historical ecological, agro-ecological, ethno-biological, and ethno-veterinary knowledge (Monteiro et al., 2011) of Amerindian indigenous populations have created and sustained anthropogenic Amazonia forests and human survival. The socio-cultural and ethno-botanical fact of anthropogenic Amazonia leads to the preliminary conclusion that Amerindian indigenous people have been living with an indigenous idea of non-anthropocentric Anthropocene for generations. They have acknowledged since millennia that human activity reshapes the agro-ecological, biological, and geographical surrounding but they have experienced that it is only through a radical interconnection – theriantropy – with the ecological, biological, and cosmological surrounding, that humans can survive.

Amerindian ontologies

Amerindian ontologies of the Amazonas exemplify best the relatedness between the human and non-human world – where the dreams of the shaman connect an animated ecological landscape with the Amerindian indigenous people. The shaman in Amerindian ontologies heals through ecstatic dream transformations and mediations of knowledge encountered through journeys to other realms of the world (Kracke, 2007). With ecological and cultural destruction, most valuable local knowledge resources and eco-cosmological worldviews disappear and have to adopt to dramatic landscape and cultural changes. These worldviews and local knowledge systems transmitted through shamanic dreams and ecological ritual practices however could be a key for finding local and global solutions for a sustainable and philanthropic global world of cultural and eco-biological diversity and mutuality.

The Yanomami

The Yanomami are a good example for the possible impact of Amerindian indigenous people on contemporary discussions around climate change and eco-biological sustainability. The Yanomami represent the largest relatively isolated indigenous community in South America. They live in the rainforests and mountains of northern Brazil and southern Venezuela. Like most indigenous groups on the continent, the Yanomami probably migrated across the Bering Straits between Asia and America some 15,000 years ago, making their way slowly down to South America. Today their total population stands at around 32,000. At over 9.6 million hectares, the Yanomami territory in Brazil is twice the size of Switzerland. In Venezuela, the Yanomami live in the 8.2 million hectare Alto Orinoco – Casiquiare Biosphere Reserve. Together, these areas form the largest forested indigenous territory in the world.

Brazilian indigenous Amazonia

The 2010 IBGE (Instituto Brasileiro de Geografia e Estatística) Census calculates the total indigenous population in Brazil to 896,917 individuals, which corresponds to approximately 0.47% of the country's total population (200.4 million). The Brazilian Amazon is home to 280,000 to 350,000 indigenous people, of which 180,000 live traditionally, heavily dependent on the ancient forest for their sustenance, spiritual, and cultural life. The 2010 IBGE Census counts 197 forest-dwelling indigenous groups, living either on reservations or in one of four national parks.

Since the beginning of the European colonisation of the Americas beginning in 1492 with Columbus, Europeans collectively killed between 70 million to 100 million indigenous people (within 80 years). In the analysis of Genocide expert David Stannard, this constitutes "the largest ongoing holocaust in the history of humanity" with 95% of indigenous people killed by European actions, 100% of indigenous lands stolen by Europeans, and with European-descent people becoming the most prosperous people on the planet (see Stannard 1992: X–XI). According to the anthropologist Darcy Ribeiro (1962), 55 indigenous populations vanished in the first half of the twentieth century alone.

In 2010, 247 peoples, speaking more than 150 different languages, were documented (IBGE, 2010). As a whole, the Indian population has been growing over the last 28 years, although some specific peoples have decreased in number and some others are even threatened with disappearance. Among the Indian peoples in Brazil listed by the Instituto Socioambiental (ISA), seven have populations between 5 and 40 individuals. According to Eduardo Viveiros de Castro, researcher and professor of anthropology at the Museu Nacional (UFRJ) and founding partner of Instituto Socioambiental (ISA), an Indian "is any member of an indigenous community, recognized by the latter as such. An indigenous community is any community founded on kinship or co-residence relations between its members, who maintain historical-cultural ties with pre-Colombian indigenous social organizations" (Viveiros de Castro, 1998).

According to the constitution, indigenous people in Brazil are considered "relatively capable" Brazilian citizens, having a childlike status which deserves particular protection from the state. In this sense, according to Instituto Socioambiental, the state follows a principle established by the old Brazilian civil code of 1916, based on the idea that Indians should be tutored by a state indigenous institution. From 1910 to 1967, it was the Serviço de Proteção ao Indio/SPI and currently it is Fundação Nacional do Indio (Funai). The aim has always been to support a full integration of indigenous communities into the national community, which means to integrate them into the majoritarian Brazilian society. The Indian Statute of the Brazilian Constitution and the rights of Brazilian Indigenous Peoples were promulgated in 1973, defining the rules on the relations of the state and Brazilian society with the indigenous communities. Even though the 1973 Statute remains in force, new approaches in the Federal Constitution of 1998 grant greater rights to indigenous peoples. The 1988 Constitution does not call for the integration of the indigenous peoples into Brazilian society anymore, ensuring them, the right to be different from the rest of the country.

Amnesty International, Cultural Survival, and Survival International have for many years raised the problem of the lack of inclusion of indigenous peoples' voices into mainstream discourses in Brazilian educational, cultural, and political life. A critical discourse on unsustainable progress and the impact of destructive development by Brazil's contemporary majoritarian society in relation to Brazil's indigenous people is lacking. The relative isolation of Brazilian indigenous people seems to have retained pre-Colombian eco-cosmological and agro-ecological knowledge systems over centuries, which could be crucial for sustainability projects in times of ecological calamities and global climate change challenges in other parts of the world.

Davi Kopenawa: Yanomami spokesman and the "Dalai Lama" of the rainforest

Davi Kopenawa is the prolific Yanomami shaman and intellectual, a contemporary global indigenous leader and environmental activist supported by Survival International. His shamanic dreams and visions on the future of the world are expressed in his testimonial auto-ethnography *The Falling Sky: Words of a Yanomami Shaman* (Kopenawa and Albert, 2013). This autobiography could be an inspiration for imagining a non-dualistic world and cosmos, where the eco-cosmological visions of the shaman hold the world by changing the worldview. In the Yanomami worldview, the shamans hold the sky by means of shamanic ecstatic rituals, and by doing so he or she heals the imbalances and disturbances between the human and non-human world.

> When they think their land is getting spoiled, the white people speak of "pollution." In our language, when sickness spreads relentlessly through the forest, we say that xawara [epidemic fumes] have seized it and that it becomes ghost.... What the white people call the whole world is being tainted because of the factories that make all their merchandise, their machines, and their motors. Though the sky and the earth are vast, their fumes eventually spread in every direction, and all are affected: humans, game, and the forest. It is true. Even the trees are sick from it. Having become ghost, they lose their leaves, they dry up and break all by themselves. The fish also die from it in the rivers' soiled waters. The white people will make the earth and the sky sick with the smoke from their minerals, oil, bombs, and atomic things. Then the winds and the storms will enter into a ghost state.
>
> (Kopenawa and Albert, 2013: 295)

As a spiritual shamanic indigenous radical critique of modernity and an indigenous apocalyptic vision, the *Falling Sky* can be read as a fundamental dismissal of global anthropocentrism, modernity, and materialism. This modernity and materialism resulting in the geological and ecological concept of Anthropocene "*a world being substantially reconfigured by human activity*" (Hamilton, 2014: 1) is facing today unprecedented humanly orchestrated ecological, social, and cultural disasters. A new knowledge based on indigenous critique of modernity and its apotheosis of Anthropocene might potentially have the power to create a new imagination of a radically interconnected world, a worldview which Anthropocene – despite its incredibly digital and global capacities – has yet not succeeded to create. "*But will an interconnected world-view arise fast enough and gain sufficient political, economic, and social traction soon enough to set humanity on a new course*?" (Hamilton, 2014: 7).

Conclusion

I am convinced that we can learn from the shamanic dream as indigenous critique and eco-cosmological worldview of indigenous peoples: only if we start learning to dream, we will start to transform. The shamanic dream can teach to transform the anthropocentric focus of Anthropocene and modernity: a transformation away from the human to the non-human or other-than-human in order to "learn from the reindeer", to "learn from the trees" and thus to tackle some of the world's most urgent problems of Anthropocene.

The shamanic dream could foster a change of thinking and could offer a frame of alternative imagination against the ethnocidal and ecocidal logic of a global neo-colonial industrial society. It could offer the possibility to create a truly post-colonial world and not a "neo-colonial" one where shamanic cultures of indigenous peoples could thrive and where they would become equal intellectual and theoretical partners in a global eco-cosmologically sustainable world.

An innovative engagement with indigenous knowledge, imagination, and use of modern technology can lead to new models of sustainability based on the care of eco-cosmological landscapes and the provision of a secured livelihood of indigenous peoples. The shamanic dream finally would transform ecocide into eco-cosmological justice for Earth and the cosmos.

Notes

1 Ethnocide means that an ethnic group is denied the right to enjoy, develop, and transmit its own culture and its own language, whether collectively or individually. This involves an extreme form of massive violation of human rights and, in particular, the right of ethnic groups to respect for their cultural identity, as established by numerous declarations, covenants, and agreements of the United Nations and its Specialized Agencies, as well as various regional intergovernmental bodies and numerous non-governmental organizations (UNESCO 1981).
2 Ecocide is defined by the Earth lawyer Peggy Higgins as "the extensive damage to, destruction of or loss of ecosystem(s) of a given territory, whether by human agency or by other causes, to such extent that peaceful enjoyment by the inhabitant of that territory has been or will be severely damaged" (Higgins 2010).
3 See www.survivalinternational.org; www.culturalsurvival.org.

References

Århem, Kaj 1996 "The cosmic food web: human-nature relatedness in the Northwest Amazon", in *Philippe Descola and Gísli Pálsson, Nature and Society. Anthropological Perspectives*, 166–184.

Balée, W., 1989. The culture of Amazonian forests. In: D. Posey and W. Balee, eds. *Resource Management in Amazonia: Indigenous and Folk Strategies: Advances in Economic Botany*. Bronx: New York Botanical Garden, 7, pp. 1–21.

Balée, W., 1993. Indigenous transformation of Amazonian forests: An example from Maranhao, Brazil. In: A.C. Taylor and P. Descola, eds. *L'Homme 33e Année, No. 126/128: La remontee de l'Amazone*, 33: 2–4, pp. 235–258. Paris: EHESS.

Beggiora, S., 2013. Teriomorfismo e licantropia in India Metamorfosi di un sistema ambientale e sociale. *La Ricerca Folklorica, No. 67/68, Ernesto de Martino: Etnografia e storia*, Aprile-Ottobre 2013. Brescia: Grafo Edizioni, pp. 259–274.

Beggiora, S., 2015. Indigenous cosmogony of Kuttia Khond. *International Quarterly for Asian Studies*, 46, pp 59–81.

Brabec de Mori, B. 2016. What makes indigenous unique? Overview of knowledge systems among the world's indigenous people. *Past-Future-2016: Seminar on the Protection of Aboriginal Wisdom; Proceedings*, Huailien: National Dong Hwa University, pp. 78–85.

Cobo, M. J. 1987. Study of the problem of discrimination against indigenous populations. http://unesdoc.unesco.org/images/0013/001356/135656M.pdf (05/2009).

Crapanzano, V., 1973. *The Hamadsha: A Study in Moroccan Ethnopsychiatry*. Oakland: University of California Press.

Descola, P., 1992. Societies of nature and the nature of society. In: A. Kuper, ed. *Conceptualizing Society*. London: Routledge.

Descola, P. and Pálsson, G. 1996. *Nature and Society: Anthropological Perspectives*. London: Routledge.

Descola, P., 2005. *Par-delà de la nature et culture*. Paris: Gallimard.

Descola, P., 2013. *Beyond Nature and Culture*. The University of Chicago Press.

Devereux, G., 1978. *Ethnopsychoanalysis*. Berkeley: University of California Press.

Eliade, M., 2005. Shamanism. In: Lindsay J., ed. *Encyclopedia of Religion*. 2nd ed., Vol. 12. Detroit: Macmillan Reference USA, p. 8269.

Garí, J., 2001. A. Biodiversity and indigenous agroecology in Amazonia: The indigenous peoples of Pataza. *Etnoecológica* 5(7), 21–37.

Guzy, L. and Kapalo, J., eds. 2017. *Marginalised and Endangered Worldviews–Comparative Studies on contemporary Eurasia, India and South America*. Bern: Lit Publisher

Harvey, G., 2005. *Animism: Respecting the Living World*. New York: Columbia University Press.

Hamilton, P., 2014. Welcome to the anthropocene. In D. Dalbotten, G. Roehrig, and P. Hamilton, eds, *Future Earth—Advancing Civic Understanding of the Anthropocene, Geophysical Monograph 203*, Ist ed. American Geophysical Union: John Wiley & Sons, pp. 1–8.

Higgins, P., 2010. *Eradicating Ecocide: Exposing the Corporate and Political Practices*. London: Arthouse Publishing

Kakar, S., 2008. *Culture and Psyche: Selected Essays*. Oxford: New Dehli.

Kapalo, J., 2014. Narratíva and kozmológia az inochentizmus ikonográfiai hagyományában. In: G. Barna and K. Povedák, eds. *Lelkiségek és Lelkiségi Mozgalmak Magyarórszáon és Kelet-Közep-Európában/Spirituality and Spiritual Movements in Hungary and Eastern Central Europe*. Szeged: SZTE BTK Néprajzi és Kulturális Antropológiai Tanszék, pp. 255–268.

Kearney, M., 1975. World view theory and study, *Annual Review of Anthropology*, 4, pp. 247–270.

Kearney, M., 1984. *Worldview. Novato*. California: Chandler and Sharpe.

Kopenawa, D. and Albert, B., 2013. *The Falling Sky Words of a Yanomami Shaman*. Cambridge: Harvard University Press.

Kracke, W. H., 2007. To dream, perchance to cure: dreaming and shamanism in a Brazilian indigenous society. In: J. Mimica, ed. *Explorations in Psychoanalytic Ethnography*, New York, NY: Berghahn Books, pp. 106–120.

Kulirani, B Francis, Kamal K. Misra, Kishor K. Basa, 2015. *Tribes and Analogous People in India: Contemporary Issues*. Delhi: Gyan Publishing House.

IBGE (Instituto Brasileiro de Geografia e Estatística), 2010. https://censo2010.ibge.gov.br.

Lee, R. B., 2006. Indigenism and Its discontents. In M. H. Kirsch, ed. *Inclusion and Exclusion in the Global Arena*. New York: Routledge, pp. 129–159.

Malinowski, B., 2014. *Argonauts of the Western Pacific*. London and New York: Routledge.

Monteiro, M.V.B., Palha, M.D.C., Braga, R.R., Schwankes K., Rodrigues, S.T., Lameira, O.A., and Bevilaqua, C.M.L., 2011. Ethnoveterinary, knowledge of the inhabitants of Marajó Island, Eastern Amazonia, Brazil. *Acta Amazonica*, 41(2), pp. 233–242.

Noll, R., Achterberg, J., Bourguignon, E., George, L., Harner, M., Honko, L., Hultkrantz, A., Krippner, S., Kiefer, C.W., Preston, R.J., Siikala, A.L., Vásquez, I.S., Barbara W., Lex, B.W. and Winkelman, M., 1985. Mental imagery cultivation as a cultural phenomenon: The role of visions in Shamanism [and comments and reply]. *Current Anthropology*, 26:4, pp. 443–461.

Padel, F., Das, S. and Roy, A., 2010. *Out of This Earth: East India Adivasis and the Aluminum Cartel*. New Delhi: Orient Blackswan

Posey, D., 1984. A preliminary report on diversified management of tropical rainforest by the Kayapo Indians of the Brazilian Amazon. *Advances in Economic Botany*, 1, pp. 112–126.

Posey, D. and W. Balee, eds., 1989. *Resource Management in Amazonia: Indigenous and Folk Strategies*. New York Botanical Garden: Advances in Economic Botany, Bronx, 7.

Ridington, R., 1971. Beaver Dreaming and Singing. In: *Anthropologica*, New Series, Vol. 13, No. 1/2, Pilot, Not Commander: Essays in Memory of Diamond Jenness, pp. 115–128.

Ribeiro, D., 1972. *The Americas and Civilization*. New York: E.P. Dutton.

Ribeiro, D., 1957. *The Tasks of the Ethnologist and the Linguist in Brazil*. UNESCO International Social Science Bulletin IX, pp. 298–307.

Ribeiro, D., 1962. Social Integration of Indigenous Populations in Brazil, *International Labour Review*, 325, pp. 325–346

Rival, L., 1999. Introductory essay on South American hunters-and-gatherers. In: R. Lee and R. Daly, eds. *The Cambridge Encyclopaedia of Hunters and Gatherers*. Cambridge: University Press, pp. 77–85.

Rival, L. 2002. *Trekking through History. The Huaorani of Amazonian Ecuador*, Columbia University Press: New York.

Rival, L. 2006. Amazonian historical ecologies. *The Journal of the Royal Anthropological Institute* 12, pp. S79–S94.

Sarma, U. K. and Barpujari, I., 2011. Eco-cosmologies and biodiversity conservation: Continuity and change among the Karbis of Assam. *The International Indigenous Policy Journal*, 2(4), pp. 1–10.

Siikala, A.L., 2004. Shamanism: Siberian and Inner Asian Shamanism. In: J. Lindsay, ed. *Encyclopedia of Religion*, Vol. 12. 2nd ed. Detroit: Macmillan/Thomson Gale, pp. 280–287.

Stannard, D., 1992. *American Holocaust: The Conquest of the New World*. Oxford, New York, Toronto: Oxford University Press.

Sutton, M. and Anderson E.N., 2004. *Introduction to Cultural Ecology*. Oxford: Berg.

Steward, J. 1946. Introduction. In: J. Steward, ed. *Handbook of South American Indians*. (Bureau of American Ethnology, Bulletin 143). Washington, DC: Government Printing Office, pp. 1–15.

Toulouze, E., 2017. *Oil Extraction and Siberian Indigenous Peoples: Reconstructing Ontology. Oral presentation at International Conference: Shamanism and Eco-cosmology - A Cross-Cultural Perspective*. Athens: Panteion University of Social and Political Sciences, 27–28 April 2017.

UNESCO, 1981. *UNESCO and the Struggle Against Ethnocide*. Paris, UNESCO: Declaration of San Jose.

Vitebsky, P., 2001, *Shamanism*. Norman, University of Oklahoma Press.

Vitebsky, P., 1995. *The Shaman. Voyages of the Soul: Trance Ecstasy and Healing from Siberia to the Amazon*. London: Duncan Baird Publishers.

Vitebsky, P., 1993. *Dialogues with the Dead: The Discussion of Mortality among the Sora of Eastern India, Cambridge Studies in Social and Cultural Anthropology*. Cambridge: Cambridge University Press.

Viveiros de Castro, E., 2015. *The Relative Native: Essays on Indigenous Conceptual Worlds*. Chicago: HAU Press.

Viveiros de Castro, E., 1998. Cosmological deixis and Amerindian perspectivism, *The Journal of the Royal Anthropological Institute*, 4:3, pp. 469–488.

Part IV

Metaphors of creativity and practice

12 Joyce's arches/arcs/arks

Portals as metaphors of transition from the antediluvian Anthropocene

Kieran Keohane

We are living in the antediluvian moment of the late modern Anthropocene, so let me begin by indicating some modalities of our predicament: first, the whole rolling, escalating militarised culture of global war that we've all been engulfed in; second, the so-called fourth industrial [digital] revolution, the Internet's infinitely expanding cyberspace, and its new generation of enormously powerful global corporations – GAFA: Google, Amazon, Facebook, and Apple; third, financialisation, and a vast transnational Market that (so we are told) is beyond human agency, that cannot be regulated by any government's fiscal apparatus, and as states and governments everywhere find their fiscal bases eroded and their regulatory powers eclipsed, we see the privatisation of public goods, from education and health, to military and policing, to entire ecosystems, fresh water, and even species' genomes. Overarching and underpinning all of these is a mutation of the symbolic order, which Dufour (2008) has called "de-symbolisation", by which he means the historical political-cultural condition in which the big Subjects and their metanarratives of legitimation that formerly reigned – once upon a time they were Gods and Kings; more recently, they have been collective nouns such as the People, the Nation, the Party; more abstractly they have included Reason, and Science: all those authorities [big] Subjects, in relation to which [little] subjects were constituted and acted 'in the name of', have been undermined, relativised, and monetised under the auspices of a new divinity, the Market: a transvaluation of all values so that "the Market" is the transcending authority that decides everything. An emancipation from previous authorities it might seem, but [neo] liberty comes at the price of the precarious identities and chronic insecurities of mal-formed subjects who suffer demands to be self-starting, self-actualising, entrepreneurial, and autonomous while they are systematically deprived of the material and symbolic resources that would support and enable them to become so; and lacking those resources they become especially prone to populist authoritarians who appear to embody and promise to restore them.

As modalities of the neo-liberal revolution, individually and severally and taken all together recursively, these developments are characterised by *limitlessness:* they are global; their structures merge into one another and their boundaries are impossible to discern; and they are un-regulated – in fact they are deliberately *de*-regulated, and insofar as de-regulation and de-symbolisation are the watchwords of late modern cultural and political economy, we can speak of the whole as that of the global

DOI: 10.4324/9781003143567-16

neo-liberal revolution and a Restoration of neo-feudal private corporate monarchs. These are modalities of the global neo-liberal revolution where warfare, corporate power, and digital communications coalesce into what Mumford (1967) called a "Megamachine," moreover a megamachine that is accelerating (Rosa, 2013).

In *The Myth of the Machine* Mumford says that his metaphor of the megamachine is in fact the *reality* of the late modern age: the coalescence of science and technology, state military and fiscal apparatus, communications and culture industries and global corporate capital constitutes what Weber, Nietzsche, and Adorno had foreseen: an iron cage of rationalised acquisitiveness, fatally nihilistic and hostile to life; Enlightenment's dialectic inverted: anti-social individualism twinned with authoritarianism, and an impending 'revenge of Nature.' Fifty years later, the accelerating megamachine grinds ahead relentlessly. The methods and mind-sets of Modernity are fully implicated in the malaises of our times, from existential threats of climate breakdown and species extinction to the insidious and pervasive dissolution of traditions, values, ideals, and holistic mythopoetic consciousness as sources of meaningfulness and models to emulate. The modern Faustian bargain comes at the price of disenchantment, fragmentation, a dearth of good models, and loss of sense of coherence that are essential to well-being and human flourishing. We are confronted with 'a yawning gulf between the problems we face and our capacity to think up workable solutions', a gulf that Evans (2018) has dubbed the "myth gap".

Given our urgent need for metaphors to enable a change of hearts and minds I turn to one of the most imaginative minds of the Modern age, James Joyce, whose work is suffused with metaphors drawn from cosmologies, theologies, and mythologies; from archaeology, quantum physics, optics, and psychology, and in Joyce I hope to trace a genealogy of the "portal" as a metaphor for imagining and enabling transition out of the antediluvian moment of the late-modern Anthropocene. Beginning with Joyce's serendipitous encounters with ancient arches in modern city spaces I trace an arc through architecture and anthropology, passing backwards and forwards between ancient and mythic portals. I will bridge Joyce's use of arches as portals of discovery with Walter Benjamin's (1999) time-travelling archaeology of Paris's shopping arcades, and with the help of Joyce and Benjamin's metaphors I hope along the way to say something about contemporary and futuristic digital portals.

For Joyce, ancient arches in modern cities were portals to memory, imagination, and the power of art to transform the subject. For Benjamin, Paris's derelict arcades could be portals to recently vanished *ur*-capitalist phantasmagorias, portals whereby contemporary time-travelers might revisit the primal scene when we first became subject to the commodity fetish, and thereby free ourselves from the spell it has cast over us. To Joyce and Benjamin, ruined arches and arcades in our contemporary cities could be portals enabling a parallax view. Cognate with the idea of "complementary dualism" developed by Ed Byrne elsewhere in this volume, parallax is the phenomenon whereby the position or direction of an object is seen to differ when viewed from different positions. The term belongs to optical physics and trigonometry, but, like Bohr's quantum mechanics' insight that an electron is simultaneously both a particle and a wave, parallax has far-reaching implications for ontology and epistemology as it reveals how seeing things from one point of view is myopic and partial, whereas binocular or multi-ocular viewing enables apperception of depth and complexity and ultimate indeterminacy that is proper to all things, and how a rational, aesthetic and moral comprehension entails appreciation of ambiguities and ambivalences, paradoxes and dialectical antitheses that do not necessarily resolve into synthesis, but they are none the less true parts of the whole. By alternating between mythic and modern perspectives, our understandings of the here and now are radically changed, opening up possibilities for imagining future horizons. Just as Benjamin's *Arcades Project* enables a 'dialectics of seeing' whereby our perspective of modernity is broadened and deepened by vignettes and illuminations, for Joyce parallax is one of the organising metaphors for *Ulysses*, in that the reader sees the world from many changing viewpoints – Stephen's, Bloom's, Molly's, as well as numerous secondary characters – the Nameless Narrator's, Gerty MacDowell's, and many others; all different, all true, in their own particular way, each contributing to our understanding of the complexity of modern life, and all provoking and requiring hermeneutical and imaginative assimilation to a whole that is greater than the sum of its parts and never graspable in its entirety. Bearing this parallax view in mind, I ask in passing, how do the transforming powers of Joyce and Benjamin's metaphorical portals fare in today's portals to the internet? But first I need to say something about metaphors, and their role in changing minds.

Metaphor and metanoia

We transcend the givenness of the world, we reach beyond, we create, and we imaginatively re-create our world by use of metaphors:

> The drive toward the formation of metaphors is the fundamental human drive, which one cannot for a single instant dispense with in thought, for one would thereby dispense with man himself ...
>
> What then is truth? A movable host of metaphors, metonymies, and anthropomorphisms: in short, a sum of human relations which have been poetically and rhetorically intensified, transferred, and embellished, and which, after long usage, seem to a people to be fixed, canonical, and binding. Truths are illusions that we have forgotten are illusions....
>
> (Nietzsche, 1873: 46–7)

Metaphors have transformative power because they help us to strain out beyond the confines of experience in the world towards what is beyond the world. Metaphors enable us to transcend the limits of both pure reason and practical reason and help "to bring reason into harmony with itself" (Kant, 2002; 49). Metaphors used in poetic, mythic, and religious language, in visual arts, music, film, dance, dramatic performance, literature, and poetry, Kant says, are

> representations of the Imagination which occasion much thought, without, however, any definite thought, *i.e.* any *concept,* being capable of being adequate to it; it consequently cannot be completely compassed and made

> intelligible by language. … Such representations of the Imagination we may call *Ideas,* partly because they at least strive after something that lies beyond the bounds of experience.
>
> (ibid)

Metaphor, from *meta* (beyond, over, across) and *pherein* (to carry, to bear, to transport), means ‘carrying over’, ‘carrying across’, ’conveying’, ‘transferring’, and in the process of carrying, conveying, and transferring, transmogrifying: metaphors transform things in surprising and seemingly magical ways so that something entirely new is fashioned out of previous forms. This understanding is encoded in our collective legacy of lore, languages, and mythologies, as Evan Boyle discusses in a chapter of this volume. Hermes, for instance, a mortal who makes himself god-like, is the spirit of metaphor and metamorphosis. Hermes carries the “spirit”, the “sense”, the “meaning” from one situation to another (Graves, 1960; Kerenyi, 1980)[1]. Hermes represents ingenuity and creativity; the spirit of Hermes animates serendipity and fortuitous mistakes that a person of wit and imagination may see as opportunities that they can make something of, especially at those times when we are stuck in a moment that we can’t get out of, when it seems that our reason has failed us, as it seems to have done at this antediluvian moment of the late-modern Anthropocene.

Metanoia is closely related to metaphor. *Metanoiein* (Gr) means ‘to change one's mind’ (from *meta-* after, beyond, and *noein* to think, from *nous* mind).[2] Metanoia means "a change in the trend and action of the whole inner nature, intellectual, affectional and moral"; a “transmutation of consciousness” (Merriam Webster; OED). Metanoia means “conversion”, that is, being “turned”, “turned around”, “turned towards” sources of illumination, ideals that lead us in a higher direction. Metanoia, a complete change of heart and mind, accords with Plato’s famous definition in *Republic* of education as the ‘turning of the soul’ (1997, 518 c–d). How is the soul turned? How does Joyce, for instance, hope to turn our souls; how was his own soul turned; and by whom? What, and who were his models? We shall come to these questions. And there is a cluster of significations associated with Metanoia as personified in Greek and Roman mythology as a spirit of repentance and regret. Metanoia is twinned with the spirit of *Kairos* – the spirit of opportunity. The spirit of Metanoia follows behind Kairos, where she laments missed opportunity: that is to say, metanoia occurs when a crisis presents us also with an opportunity but we fail to see it and to seize it at the crucial moment, and we have a change of heart and mind only when it is already too late, when the opportunity to do the right thing has already passed (Myers, 2011). Perhaps this is where we are at in the late modern antediluvian moment of the Anthropocene, that we have missed the opportunity to turn things around? But this is exactly the sort of moment when Hermes might show up, as he does for James Joyce, in the form of a series of serendipitous and fortuitous mistakes.

James Joyce’s portals of discovery

> A man of genius makes no mistakes. His errors are volitional and are portals of discovery.
>
> (Joyce, 1990; 190)

James Joyce landed in Pula by mistake, but his error led to a vital portal of discovery. Having eloped, James and Nora borrowed their way from Dublin through London and Paris to Zurich only to find that the teaching position he had been told was at the Berlitz language school there turned out to be at Trieste. On the way to Trieste, Joyce misread the midnight train-stop and got off at Ljubljana, where he and Nora slept the night in a park. In Trieste, next morning Joyce magnanimously offered to translate in a multilingual dispute involving locals and some English sailors, but the academic Italian he had learned in Dublin only added to the confusion of the Triestine babel and the police misrecognised Joyce not as a mediator but as a protagonist, which landed him in jail for the afternoon. Joyce eventually made it to the Berlitz School, only to be told that the teaching position at Trieste was actually in their newly opened branch in Pula, 120 miles down the Istrian coast. Joyce had left Dublin to fly free of the nets of Church, Empire, and Nation only to alight in a city where, he complained, there were 'faded uniforms everywhere', and one of his first students was a naval officer who became Admiral Horthy the Hungarian dictator. But the series of errors that led Joyce to Pula was serendipitous, for Pula was once a Roman city, and among its Classical ruins, directly at the front door of the Berlitz school, outside his classroom window, around the corner from his apartment, facing a trattoria where a life-size bronze statue of Joyce sits today gazing at it, was the *Arco Sergius* (the Arch of the Sergii).

Michelangelo and Dante had both travelled to Pula specifically to see the Arch of the Sergii, and Joyce found it inspirational too. Though Roman, built around 30 BC as one of the gateways into the city to commemorate a patrician family, the Arch of the Sergii is decorated with Hellenistic motifs, because Pula's mythopoeic origins are Greek. "They calmed their oars on the river of Illyria, By the gravestone of the blond Harmony-Serpent, And founded a city: a Greek would say - a 'City of Fugitives', But in their tongue they called it Pula" (Alexandrian librarian Callimachus of Cyrene (320–240 BC). Pula [Pola/Polis] was the city of fugitives: Jason and the Argonauts took refuge there after Poseidon had pummelled their protector-goddess Athena and swamped their ship. Like the mythical Greeks, Joyce and Nora were refugees too, from the too-real neo-Olympian powers of Empire, Church, and Nation they had washed up in Pula. Joyce was just twenty-two; Nora and he were in the early days of their relationship; honeymooning; starting a first professional appointment; setting up their first household together; falling in love with one another, squabbling and falling out with each other; variously unhappy and joyful. Nora became pregnant, and Joyce too conceived the literary children of his imagination. Looking awry at the modern city of Pula through the ancient Arch of the Sergii Joyce caught a glimpse of the Argonauts – Jason (the navigator), Hercules (the culture-maker), Orpheus (who's musical poetry had transformative powers) – and from that epiphany of parallax, the portal through which myth and modernity altered the perspectives of one another, James Joyce, formerly Stephen Hero, began to transition and styled himself thereafter as Stephen Daedalus, architect of the Labyrinth as an ingenious maze of passages that tamed the Minotaur, enabling a transition to Civilisation. And having glimpsed him among the Greeks through the Arch of the Sergii, James Stephen Daedalus Joyce invented a metaphor and model for contemporary fugitives from the wrath of the Gods that our hubris has incurred, Leopold Bloom the modern Ulysses.

When Joyce eventually made his way back to Trieste one of his favourite places became the *Trattoria di Trionfe* at the *Arcoriccardo*. The Arco Riccardo: one etymology links the name of the triumphal arch to Richard the Lionheart who passed through Trieste during the Crusades; but the Arch is a thousand years older than that. Another etymology suggests that it is *Arco di Cardo*, meaning the portal to the heart of the city. "Cardo" – the "heart" of the city meant more than the forum; specifically "cardo" meant the "sacred" heart, and not in the Christian sense but in the much older pagan sense of the shrines and temples of the city's' deities, meaning that the Arco di Cardo was the portal into the sacred core of ideas and ideals. The *Arcoriccardo* is located mid-way on Joyce's route between his and Nora's apartment in a respectable modern piazza, and the portal of his descent into the *cittia vecchia*, the old city centre, thriving and teeming with an inspiring mix of vice and virtue; and, inspired, it was the arch through which "See, the conquering hero comes." (Joyce, 1990: 264)

During this time Joyce and Nora lived in Piazza Vico, another serendipitous coincidence, appreciated by Joyce as it is named for one of his most important sources, Giambattista Vico. It was in Trieste that Joyce discovered Vico, again by happy chance, when a colleague at work gave him a gift of Vico's *Scienza Neuvo*. The name Vico -*vico/via*- means "way", "passage", as in Greek where "road" is *hodos*, from which we have "method", and in Vico Joyce found his "way", the two principles of his artistic method: at the level of form, the philosophy of history in the metaphor of *ricorso;* and at the level of content, Vico's mythology showing how all of the world's ideas and meanings are encoded as a treasury of metaphors. With these two principles Joyce takes also the formula of memory and imagination: "Imagination is simply the resurfacing of recollections, and ingenuity is simply the elaboration of things remembered" (Vico, 1999: 699).[3]

In 1906, still unsettled financially and artistically, Joyce and Nora left Trieste to take up a better-paying position with a bank in Rome. But Joyce disliked Rome; the cafes of Trieste were more beautiful, and their people were more interesting, he said. Multicultural, polylingual Trieste was alive, whereas Rome was 'a man who makes his living by charging people to see his grandmother's corpse.' (Joyce, in Ellmann 1982; 225).

On the basis of this disparaging comment, it is commonly thought that Rome wasn't important for Joyce, but that's not the case at all, as his letters to Stanislaus back in Trieste reveal (Ellmann, 1975). While in Rome Joyce was avidly reading the work of historian and sociologist Guglielmo Ferrero, particularly *L'Europa Giovane* (*Young Europe*) and *Grandezza e Decadenza de Roma (The Greatness and Decline of Rome).* The Emperor Constantine figures prominently in Ferrero's accounts. A flawed hero, like Ulysses and Bloom, Constantine was a wily maneuverer trying to hold together a crumbling empire besieged by external enemies and bedeviled by internal schisms. Of all the challenges Constantine faced the most subtle and ultimately the fatal one was Rome's lack of a moral centre. During Constantine's era, Rome was internally riven by bitter and violent conflicts about heresies of one sort or another. These many individually trivial conflicts that seem to have been "about nothing", Ferrero says, were in fact about Nothing! By this Ferrero means that the time during which Constantine ended its suppression and promoted Christianity to become a common religion across the Empire was an interregnum wherein numerous pagan cults struggled for hegemony with Christianity, which itself was internally split among various sects. In this schismatic context while there was no agreement on the Name of the Father and the mystery of the divine Trinity there could be no common foundation and no horizon of transcendental authority: instead there was *anomie* [no Name] Nothing! The apotheosis of Constantine's Rome foreshadowed a second coming of a dark age, as Yeats (1920) represents the apotheosis of Modernity: 'Things fall apart/ the centre cannot hold.' In Constantine's time that [unstable, contested] centre of the temporal and spiritual realms was God the Father, Christ the Messiah, and the unity of the Holy Spirit. In Modernity the equivalent holy trinity had been Reason, the Scientific Method, and the Spirit of Enlightenment, a modern holy trinity that has been becoming undone, and as in Rome it cannot be put together again. This is a central theme of *Finnegans Wake*: after the shattering thunderwords, the fallen Humpty Dumpties cannot be put together again: not by all the king's horses and all the king's men, and not by the senile demented evangelists, Mamalujo, professors of whatever have been the hitherto prevailing orthodoxies. In the ruins of Rome, Joyce could already foresee the future fallen ruins of Enlightenment and Modernity and the rise of the "feary fathers" – Mussolini, Franco, Stalin, Hitler, and their present-day revenants; and Joyce could see the need for a good model to emulate, Bloom, and a hopeful philosophy of history represented by eternally recurring Anna Livia Plurabelle.

The extraordinary portal through which Joyce could see our future-present was a parallax view through the Arch of Constantine. Joyce, perpetually walking, left his apartment on via Fratella, walked 50 metres onto via del Corso, turned left; passed between the bank where he worked and the Italian Parliament; straight ahead for 500 metres; skirted around the colossal monument [then under construction] to Vittorio Emanuele II (the king on his high horse; impostures of "*patria*" and the glory of Rome's second coming).[4] Joyce walked on through the Forum, the ruins of a collapsed civilisation 'like an old cemetery with broken columns of temples and slabs' and he came directly to the Colosseum[5] and the Arch of Constantine.

Through the portal of the Arch of Constantine Joyce got a parallax view: east to Byzantium, and west to Ireland and to Kells. Constantine's name resonated with Joyce because of his good friend Constantine Curran. Generous and supportive, intelligent and fair-minded, Curran maintained a lifelong correspondence with Joyce; he took care of Joyce's father's funeral and took in Lucia Joyce when she became mentally ill. But while Curran was an essential reference point of constancy and the ideal of friendship for Joyce, nowhere does Constantine Curran appear explicitly; he is a silent partner. Similarly with the Emperor Constantine. The Arch of Constantine commemorates Constantine's victory over his predecessor Maxentius, and his reunification of a fragmenting empire. Like all of Rome's triumphal arches, it was erected as a portal to the city on which master signifiers of the big Subject were 'set in stone' literally and figuratively. The arch signified the Emperor himself as big Subject, as a living divinity, and the ruling powers of his Empire that integrated all of its subjects under him. All who entered Rome, passing under the Arch of Constantine, became subject to the powers that it symbolically represented: the military might of the historical hyperpower, a gold standard currency embossed with Constantine's head, and a new monotheistic universal religion. But unlike other Caesars before him Constantine inscribes his name on the arch not as a living god, but as ruling by *Instinctus Divinitatis* – through divine [Christian] inspiration. Constantine's Arch stands adjacent to the Colosseum because among many other reforms Constantine closed down gladiatorial combat and other gory spectacles, and proclaimed a new golden age for Rome, founded on benevolence and justice. The triumphal arch represents the triumph of these ideals as master signifiers of Constantine as big Subject, and as commemorative arch it reminds subjects to remember a metanarrative of Rome of which these master signifiers are *points de capiton.* Constantine's Arch is a pastiche of motifs and materials from earlier commemorative arches, to Trajan, Hadrian, Aurilius, and Maxentius, whose monuments were demolished to make a place for Constantine to impress his mark. So in spite of the claims it makes to the coherence and unity of Rome, like the Arch of the Sergii's traces of Greece and the Arco Riccardo's semantic overdeterminations the Arch of Constantine is what Walter Benjamin would call a palimpsest, a text on which legacies of the past, erased by Constantine, are still visible as traces in a collage of *spolia* from earlier iterations of Rome, for at the time of Constantine Rome was in the throes of seismic transitions[6] and, ironically, despite of his name connoting the virtue of constancy, "Constantine was a restless figure typical of an age of transition" (Ferrero, 2019; 428). Constantine's legions went to battle with "PX" emblazoned on their shields and carried as their ensign. "PX" is "*Chi Rho*" the first letters of the word Christ in Greek, and *Chi Rho* is the most lavishly decorated folio of the *Book of Kells.* This would have resonated powerfully with Joyce, for at this time his most treasured possession (and he had very few!) was a facsimile copy of that iconic illuminated manuscript:

> In all the places I have been to, Rome, Zurich, Trieste, I have taken it about with me and I have pored over its workmanship for hours. It is the most purely Irish thing we have, and some of the big initial letters which swing right across a page have the essential quality of a chapter of *Ulysses*. Indeed you can compare much of my work to the intricate illuminations.
>
> (Joyce, in Ellmann, 1982; 545)

The monograph page of the Gospel of Matthew in the Book of Kells, described in the facsimile copy as 'the most elaborate specimen of calligraphy ever executed', as Joyce remarks, consists almost entirely of the initials of Christ's name, the symbol "PX" "*Chi-Rho*" (kai-roe). Resonating with the Greek "*kairos*", it means 'the right time'; in rhetoric it means the opportune moment to turn a discourse, to make a transition. Christ is messiah, teacher, who at the right moment makes a rhetorical intervention, announcing the "good news" that turns the soul and enables a transition from a Rome founded on cruelty and sacrificial violence towards a new Rome founded on charisma, grace, the gift. Matthew's Gospel is the first book of the New Testament, the genealogy of Christ, his teachings, and parables. Crucially, Matthew records the Sermon on the Mount, which contains all in one direct speech the essential tenets of Christianity's new Covenant – a covenant of *philia* and *agape*, between God and man with Christ as mediator; a covenant transacted on both the spiritual and temporal planes simultaneously, with Christ as its living embodiment –the "Ark" of the New Covenant. This is the legacy of Christianity that underwrites and guarantees all subsequent Social Contracts and New Deals in Western Civilisation:

> For the normative self-understanding of modernity, Christianity has functioned as more than just a precursor or catalyst. Universalistic egalitarianism, from which sprang the ideals of freedom and a collective life in solidarity, the autonomous conduct of life and emancipation, the individual morality of conscience, human rights and democracy, is the direct legacy of the Judaic ethic of justice and the Christian ethic of love. This legacy, substantially unchanged, has been the object of a continual critical reappropriation and reinterpretation. Up to this very day there is no alternative to it.
>
> (Habermas, in Calhoun, 2013; 207)

The *Book of Kells* was used in ceremony as a sacred talisman to ritually achieve metanoia, a turning of the soul. Held aloft, the Light of the Word was revealed in and by the illuminated text. A pilgrim during the Dark Ages travelled to Iona or to Kells entered the monastery through an escutcheoned archway, stood under the nave and chancel, its arched vault representing Heaven, and under these auspices, within this ceremonial space the illuminated Book of Kells was ritually displayed as the divine Logos: 'This is the Word of the Lord!': the order and meaningfulness of the cosmos revealed, enunciated, centered. The Logos, Derrida (1991, 32) says, has symbolic surplus, a metaphysics of presence: the illuminated Word is the Light of God in the world; in the illuminated manuscript the pilgrim gets a glimpse of the face of God, infinite and absolute, awesome and sublime. By the power of the Word, revealed in ritual, the pilgrim undergoes a rite of passage effecting metanoia; he is transformed and transfixed, as an individual, and as a member of a moral community of believers, he is integrated into the Church. He has become subject to the power of the Logos; he is stamped by the experience, and with his soul turned he transitions to a new life, exiting by a plain door.

The Book of Kells used in ritual is a quintessential instance of what Benjamin calls a *denkbild* – a "thought-image", a metaphor that effects metanoia by "establish[ing] a connection which is sensually perceived in its immediacy and requires no interpretation" (Arendt, in Benjamin, 1992; 19). Because it conveys cognition and establishes correspondences between the most remote things, metaphors are the means by which the oneness of the world is poetically brought about. The linguistic cognitive "transference" effected by metaphors "enables us to give material form to the invisible … and thus to render it capable of being experienced" (ibid).[7]

Joyce tells us that he pored for hours over the workmanship of the Book of Kells[8], so he knew that hidden in its intricacies and interstices were strange and peculiar signs: zoomorphic, playful, profane, pagan, and irreverent; and that the Book had plentiful errors and mistakes too, left un-corrected. And as Chaucer's *Canterbury Tales* was another of his influences, Joyce knew that pilgrimages were not entirely solemn and po-faced affairs; they were polysemic and ambiguous, undertaken often as a quest for excitement and adventure, in a spirit of play as well as piety. Attuning with one ear to Vico's poetic counterpoint to the monotone of Descartes' methodologism, with his other ear to Bohr's quarky dissonance in Newton's clockwork cosmos; with one parallax eye glancing at past and present flooding and ebbing through the ruins of Roman arches on modern city streets, and with his other eye focussed on the Eternal as represented in the Book of Kells, Joyce asks us "Do you hear what I'm seeing…?" (Joyce, 1995; 193:10): The Logos is overdetermined, ambiguous, and ambivalent; the Fall from Grace is a fall into language, into polysemy and paradox; a Fall that makes us human; a fall that may soon turn out to be fatal, but one that for the time being, in the here and now, means that we the authors of our own fate.

In *Finnegans Wake*, Joyce dwells on the *Book of Kells'* overdeterminations, and on its errors and ellipses and incompletions. The Word is simultaneously 'the empty word and the full word' (Lacan, 1994). The empty word – incomplete, with mistakes left uncorrected, and moreover, rather than erasing and correcting them as the scribes could have done, instead they left them as they were and even marked them with red dots, as if to draw our attention to them:

> Look at this prepronominal funferal, engraved and retouched and edgewiped and pudden-padded, very like a whale's egg farced with pemmican, as were it sentenced to be nuzzled over a full trillion times for ever and a night till his noddle sink or swim by that ideal reader suffering from an ideal insomnia: all those red raddled obeli cayennepeppercast over the text, calling unnecessary attention to errors, omissions, repetitions and misalignments….
>
> (Joyce, 1995; 120)

These errors, omissions, repetitions, and misalignments – "mistakes" – are themselves portals of discovery, in this case portals to discovering the fullness of the word, its overdeterminations. Do the mistakes mean, for instance, that a human scribe is inadequate to the task of representing the Logos; that the Absolute, being unrepresentable, is, appropriately and necessarily mis-represented? Yes, it may mean this; and equally it suggests that the Absolute is not absolute, that even the Logos does not have 'the last word'; that it is incomplete, in error, lacking; leaving more to the world than can be said, even by God; a surplus, excess, that remains yet to be said, by man: humankind is the source of the divine logos, not the other way around. Hermes and Christ are both hybrid mythic embodiments of god and man, and as mythic embodiments they are arks containing a treasure trove of metaphors representing 'knowledge of things both human and divine" (Vico, 1999) an inexhaustible resource for our reflexive critical self-understanding and imaginative re-invention that can be 'nuzzled over a full trillion times' by 'an ideal reader suffering from an ideal insomnia' (Joyce, 1995; 120).

The parallax power of portals

Passing through the portal, to come under the arch, is to become subject to the sign of the big Subject. The arch is built to impress; it impresses; it stamps! Portals, whether the Arch of the Sergi asserting the power of the ruling Roman Patrician family (though its Greek motifs also indicating that the Sergi were only Johnnies-come-lately) or the Arco di Cardo as portal to the city's sacred precinct and its ruling deities (but the immortals long dead, usurped by lively trading houses, taverns, and bordellos) or Constantine's triumphal arch, with his three coalescing metanarratives and their master-signifiers of Rome as megamachine – the stream of gold circulating between Babylon and Bath and the disciplined Legions with PX on their ensigns proclaiming that 'Christ and Caesar go hand in glove' – is Rome at its apotheosis (but simultaneously at the cusp of terminal decadence). What all of Joyce's arches have in common is overdetermination and ambiguity: their signs may have multiple meanings, depending upon the perspective of the reader, context, and interpretation. Because the symbolic order is always over-determined, incomplete, flawed, lacking, there are always interstices in which to compose an alternative text, even in a coded and algorithmised hypermodernity wherein it seems, to paraphrase Horkheimer and Adorno (1972) that 'the fully digitised Earth radiates disaster triumphant.'

Joyce discovered his first arch in Pula, his second in Trieste, and a third arch in Rome, and these arches were for Joyce inter-dimensional portals though which he could, by taking a parallax view of the flood and ebb of history, transition to a hypertextual fantasy space so that he might 'sing of all that is past, or passing, or to come' (Yeats, 1928). Walter Benjamin's *Arcades Project* similarly used the dilapidated Parisian passages as portals in the heart of the modern city to foresee future ruination, but Joyce (in keeping with his name – Joyce/"joy"/*freud/jouissance*– and thus always attuned to Eros) is more hopeful. All of human life flows through these arches, standing as they do, and have done for ages, on main public thoroughfares.[9] Passage through the arch, to the symbolic arc of language, to the ark of shared ideas is an inter-dimensional portal where metanoia, transfiguration, and transformation can take place. Joyce explicitly intended *Ulysses* and the *Wake* to have such effects – that their over-determinations, portmanteaus, and metaphors would be as an Ark containing a treasure trove of ideas, encoded in metaphors, enabling us to articulate a new covenant that might sustain civilisation through a modern Dark Age: *Ulysses* was written during WW1, and the *Wake* was published on the eve of destruction in 1939. The reader reads, but, through reading, the book re-writes the reader: in the reciprocal exchange between text and reader the reader becomes subject to the influence of the text, and by the power of metaphors the reader undergoes metanoia.[10]

Joyce's education and his model for how to live the good life

Educare means to cultivate, to lead forth, to draw out from within; and wisdom is derived from *vis* in "vision" and *dom* meaning judgment and authority. When we consider the transition and transformation that climate breakdown requires of us, we realise that it is not just innovation in economy and technology that is at issue, but more fundamentally a revitalisation of our political, cultural, and moral institutions. Our individual and collective abilities to be self-reflexive, innovative, and creative, to adapt to change and to reinvent our society and our economy to face the challenges of climate breakdown will come primarily from vision and the exercise of judgement based on good authority; guided by the light of higher values and ideals and inspired by good models, for education is mimetic: we learn by imitation. Education is concerned with relations of influence, and thus with models that inspire us, and that we emulate. Joyce provides us with a model of human flourishing in Bloom, but who was the model for Bloom, and how did Joyce emulate him?

In Trieste, Joyce made a good friend in one of his English students, Ettore Schmitz (Italo Svevo). Schmitz too was a writer, among many other things – a businessman, a raconteur, a flaneur, a husband, and father. Joyce read to Schmitz from working draft of "The Dead", one of the leitmotifs of which is Gabriel Conroy's "absence of mind". Gabriel Conroy is not attuned or clued-in to the world around him, either at the level of the cultural and political issues of the day, or at the level of his wife's inner life. Conroy – "*con*", "with", and "*roi*" "king", "Gabriel with-the-king" is a familiar contemporary personality type: he fancies himself to be a modern, progressive, cosmopolitan, rational, man of the world, but actually he is a conformist, his interests are narcissistic, his views are myopic and banal. A creature of the herd, Conroy lives by a morality borrowed mimetically from prevailing conventions, unwilling and unable to attune appropriately either to his wife's or to his country's needs, and so, at the end of the story Gabriel's soul swoons away like 'the snow falling faintly through the universe and faintly falling, like the descent of their last end, upon all the living and the dead.'

Reciprocally, Schmitz showed Joyce two of his own unpublished books – *Una Vita* and *Senilita*. Both feature a man who imagines himself to know what's going on around him, but in actuality he is a witless dope. He mis-reads social situations and people's motives; he fails to see and to grasp opportunities; he is "inept", in the sense that he doesn't have his wits about him, so to speak, and so he adapts poorly to changing currents and contexts, and, like Gabriel Conroy, mimetically aping conventional morality and prevailing opinions and monocular viewpoints and lacking strong guiding ideals that would enable autonomous moral transcendence and ability to push back against the flow of events, he is snowed under and drifts away in the currents of history.

Both Joyce and Schmitz are concerned with how modern people generally suffer from orthodox frames of thinking and understanding, conventional politics and morality, solipsistic isolism and inept maladaptation to the demands of changing cultural, political, psychological, and moral contexts, and that this willing conformity to the flow of prevailing discourse, an inability and unwillingness to turn around is one of the pathogenic social currents of our time. Both Schmitz's books end unhappily, *Una Vita* in suicide, and *Senilita* ends as the title suggests, the protagonist ageing unto banal demise. Joyce loved Schmitz's books, and later, when he was in a position to do so, he promoted Schmitz strongly in Paris, and Schmitz became famous in his own right. In the meantime, Joyce and Schmitz became firm friends, and Bloom is an ideal-type representation of the qualities of character of both of them, assimilated from the experience of their friendship and aesthetically – intellectually sublimated into the form of a person who is especially adept at reading and interpreting, engaging with, and responding strongly to the moral-practical complexities of the milieux of modern life; Bloom is a man who is the very model of "presence of mind".

By creating Bloom, Joyce is reaching for something beyond Art.[11] The younger Joyce, Stephen of *A Portrait of the Artist as a Young Man,* is very much in thrall to the Ideal, he is an Icarus at risk of crashing and burning; whereas Joyce, Daedelus, as he becomes in Trieste through his friendship with Schmitz, flies closer to the earth, and through and in Bloom Joyce develops a model – a living human "real" model that we may emulate if we wish to lead a good [Modern, and beyond modern] life, just as the life of Christ is a model to be imitated if one seeks redemption. It is quite striking that the great literary critic, mythologist, and philosophical anthropologist Rene Girard doesn't engage with Joyce at all. He doesn't, because Joyce's Bloom doesn't fit Girard's powerful, but too-negative frame. For Girard any model that isn't an "external" mediator, exalted to the plane of the Ideal – gods, saints, nobility – but belongs rather to a shared egalitarian horizon of fellow human beings, citizens, and friends "in real life" – can only generate a downward spiral of mimetic envy and violence. For Girard, without the mediating power of an exalted model in a relation of binary polarity modern people relate to one another only as rivals and enemies, and the only solutions are either the scapegoat mechanism or/and a Restoration.[12]

Joyce and Schmitz belong to the same modern egalitarian plane of being: they have similar backgrounds, similar aspirations, similar relations to their respective religions, similar relationships with their respective partners and families, and so on. There are differences between them for sure, but these are matters of small degrees: not "binary antagonisms" but "complementary dualisms" (Byrne, in this volume), and talking through their differences and similarities is how they become friends, so that rather than becoming rivals who are envious of one another, Joyce and Schmitz both benefit mutually and reciprocally from their relationship. Joyce personally – especially younger Stephen as we meet him in *A Portrait* – is terribly prone to envy and a sense of betrayal by rivals – Gogarty and Cosgrave among a great many others, but through his growing friendship with the genial, avuncular, and magnanimous Schmitz Joyce becomes less austere and more fully human, while not losing the conceit of genius that was essential for his Art; and reciprocally Joyce was hugely encouraging and fortifying to Schmitz who as a writer was modest and self-effacing almost to a fault. Bloom emerges as the quintessential artistic representation of a form of life that shows how finding a good model is possible after the modern Fall – not only "shows" but actually "is" a living model to be emulated. The story of Joyce and Schmitz and Bloom as the literary child of their friendship offers a strong and hopeful alternative to Girard whose only way forwards is a second coming of Christ (or more likely Yeats's "second coming" of a demonic "rough beast").[13]

Maieutics, mimesis, and educating for transition

Education is *maieutics* – midwifery (Socrates) and *mimesis* – imitation (Plato) and Bloom as literary child of the friendship between Joyce and Schmitz represents a synthesis of that dialectic; which leads us to consider Joyce as a teacher. It is mostly overlooked that Joyce was a teacher, glossed as merely how he made a living while he was working on his Art, but that account underplays something that was very important to Joyce, especially in Trieste, during his most creative period. Teaching was essential to Joyce's vocation, and it is not insignificant that Schmitz' and Joyce's friendship was first sparked in a teaching session, and thereafter, and again not insignificantly, many of their "classes" took place in the city streets, where Joyce and Schmitz walked and talked for hours. A contemporary witness describes them "clinging to ropes fixed along the steep side streets under the *bora*'s blast, as if they were climbers roped together, and talking incessantly' (Price, 2016; 74). What belongs particularly to being a language teacher is that it involves guiding someone through a transition into a shared symbolic order, the same arch of language and meaning, and insofar as 'to imagine a language is to imagine a form of life' (Wittgenstein, 1994) both teacher and student are mutually transformed by the experience. So, here we have the arch, the arc, and the ark, their overdetermination as a portal into a shared imaginative structure, a passage between mythopoetic, historical, and modern worlds, a metaphoric zone of metanoia and transition. With Joyce, the commemorative arch becomes the portal into the symbolic order, the arc of language, the arc of history, and the arc of individual life – life that can be mimetically emulated after a good model, a model, Bloom, drawn from real life. Taken all together they become through Joyce's work as both Artist and as teacher an ark of a modern covenant that might carry us safely through the coming flood.

If we hope to turn things around, to transform, and to transition away from the antediluvian moment of the late modern Anthropocene, it will be by an education, using metaphors and good models to give new meanings to the key terms of the prevailing language games so that the trajectory of the narrative of Progress ['Progress through Technology' -*Vorsprung durch Teknik*, the keynote ideology of Enlightenment and Capitalism condensed into an advertising slogan!] is turned-around, effecting a metanoia that is at once subjective – psychological- and collective – sociological-historical, and moral-political, a message that resonates so that it "sells" as well as Audi's does.[14] Such use of metaphor to bring about a transformative metanoia is what Benjamin has in mind when he says somewhere that his *denkbilder* are as tools in the hands of a skilled engineer on a runaway engine, by means of which he need make only a small adjustment that will change the direction of the huge machine.[15]

Contemporaneously with Joyce, Walter Benjamin began his *Passagenwerk* – the "Arcades Project", an archaeology of the present conducted within the derelict covered passages of Paris. Joyce's arches were inter-dimensional portals in the heart of the city through which he could transition imaginatively between past, present, and future. For Benjamin similarly the Paris arcades were phantasmagoric dream-machines that originally served as theatres for a rite of passage into consumerism as a form of life, but the now derelict arcades could become inter-dimensional portals through which we could escape from that dreamworld. On entering the portal of the arcades people came under the influence of the master signifier of "Progress" and were impressed by the trade mark[16] of fashion [that] 'prescribes the ritual by which the commodity fetish wishes to be worshipped'

(Benjamin, 1999; 153). Shopping in the arcades whereby people encountered the commodity fetish, displayed and illuminated in "*panoramas*", "*spectacles*", and the "*theatre des varieties*" became the *rite de passage* whereby people became transformed and transfixed as subjects of modern consumerism. However, some people, a minority, whom Benjamin hoped could serve as models, *flaneurs* and *flaneuses,* while immersed in the crowd, cultivated a blasé outlook, a cool detachment and intellectual distance, their knowing irony disguising skeptical agnosticism and even apostasy. Benjamin's hope was that Paris's derelict arcades could be portals to a parallax view of recently vanished *ur*-capitalist beds of conception, portals whereby contemporary people, as though they were time-travelers, might revisit the primal scene when we first became subject to the commodity fetish, and, seeing Noah naked and drunk as it were, "begin to recognize the monuments of the bourgeoise as ruins even before they have crumbled" (Benjamin, 1999; 14).

Conclusion: At the vestibule of hell

To bring this reflection on metaphors of transition in the *Age of Total Capitalism* and its divine Market (Dufour 2008) to something of a close, let us return again to Joyce, and that other key source of his inspiration, Dante's *Divine Commedia* (2008).

Outside the Vestibule of Hell, Dante says:

> ...I saw a flag
> Flapping wildly as it was carried forward,
> As if it was not fit to rest a moment;
>
> Behind it came a huge torrent of people;
> So many that I never should have thought
> Death had been able to undo so many.
>
> ...
> That calamitous crowd, who never were alive,
> Were naked, and their skins blown with the bites
> Of swarms of wasps and hornets following them.

T.S. Elliott borrows these lines as his metaphor for the plight of the modern subject in *The Waste Land*, that form of life – which is so characteristically our own – of those uncommitted and poor in spirit, who, having no faith, belief, or commitment to any big Subject, those who, not being subject to any master signifier – de-symbolisation having deprived them of the metanarratives that are necessary for subject-formation and thus they never been fully alive as autonomous subjects – are compelled to chase after every flapping ensign; a futile pursuit, hunting for meaningfulness in a succession of fleeting empty signifiers, all the while chased by swarms of stinging hornets. Dante gives us the metaphor for the condition of the subject of de-symbolisation, the precarious subject of the neo-liberal revolution, the depressive-anxiety-ridden, isolist-narcissist-protopsychotic of the digital and social media revolution. We are at the Vestibule of Hell; and if we follow Dante, the only way for us to go may be to enter the portal and follow the passage all the way through to the other side, down through the nine circles of Hell, to meet Satan himself at the centre of the world, and then, having faced the consequences of our sins, to progressively work our way back up through the levels of Purgatory, where crimes of speculation, avarice, gluttony, fraud, treachery, and false leadership are systematically purged, until we reach at last the shore of a sustainable Paradise. We know the way that we have to go if we are to get out of the mess we have made, but over the portal at the Vestibule of Hell hangs an ominous sign: "Abandon Hope All Ye That Enter Here."[17]

'Abandon hope, all ye that enter here' shows the seriousness of the task confronting us at this antediluvian moment of the late modern Anthropocene. We will need courage and hope to embark on the journey before us, and, thankfully, neither Dante nor Joyce were ever short of these two cardinal virtues. When Yeats met the young Joyce he found him to be self-confident to the point of conceit, but what really impressed Yeats was Joyce's "joyous vitality" (Ellmann, 1982; 101) and it is the same 'holy spirit of joy' that sustained Dante too in his lonely exile from his home in Florence and his lost love Beatrice. Dante and Joyce never despaired, never lost hope. In the structure of *Ulysses* Joyce took Dante's infernal landscape and laid it over 1904 Dublin creating a double-layer of metempsychosis for his characters. Leopold Bloom encounters all nine levels of Dante's hell (in order), crosses four rivers, and emerges from the other side with a renewed passion for life. (Devine, 2014).

"In Dante", Joyce says, "dwells the whole spirit of the Renaissance. I love Dante almost as much as the Bible. He is my spiritual food" (Joyce, in Ellmann, 1982; 218). Dante is the quintessential voice of transition from the darkness of the Middle Ages to the hopefulness of Renaissance, and the medium of this transition is the vernacular language of everyday life. This is what "*comedia*" of the *Divine Comedy* means: the style is low-brow; written in the everyday language of common life, and intended to be educational, enabling ordinary people to self-reflexively grasp and understand their predicament and thereby enable a transition from perdition to salvation for the community as a whole.

Dante describes the mode of *Divine Comedy* as "poetic, fictive, descriptive, digressive, [and] transumptive," (Tardif, 2015; 244). To "transume" is an archaic verb that means to transfer meaning, to convert, to metamorphose, to transubstantiate. So it is from Dante as much as Vico that Joyce learns the use of myth and metaphor to bring about metanoia. Dante teaches Joyce how to harness even those authorities that Joyce repudiates, such as Catholic dogma, nationalist chauvinism, and the myopias and partial perspectives of histories, sciences, expert discourses, and everyman's opinions; they all become grist to his mill. Joyce works with and within the traditions and inheritances of Religion, Science, Art, as well as politics, popular opinion, and the banal contents of everyday life. More than simply influencing him, Dante teaches Joyce how to be influenced, and how to be influential: by working with the repertoire of cultural inheritance of ideas and metaphors and giving them new usages to help us to understand and to transform our changing circumstances, just as "Dante" becomes "d'aunty" [the god-fearing kindly aunt who sustains the chaotically self-destructive Joyce family], becomes the Ondt [the prolific ant who shows us how to live collaboratively, sustainably], and Anna Livia Plurabelle, the eternally recurring, always beautiful goddess Anu, the river of life flowing through Dublin's many arched bridges; flooding by the docklands' headquarters of GAFA – Google, Amazon, Facebook, Apple, all those tax-dodging dead-beat dads, would-be fathers purporting to be the new big Subjects; to the port, into the sea, where everything is dissolved, and where everything may begin-again. The new metaphors that we looking for do not need to come into the world out of nowhere; we can find them already close to hand, by working over what we have known all along; for "imagination is nothing but the working over of what is remembered; imagination is memory".[18]

Notes

1 One of Hermes many roles is "psychopomp", the guide who conveys souls between the realms of the living and the dead.

2 The corona of shades of meaning in the semantic field that bear family resemblances to metanoia include "transmutation", "transfiguration", "metamorphosis", and "metempsychosis" (in the sense of being taken over, becoming possessed and animated by powerful spirits); transfiguration – a complete change of form into a higher and more beautiful state (Merriam-Webster; OED). What metanoia has come to mean today has a long genealogy, from Philology, through Theology, to Psychology, and to Pedagogy; a genealogy showing a progressive deepening of meaning, from an original usage signifying a relatively superficial change of mind, as in "to change one's opinion" about a particular matter or point of view, to metanoia used "to express that mighty change in mind, heart, and life wrought by the Spirit of God".

3 "Imagination is nothing but the working over of what is remembered" (Joyce, in Ellmann, 1982; 661).

4 Joyce satirically juxtaposes the Savoy monarchy's bombastic monumentalism with his co-workers endless chatter about their *callo, coglioni,* and *culo*.

5 There Joyce has an epiphany. Some English tourists recite Byron's lines:
"Whowail stands the Colosseum Rawhm shall stand/When falls the Colosseum Rawhm sh'll fall/And when Rawhm falls the world sh'll fall – But adding cheerfully Kemlong, 'ere's the way aht". Blissfully unaware that their own Empire, at the apex of modern civilization, is on the cusp of ruination too.

6 Constantine was an Illyrian, from Istria, the liminal region to which Pula and Trieste belonged, and under Constantine the Empire was re-united, but simultaneously decentring to Byzantium, which, re-named for Constantine became Constantinople. Constantine's dominion stretched from Asia and North Africa to Hadrian's Wall, encompassing most of what is now the EU, and a great deal more besides. Constantine established the gold standard *soldus* two millennia before the euro, facilitating trade across languages and cultures through four and a half million square kms for a thousand years.

7 Consider the difference – and also the continuity – between the Book of Kells and the smartphone: its illuminated apps giving unlimited access to profane mysteries, held aloft on a selfie-stick to frame the soul in narcissistic splendor, uploaded temporarily, stored on the cloud in an infinite abyss of self-referentiality; the ubiquitous selfie is the banal ritual of dis-integration from community and society and self-alienation from transcending ideals of the common good; it cultivates an isolist, solipsistic "inability or reluctance to imagine what others are thinking, or, as Kant once said, "to think from the standpoint of everyone else" is eclipsed by the "elevation of possessive individualism as the only value that matters" (Giroux, 2015; 155).

8 Joyce's facsimile copy of the Book of Kells has an Introduction that celebrates "its weird and commanding beauty; its subdued and goldless colouring; the baffling

intricacy of its fearless designs; the clean, unwavering sweep of rounded spiral; the creeping undulations of serpentine forms, that writhe in artistic profusion throughout the mazes of its decorations; the strong and legible minuscule of its text; the quaintness of its striking portraiture; the unwearied reverence and patient labour that brought it into being" (Sullivan, 1914)

9 Which is to say that through them all flows Proteus. One of the leitmotifs of *Ulysses* [later, in the *Wake,* Proteus undergoes a gender transition, becoming Anna Livia Plurabelle] Proteus, the "Old Man of the Sea" has the gift of prophecy, but he changes forms bewilderingly, from a lion, into a snake, into a leopard, then into running water, then into a tree. But if he is grasped and held in spite of his metempsychosis, he will give prophecy. Consider this in the context of Cambridge Analyticas' grasping the psychological profiles of 75 million people from Facebook's stream of data so that their political inclinations can be rendered and predicted, and then micro-targeted and manipulated by orphic spindoctors and hermetic codeweavers. Are there limits to such endeavours? Hermes, a Trickster figure, typically slips away quietly from the chaos, leaving the Robert Meullers of the world to reconstruct the scene of the crime!

10 Consider how Joyce's aesthetic-political emancipatory principle of parallax and recursive reciprocity between writer, text, and reader wherein the reader is re-reformed by the influence of the text is appropriated, instrumentalized, commercialized, and weaponized as the new social media engines' dopamine-driven bio-feedback loops fuelling proliferating solipsistic echo-chambers that can be manipulated at will with alternative facts and fake news.

11 I am indebted to Kieran Bonner of the University of Waterloo for much of the following.

12 Which is why deep-conservatives, neo-reactionaries, and Silicon Valley billionaires all love Girard!

13 A metaphor that could be a radical Humanist T shirt slogan rather than an epitaph for a form of life that might survive the late modern Anthropocene, and echoing Ecology's principle diversity is the key to sustainability, would be "Let a thousand Blooms flower!"

14 Hermes as God of metaphor, spirit of commerce and politics may help us find a way of getting the message across (Hyde, 1998).

15 And to update Benjamin's analog metaphor so as to attune with our digitized world, we may speak of a meme – an idea, behaviour, or style that spreads from person to person. A meme acts as a hermetic device for carrying ideas, symbols, or practices that can be transmitted from one mind to another through writing, speech, gestures, rituals, or other imitable phenomena. In a previous and still prescient register full of similar resonances, Donna Haraway (1991) uses the notion of a "viral vector" in the cyborg world of C3I [the algorithms and codes of command-control-communications artificial intelligence integrating five-dimensional contemporary warfare – land, sea, air, space, and psych-ops]. Digital-biological interfaces effected by neuro-genetic and nano-engineering need not be only one-way circuits, Haraway says, but can be conduits for aesthetic images and metaphors as Internet memes, viral hermetic keys that we may send back through the matrix in such a way as to re-code it. Against the pathogenic "inscription of a digital rhythm on the social body" Berardi identifies already a dialectical interior counter-rhythm in terms of a "rhizomatic poetics of insurrection", "of slowness, withdrawal…the autonomization of the collective body and soul from the exploitation of speed and competition" (Berardi, 2012; 29; 68).

16 Trade "mark" is "*merc*" from Mercury, the Roman Hermes.

17 …and yet there is hope: a generation have already abandoned Facebook and are treating others like Instagram much more circumspectly. Like the Paris Arcades these portals to phantasmagorias of the world wide web were fun at first, and even became for many something like a new religion, the online social network functioning as a sort of church; but a church of self-worship, wherein a sense of community was paid for by cathartic rituals of scapegoating and sacrificial violence, trollings, abasements and humiliations, epidemics of anxiety-depression often culminating in suicide.

Furthermore, as investigations into the role of Facebook and other social media bring to light wholesale manipulation, whether by Republicans or Russians in global cyber psychop wars that makes cultural dupes of everyone, the digital portals to the Internet begin to resemble already the Paris Arcades as Benjamin found them, deserted and decrepit…

18 Joyce, in Ellmann 1982, 661.

References

Benjamin, W. (1992) *Illuminations*. New York: Fontana.

Benjamin, W. and Tiedemann, R. (1999) "Paris: Capital of the Nineteenth Century" in *The Arcades Project*. Cambridge MA: Belknap-Harvard University Press.

Berardi, F. (2012) *The Uprising: On Poetry and Finance*. Los Angeles: Semiotext(e)

Calhoun, C. et al. Eds. (2013) *Habermas and Religion*. Cambridge: Polity Press .

Dante (2008) *The Divine Comedy*. Oxford: Oxford University Press.

Derrida, J. (1991) *A Derrida Reader: Between the Blinds*. New York: Columbia University Press.

Devine, B. (2014) "*Bloom's Inferno: James Joyce's Hidden Dantean Landscape in the "Hades" Episode of Ulysses*" *American Conference of Irish Studies*, University of Miami, Coral Gables, Miami, Florida.

Dufour, D.-R. (2008) *The Art of Shrinking Heads: on the New Servitude of the Liberated in an Age of Total Capitalism*. Cambridge: Polity Press.

Ellmann, R. (1975) *Selected Letters of James Joyce*. New York: Viking Press.

Ellmann, R. (1982) *James Joyce*. Oxford: Oxford University Press.

Ellmann, R. (2003) *Selected Letters of James Joyce*. London: Faber & Faber

Evans, A. (2018) *The Myth Gap: What Happens when Evidence and Arguments Aren't Enough*. Cornwall: Eden Books.

Ferrero, G. (2019) [1898] *L'Europa Giovane*. Delhi: Pranava Books.

Ferrero, G. (2013) [1907] *The Greatness and Decline of Rome*. Los Angeles: Hardpress.

Giroux, H (2015) "Selfie Culture in the Age of Corporate and State Surveillance", *Third Text*, 29:3, 155–164.

Graves, R. (1960) *The Greek Myths Vols 1 and 2*. London: Penguin.

Haraway, D. (1991) [1984) 'A Cyborg Manifesto' in *Simians, Cyborgs and Women: The Reinvention of Nature*. London: Free Association Books.

Horkheimer, M. and Adorno, T. (1972) *Dialectic of Enlightenment*. New York: Herder & Herder.

Hyde, L. (1998) *Trickster Makes this World: Mischief, Myth and Art*. New York: Farrar, Straus & Giroux.

Joyce, J. (1995) [1939] *Finnegans Wake*. London: Picador.

Joyce, J. (1990) [1922] *Ulysses*. New York: Vintage.

Kant, I. (2002) [1875] *Groundwork for the Metaphysics of Morals*. Oxford: Oxford University Press.

Kerenyi, C. (1980) *The Gods of the Greeks*. London: Thames & Hudson.

Lacan, J. (1994) *Speech and Language in Psychoanalysis*. Baltimore: Johns Hopkins University Press.

Mumford, L. (1967) *The Myth of the Machine*. New York: Harcourt.

Myers, K.A. (2011) "Metanoia and the Transformation of Opportunity" *Rhetoric Society Quarterly* Vol 41 (1) 1–18.

Nietzsche, F. (1994) [1873] 'On Truth and Lies in a Non-Moral Sense' in W. Kaufman Ed., *The Portable Nietzsche*. London: Viking.

Plato (1997) *Republic*. Translated by D.J. Vaughan and J. L. Davies. London: Wordsworth Editions

Price, S. (2016) *James Joyce and Italo Svevo: The Story of a Friendship*. Bantry: Somerville Press

Rosa, H. (2013) *Social Acceleration: A New Theory of Modernity*. New York: Columbia University Press.

Sullivan, E. (1914) *The Book of Kells*. London: Studio Publications

Tardif, S. (2015) "Joyce's Dantean piety, or the survival of acceptable ideas" in *Dante and The Christian Imagination*. Ed. Domenico Pietropaolo and Jenna Sunkenberg. Ottawa: Legas Publishing.

Vico, G. (1999) [1744] *New Science*. London: Penguin.

Wittgenstein, L. (1994) [1958] *Philosophical Investigations*. London: Blackwell.

Yeats, W.B. (1920) 'The Second Coming' in *Michael Robartes and the Dancer*. Dublin: Cuala Press

Yeats, W.B. (1928) 'Sailing to Byzantium' in *The Tower*. Dublin: Cuala Press.

13 Patterns of interference

The ethics of diffraction in Mike McCormack's solar bones

Maureen O'Connor

Introduction

Few social problems or political conflicts can be considered without taking into account the environmental dimension, especially the fact of diminishing resources along with the suffering, forced migration, and violent conflict to which these changing conditions give rise. The humanities are central to understanding the issues at the core of such crises, including the ways in which the environment is represented and how the role of the human in nature has been theorised and understood. The field of environmental humanities addresses this urgent and ethical need. Ecocritical theory was first developed in literature departments in North American universities in the 1990s. In the ensuing decades, ecocriticism as theoretical praxis has extended well beyond literature and beyond the humanities, though its literary origins remain influential, if not always acknowledged, due to the power of narrative, recognised by environmental theorists across disciplines. For example, in outlining the problems of enshrining dangerous nature/culture dualism in the increasingly cited idea of the "Anthropocene", historian Jason Moore locates the argument's influence in its "storytelling power" (2016: 82). Biologist-philosopher Donna Haraway shares Moore's reservations about the limitations of the human-centring implicit in presenting the Anthropocene as a story about the dynamic "ongoingness" of being, arguing that the "practice of thinking [...] must be thinking with: storytelling. It matters what thoughts think thoughts; it matters what stories tell stories" (2016, 42).

Recent environmental commentators, including Moore and Haraway, have offered a less hopeless narrative than the teleological Anthropocene plot, shifting the focus away from fatalistic humanity-blaming scenarios about imminent environmental collapse, proven to be counterproductive. They suggest more positive engagements with the threat posed by climate change; in the words of Haraway, "I want to make a critical and joyful fuss about these matters [...] and the only way I know to do this is in generative joy, terror, and collective thinking" (2016: 34). One source of generative and joyful collective thinking with promise for understanding the place of humans in the world is the work of ecofeminist scholars, particularly in the field of new materialism, which can be both illuminating and inspiring. Haraway, Karen Barad, Stacy Alaimo, Serenella Iovino, and Serpil Opperman, who often write together, Jane Bennett, Iris Van der Tuin, and Rosa

DOI: 10.4324/9781003143567-17

Braidotti are some of the important contributors to this field. The material turn in ecological humanities seeks to destabilise conventional notions of subjectivity, to argue against the idea of independent entities, to understand ourselves and everything around us, seen and unseen, as phenomena in an ongoing process of "becoming" through relationships and interactions with other phenomena. An historically entrenched perception of radical discontinuity between the human and the nonhuman has had disastrous consequences for the environment and all the planet's inhabitants. Metaphors have been instrumental in the arts and sciences in establishing and perpetuating the idea of objective, scientific "truth" regarding the natural world and natural phenomena. The figure of the mirror that science holds up to its subjects in order to deliver unproblematic, unbiased reflections is one such metaphor. These metaphors have both relied on and created the dangerous concept of "nature" as something unchanging and ahistorical, as constitutively "other" to the human realm of culture.

The metaphor of diffraction

In her 1992 book, *Promises of Monsters,* Haraway first proposed another metaphor for the relationship between the observing human consciousness and the finally unknowable world around us: diffraction, a different approach to understanding physical phenomena, one that accepts the incomplete nature of human knowledge, that recognises and even embraces the unstable, the plural, and the partial. Tracing the origins of the deployment of this metaphor and its application in critical humanities, Birgit Mara Kaiser and Kathrin Thiele have summarised its significance: "Drawing on physical optics, where it describes the interference pattern of diffracting light rays, Haraway adopted diffraction to move our images of difference/s from oppositional to differential, from static to productive, and our ideas of scientific knowledge from reflective, disinterested judgment to mattering, embedded involvement" (2014: 165). As a metaphor for thinking our place in the world, diffraction extends beyond its origins in science. As Barad has observed, "Diffraction owes as much to a thick legacy of feminist theorising about difference as it does to physics" (2014: 168), and "attends to the fact that boundary production between disciplines is itself a material-discursive practice" (2007: 90). As a metaphor, then, diffraction, by definition, cannot be restricted to its "original" disciplinary context. According to Haraway, "Diffraction patterns record the history of interaction, interference, reinforcement, difference. Diffraction is about heterogeneous history, not about originals. Unlike reflections, diffractions do not displace the same elsewhere, in more or less distorted form, thereby giving rise to industries of metaphysics" (1997: 273).

Enthusiastically adopted by other thinkers like Barad and van der Tuin, this metaphor can keep us – literally – grounded, grappling with the material, rather than retreating into transcendent, invisible positions of objective authority. Diffractive reading asks that we truly inhabit our critical positions as well as our environments. Barad argues that "Diffractive readings bring inventive provocations; they are good to think with. They are respectful, detailed, ethical engagements" (2012a: 50). The 2016 prize-winning novel, *Solar Bones*, by Irish writer

Mike McCormack, not only directly addresses the ecological destruction brought about by late capitalism in twenty-first century Ireland, but also, in the text's innovative form, enacts the crisis through a diffractive aesthetics of fragmentation and heterogeneity that reveals unsuspected but vital continuities and connections, an example of "respectful, detailed ethical engagement" with our place in the environment as well as our responsibilities.

Mike McCormack's *Solar Bones*

Solar Bones is narrated in first-person by a middle-aged family man, Marcus Conway, in a kitchen just outside of the village of Louisberg in County Mayo, where he waits for his wife to return from work. The novel is a single, unbroken sentence, unostentatious in its structural virtuosity, appropriately so, considering Marcus is an engineer obsessed with unseen but vital supporting structures and systems. Textual rhythms and pauses are achieved by apparently random, but ingeniously crafted line breaks so that the prose is delivered, as Martin Ryker describes it, "as if spoken by a friend across the table" (2018). The flexibility of this prose style, the openness of its interrogative mode, and its avoidance of even the benign authority of the full stop are formal qualities inextricable from the environmental themes of the novel that trace the enmeshed nature of all phenomena. In the first few pages, Marcus, who is scanning the newspapers, refers to an "environmental campaigner who has been on a hunger strike" against the Corrib gas pipeline (2017: 10) and a story about "an asbestos conversion plant which will form part of a massive toxic dump to process industrial and medical waste" (2017: 12). This last story prompts a childhood memory of Marcus's father working on the construction site of an acrylic yarn factory where "the manufacturing process would utilise a highly toxic compound [...] a chemical that would have to be transported overland in the middle of the night under security escort, shipped in double-hulled, crash proof containers", a recollection suffused with a "credible apocalyptic glow", in the light of which his father appears "fearless, heroic" (2017: 13). As it unfolds, the novel dedicates much of its energies (and the narrator's) to attending to Mairéad, Marcus's wife, who becomes stricken with cryptosporidiosis, a water-borne disease caused by another environmental disaster, the fecal contamination of Galway's water supply.

The novel's narrative voice is not especially authoritative, nor, despite being first person, does it comfortably occupy a stable subject position. There is no plot; the text proceeds associatively, dissolving and resolving by turns, "swept up in that sort of reverie which has only a tangential connection to what you were thinking of" (McCormack, 2017: 15). Movement through the novel is horizontal, rhizomatic in offering multiple entry and exit points. The grammatical structure relies on juxtaposition and parataxis; no subject, or train of thought, or even phrase is subordinate to any other. In place of linearity, the text is patterned by interference; the speech, thoughts, and actions of others repeatedly, and productively, interrupt and distract the narrator, as do his own thoughts and recollections. Some of these effects emerge from Marcus's own radical ontological indeterminacy: he is a ghost returning to the family home on All Souls' Day, though he does not realise this until the last pages when he recalls the day of his death months earlier. This layer of "unreality" allows for a narrative that is not traditionally representational, a narrative that both requires and performs a diffractive reading, which "denote[s] a more critical and difference-attentive mode of consciousness and thought" (Geerts and van der Tuin, 2016). In her book, *Meeting the Universe Halfway: Quantum Physics and the Entanglement of Matter and Meaning*, Barad, whose training is in quantum physics, reveals, through diffraction, "the entangled structure of the changing and contingent ontology of the world, including the ontology of knowing" (2007: 73). Reading phenomena diffractively is at the centre of Barad's account of agential realism, which posits that "knowing, thinking, measuring, theorizing, and observing are material practices of intra-acting within and as part of the world" (2007: 90). As a structural engineer, Marcus has been keenly involved in measuring and observing all of his working life, appearing to think more deeply about the implications of these actions than many of his peers. In Barad's relational otology and epistemology, "all bodies, not merely human bodies, come to matter through the world's performativity—its iterative inter-activity" (2012b: 32). Marcus's intense interest in the structure of all phenomena, human and nonhuman, animate and inanimate, imagined and experienced, provides a compelling literary instance of Barad's diffractive agential realism, with Marcus practising a kind of diffractive methodology himself, in his attention to "fine-grained details" (2007: 90).

Science and literature

Suggesting connections between science and literature is nothing new, though such a link is rarely noted in discussions of *Solar Bones*, the important exception being Sharae Deckard's praise for the novel's foregrounding of subjects which are often absent or obscure in Irish literary fiction: politics, economy, science (2016). Deckard considers *Solar Bones* to be the "fullest expression" of McCormack's career-long preoccupation "with the way science and technology shape everyday life" (2016). Most reviews have made comparisons with the work of Irish Modernist James Joyce, writing and publishing a hundred years earlier. Joyce was composing his late work when quantum mechanics and quantum physics were first theorised, developments of which Joyce was aware and to which he refers in his correspondence and fiction.[1] Andrzej Duszenko has demonstrated that, by the time Joyce was writing his last novel, his working method had become more "scientific":

> the development of a philosophical interpretation of quantum phenomena coincided with the writing of *Finnegans Wake* […]. Joyce, submerged in his new work and as always sensitive to anything that could support his ideas, found quantum physics fitting for inclusion in his project. In *Finnegans Wake* he made use not only of the ontological and epistemological implications of particle physics, but he also enriched his work with numerous references and allusions to specific elements of the subatomic realm that twentieth-century physics finally managed to uncover.
>
> (1994: 272)

Marcus is a revenant haunting his own life, while the text is haunted by the spectre of Joyce, including a largely unacknowledged indebtedness to theories from sciences and other discourses. Haunting inheres in diffraction, as Barad explains, when describing the way experiments with diffraction undo our perceptions of separation, of hard edges and outlines: "Darkness is not mere absence, but rather an abundance. Indeed, darkness is not light's expelled other, for it haunts its own interior. Diffraction queers binaries and calls out for a rethinking of the notions of identity and difference" (2014: 171). The hauntings in the novel include, then, not only Marcus's spectral appearance in the family home and the ghosts of Irish writing past, but also alternative, "queer" visions, different from those conventionally allowed to be imagined, of identity, being, time, and even difference itself.

A diffractive methodology moves away from "familiar habits and seductions of representationalism" (Barad, 2007: 88), which could also describe Modernist formal experimentation, but, as Ryker notes of *Solar Bones*, the novel is a "wonderfully original, distinctly contemporary book with a debt to modernism but up to something all its own." Barad's formulation of diffractive methodology demands an analytic practice that is not "a self-referential glance back at oneself" that "reflect[s] the world from outside", but requires instead "understanding the world from within and as part of it" (2007: 88). *Solar Bones* is as preoccupied with the partial and fragmented as earlier Modernist work. However, while the subjective artist-self treated heterogeneity as threatening material to be hastily assembled into a new whole, as when T.S. Eliot famously referred to "these fragments I have shored against my ruins", McCormack's text emerges as part of a larger, constant re-configuring of the world. The novel is, from the outset, aware of the process of world-making as an ongoing, dynamic phenomenon, reliant on multiple actors, human and nonhuman, all "materially engaging as part of the world in giving it specific material form" (Barad, 2007: 91). Marcus begins by describing himself as:

> here
> standing in the kitchen
> hearing this bell
> snag my heart and
> draw the whole world into
> being here
>
> (McCormack, 2017: 1–2)

The environmental implications of such an understanding of the world as process are clear, as Haraway has observed: "A common liveable world must be composed, bit by bit, or not at all. What used to be called nature has erupted into ordinary human affairs, and vice versa, in such a way and with such permanence as to change fundamentally means and prospects for going on, including going on at all" (2016: 45). Marcus hears the bell contributing to the never-ending re-configuration of the parish and the village of Louisburg, and, therefore, to the whole world's constant renewal, which has been in process since the "beginning of time":

> drawing up the world again
> mountains, rivers and lakes
> acres, roods and perches
> animal, mineral, vegetable
> covenant, cross and crown
> the given world with
> all its history to brace myself while
> standing here in the kitchen
>
> (2017: 3)

The theme of re-constitution as dynamic and ongoing necessarily challenges stable subject-object dichotomies. The novel does not posit the universe as random disorder for which we can abdicate responsibility, but a protean assemblage in which every part, from the microscopic to the vast, plays a role and has a connection, or connections, to every other part, if of varying intensities and proximities.

Time and space

The "luminous bones" of the title refer, inter alia, to sections of a disassembled wind turbine, a symbol to Marcus of a failure of imagination, "the world forfeiting one of its better ideas" (2017: 25). It is telling that this defeat in the struggle to imagine a new and better world incites in Marcus, "a renewal of that same old anxiety I had experienced as a nine-year-old" (2017: 25), as he recalls his father dismantling a tractor engine. For Marcus, the image evokes "the beginning of the world, the chaotic genesis which drew it together and assembled it from disparate parts" (2017: 26). Marcus speaks of himself as similarly susceptible of reassemblage: "my own heartbeat suspended in mid-air, nothing but a fat systolic contraction of the light, waiting for the dawn and the sun to shine upon it so that I might coalesce around it once more flesh and blood/time and again" (2017: 150). When he says this, Marcus does not yet realise that he is a revenant whose "coalescence" does not bear the same relationship to time and space as did his mortal form. His anomalous status affords a kind of diffractive perspective:

> Diffractions are untimely. Time is out of joint; it is diffracted, broken apart in different directions, non-contemporaneous with itself. Each moment is an infinite multiplicity. 'Now' is not an infinitesimal slice, but an infinitely rich condensed node in a changing field diffracted across spacetime in its ongoing iterative repatterning.
>
> (Barad, 2014: 169)

In this text, the extreme gothic version of "loss" of self and identity through death enables ethical insights into the constitutive patterns and connections among all phenomena, insights that decentre the human subject, including conventional androcentric understanding of space and time. In a diffractive reading, space and time "are intra-actively produced in the making of phenomena; neither space nor time exist as determinate givens, as universals, outside of phenomena" (Barad,

2010: 260–1). In its representation of time and space as collaborative processes, the novel also acknowledges the historicity of all matter, with implications for the self's contingent plurality:

> all things are out of sync and kilter, things as themselves but slightly different from themselves, every edge and outline blurred or warped and each passing moment belated, lagging a single beat behind its proper measure, the here-and-now beside itself, slightly off by a degree as in
> a kind of waking dream in which all things come adrift in their own anxiety
> (McCormack, 2017: 91)

"Things" described here have anxiety and lack clear outlines and self-similarity, things that include the narrator, whose altered relationship to time and space increases his sensitivity to his status as a phenomenon among phenomena. Barad observes of diffractive reading that "'Past' and 'Present' are iteratively reconfigured and enfolded through the world's ongoing intra-activity [...]. Phenomena are not located in space and time; rather phenomena are material entanglements enfolded and threaded through the spacetimemattering of the universe" (2010: 261).

Marcus becomes most acutely aware of the enfolded nature of all phenomena when his wife falls desperately, frighteningly ill, and the very house seems to pulse with her fever, an uncanny experience of dissolved boundaries and distributed agency that initially reminds Marcus of another crisis decades earlier when Mairéad left the family home for seven weeks, while pregnant with their first child after discovering her husband's infidelity. That earlier experience of unravelling loneliness and loss made household objects also seem poised for flight: "everything slowly shifting through the house as if they had a meeting to keep somewhere else, possibly in some higher realm where all this chaos would resolve into a refined harmony which had no need of my hand or intervention" (2017: 166). Marcus abnegates any possibility of authoritative consciousness, of acting as a powerful, ordering subject. This is a point in the text when Marcus's own ghostliness begins to make itself felt, though not fully understood, through a growing understanding that his sensations don't "belong" to him and never have. The sound of the "interference" on the radio that won't quite resolve into intelligible sounds (2017: 187) echoes the "interference" he had been experiencing whenever he struggled to communicate through Skype with his backpacking son thousands of miles distant (2017: 149). Ideas of "here" and "there," "then" and "now" grow less secure. Marcus feels "all things, myself included, suspended in a kind of stalled duration, an infinity extended moment spinning like an unmeshed gear [...] /time itself could decay here, lapse completely" (2017: 230). Not yet fully aware of his own altered state, he nonetheless returns repeatedly to his wife's earlier illness, *her* brush with mortality. He remembers a protest piece by his daughter, Agnes, an artist whose work combines activism and performance, staged in Galway, the source of the cryptosporidium parasite that infected Mairéad and hundreds of others in 2007. Marcus sees the city itself is a suffering body and consciousness during the outbreak, when it "began to inhabit a kind of dreamtime when its past and future unfolded simultaneously, a while city dreaming itself [...] every part of it twitched between its real existence and its own dream-life where it morphed through all the changes of itself, its history unfolding in an ongoing delirium" (2017: 233).

Environmental/personal crises

This picture of Galway parallels an earlier description of Marcus's feelings of helplessness and metaphysical infection in the face of his wife's serious illness: "drifting in that state between sleep and waking it is easy to believe that I inhabit a monochrome X-ray world from which I might have evaporated, flesh and bone gone, eaten up, not by any physical rot or wasting but by some metaphysical virus which devours and leaves nothing of me behind" (2017: 150). Galway is given a momentary embodied realism, as fleeting and flickering as Marcus's own. He repeatedly grants a kind of subjectivity and agency to nonhuman entities, especially cities and buildings, but also other environmental elements such as "mountains, rivers and lakes," one of the novel's incantatory refrains. In a moment of unconscious irony, Marcus refers to his house, a "living pulsing thing", as "flickering" with its own "ghost neurology" (2017: 150). His expansive and flexible understanding of identity is a diffracted one. According to Barad, "*Identity is a phenomenal matter; it is not an individual affair. Identity is multiple within itself, or rather, identify is diffracted through itself*" (2012b: 32; italics in original). This shared and distributed model of subjectivity and agency is not simply or naively assuasive but implies shared suffering, a significant moral element in the novel. As Barad observes, "The existence of indeterminacies does not mean that there are no facts, no histories, no bleeding—on the contrary, indeterminacies are constitutive of the very materiality of being, and some of us live with our pain, pleasure, and also political courage" (2014: 177). If health is traditionally associated with "wholeness", it is fitting that Marcus often uses metaphors of illness and suffering to articulate connection across phenomena, as when the house pulses in sympathy with its ailing inhabitant, and Marcus's thoughts drift to avian flu: "God's creatures bound together in common suffering, our aches and pains one and the same as those of the duck and the turkey and the chicken" (2017: 174–5). This entangled existence is messy and troublesome. Haraway refers to the need to "build attachment sites and tie sticky knots to bind intra-acting critters, including people, together in the kinds of response and regard that change the subject—and the object" (2007: 287).

The intra-acting "critters" recognised in the novel, for good or ill, include viruses and parasites. Cryptosporidium is described sympathetically as "circl[ing] back to its source to find its proper home where it settled in for its evolutionary span , rising through every degree of refinement [...] we would be host to this new life form" (2017: 125), an acknowledgement of the uncanny multiplicity and dependency of human bodies.[2] The novel constantly returns to Mairéad's illness, to descriptions of the sweat, vomit, and diarrhoea that confront Marcus with the unstable contours and borders of the human body. Mairéad fills the sickroom "with a stench beyond what was human, as if her very soul was being drawn from her body, out through the pores of her skin" (2017: 122). Another bodily excretion undermining dangerous illusions of human singularity and separateness is deployed by Agnes in her first solo exhibition, transcriptions of court reports detailing violence, abuse, and crime from local newspapers, written in the artist's own blood. Marcus is unnerved by the experience, as his child's blood becomes

> a red mist that suffused the weak evening light [...] a light so finely emulsified that we might take it into our very pores and swell on it, so that even if the crowd broke up the continuity of the space there was no doubting that the light served to make everyone a part of a unified whole [...] Agnes's blood was now our common element, the medium in which we stood and breathed (2017: 46)

While Marcus is overcome to the point of having to leave the exhibition, he finds himself reflecting later on the sense of connection engendered by the event when listening to country music, relishing the melancholic knowledge "that we are all part of the world's heartache, its loss and disappointment mapped out in the songs" (2017: 75).

The series of revelations Marcus experiences over the course of the novel regarding phenomenal "entanglement" and commonality are in contrast to the "stable and unified" world of his "childish imagination" (2017: 23), first challenged, paradoxically, by an intended demonstration of his father's mastery. Brooding over images of collapse and decay – domestic, economic, political, environmental – relayed by the headlines of the papers on the kitchen table, Marcus acknowledges "that smidgeon of chaos which brings the whole thing down around itself", and tips the balance of the "gravitational pull we feel in everything around us now, the instability which thrills everywhere like a fever" (2017: 16). Blaming his "engineer's mind" for a tendency to obsess about collapse (2017: 17), particularly his father's disassembling of large machines in order to put them back together. Marcus concludes that his father's impulse grew not out of doubt as to "any fault or redundancy in the constructs themselves, but because there was in him that need to know how these things held together so that he could be assured his faith in them was well placed" (2017: 17). Marcus has not inherited the faith of his father.

One particular memory of a Massey Ferguson tractor that was burning oil has a lasting effect on Marcus. If the "fever" of social instability and the "glossy and hard" zeroes of the economic crash, "so given to viral increase" (2017: 9), anticipate Mairead's physical torments, the "ailing" tractor is similarly, in the frightened young Marcus's eyes, stricken with a

> viral malfunction likely to spread from the machine itself and infect the world's wider mechanism, [...]
> my fear only deepened as I recoiled at the thought that something so complex and highly achieved as this tractor could prove so vulnerable, so easily collapsed and taken apart [...]
> this may have been my first moment of anxious worry about the world, the first instance of my mind spiralling beyond the immediate environs of

hearth, home and parish, towards
the wider world beyond
way beyond
since looking at those engine parts spread across the floor my imagination took fright and soared to some wider, cataclysmic conclusion about how the universe itself was bolted and screwed together.

(2017: 21–2)

This anxious, frightened "spiralling" can be understood as an undoing of the self in a moment of apprehension of Marcus's entanglement with the cosmos, "the wider world beyond/way beyond."

Living with the chaos

As Deckard observes, Marcus goes on to be a "dismal" failure at "the strategic world-building game" (2016). Even though he works in a field that "concerned itself with scale and accuracy, mapping and surveying so that the grid of reason and progress could be laid across the earth," and lives a professional life "governed by calculation," he never comes to feel "so accurately placed as [his] father" (McCormack, 2017: 105). In place of the grid of "reason and progress" the novel proposes relational, diffractive mappings, which "are not rationally made, because the production of diffraction comes from elsewhere" (van der Tuin, 2014: 236). Marcus's loss of faith in the rational is first apparent when he moves from his contemplation of the association between Ireland's (and the world's) economic collapse and his father's dismantling of machines, to connect people's unwillingness to assume and anticipate disaster with the loss of "that brute instinct for catastrophe" that is "now buried too deep beneath reason and manners to register" (2017: 16). Marcus retains that instinct; his own attempts to find reassurance in the steady comforts of reason settle "instead into a giddy series of doubts, an unstable lattice of questions so far withholding any promise I might inherit/my father's ability to comprehend the whole picture" (2017: 105–6). Conceiving the universe as a homogenous, uninterrupted "whole picture" eludes Marcus. Though a source of anxiety and insecurity, his thinking emerges as more ethically responsive, more accepting of the partial and fragmented, than the patriarchal model of omniscience and control he fails to duplicate. Marcus is "seeing and thinking diffractively [...which] implies a self-accountable, critical, and responsible engagement with the world" (Geerts and van der Tuin, 2016). Marcus cannot confidently assume the self-satisfied role of "fearless, heroic" patriarch.

In Marcus, McCormack has created a largely unremarkable everyman who takes pride in his professional achievements while recognising his failures, who has betrayed and cherished those close to him, who has both regrets and joys to recall in looking over his life. How does such a character, such a novel, make any kind of difference in the world, persuade readers to reconsider their own life patterns, and thereby inspire transformative change? For one thing, Marcus is capable of insights with far-reaching implications for those who read the novel and identify with his everyday strengths and weaknesses. In addition, what is "most powerful about the book," Deckard argues, is that the pain of Marcus's realisations "is always held in tension with his bright imagination of a world made and constructed for the better" (2016). Conceiving alternatives, imagining possible new worlds, the first step to achieving transformative change, is the vital, constructive work of literature, the essential counter to market-driven destruction. In arguing the need for stories, Haraway insists that "Another world is not only urgently needed, it is possible, but not if we are ensorcelled in despair, cynicism, or optimism, and the belief/disbelief discourse of Progress" (2016: 54). Through its narrator's musings, *Solar Bones* draws painful, miraculous, comforting, disquieting connections among ostensibly unrelated phenomena, making the novel an example of what Haraway refers to as "the systemic stories of the linked metabolisms, articulations and coproductions [...] of economies and ecologies, of histories and human and nonhuman critters" (2016: 52). The novel exposes the fiction of "us" versus "them", the unworkable approach adopted by millennium development goals, finally unsustainable in promoting sustainable development.

As a visiting spirit, Marcus challenges every dichotomy, including those that distinguish here from there, now from then, and creating such a narrator is an ethical choice with environmental implications that the metaphor of diffraction can help elucidate:

> To address the past (and future), to speak with ghosts, is not to entertain or reconstruct some narrative of the way it was, but to respond, to be responsible, to take responsibility for that which we inherit (from the past and the future), for the entangled relationalities of inheritance that "we" are, to acknowledge and be responsive to the noncontemporaneity of the present, to put oneself at risk, to risk oneself (which is never one or self), to open oneself up to indeterminacy in moving towards what is to-come.
>
> (Barad, 2010: 264)

The novel's present is non-contemporaneous with itself as the narration stitches back and forth in time. Marcus's final memory is of one of opening to indeterminacy, being "cast out beyond darkness into that vast unbroken commonage of space and time [...] in which there are no markings or contours to steer by" (2017: 265).

Hauntings, interruptions

Throughout the text, Marcus's ghostly embodiment is equivocal and flickering; he is, to himself, "the surprise, the interruption, by the stranger (within) re-turning unannounced" (Barad, 2014: 178). His earthly life has been interrupted, as has his "afterlife"; neither is a frictionless, self-sustaining whole. Every ghost returns with a message, a warning. The purpose of Marcus's return has been to map out the patterns of "interference", creative and destructive, that world and re-world the cosmos, with or without him. In the novel's last pages, when Marcus is remembering the events leading up to his death and "relives" his final physical undoing, he continues to understand himself as entangled with fellow phenomena: "animal, mineral, vegetable/father, husband, citizen" (2017: 264–5). Even in the final moments, happening simultaneously in the past and in the present, Marcus retains a "capacity for imagining and caring for other worlds, both those that exist precariously now [...] and those we need to bring into being in alliance with other critters" (Haraway. 2016: 53). His memories pay detailed attention to "the interference patterns on the recording films of our lives and bodies" (Haraway, 1997: 16). Marcus's focus has been largely on bodies other than his own, until the final pages, when he returns to the "here" of the first page, a recursive gesture that refuses progress, teleology:

> being here as this electrical interval held within its circumference of flesh and bone, the full sense of myself to myself as a kind of bounded harmonic, a bouquet of rhythms meshing into one over-emergent melody which homes me within the wider rhythms of the world, the horizontal melody of the cosmos, the celestial harmonic which inscribes me against the biggest magnitudes, the furthest edge of the universe.
>
> (2017: 231)

The narrator, like every other element of creation, is not finally constrained by its ostensible structure – in Marcus's case, "flesh and bone" – but is a kind of vibration, contributing to the melody that reaches to "the furthest edge of the universe".

Conclusion

McCormack has created a white, middle-class male narrator for this novel, not in order to ignore or erase alternative voices, but to address a particular audience that may choose to comfortably position itself as somehow not *as* responsible for counter-implicated social and environmental crises as either powerful world-stage actors or especially affected minority populations. In this, McCormack is practising what Karen Thiele argues is a "different ethicality" that aspires to responsibility, but is careful "not to either appropriate otherness or sameness or patronize others": "Affirming that there will never be an innocent starting point for any ethico-political quest, because 'we' are always/already entangled with-in everything; and yet that this primary implicatedness is not bound to melancholy or resignation, which for too long has been preventing us to think-practice difference(s) that really might make a difference" (2014: 213). Diffraction, in this context, allows us to read insights by "attending to and responding to the details and specificities of relations of difference and how they matter" (Barad, 2007: 71). McCormack's protagonist lives in the same place where the author himself grew up and where he continues to live, a narrative choice that is emphatically local and intimate, so that the text in more ways than one "spoken by a friend across the table", rendering immediate the message of the need for transformative change. Marcus Conroy is ostensibly blandly indistinguishable from any other Western, middle-class professional and family man, and he suffers the ordinary fate of all phenomena, death, though even that "ending" is not a conclusive, melancholy transformation. It is both terrifying and invigorating, like the collapse of a star. All entities, from an infectious parasite to "mountains, rivers and lakes" to suffering metropolises, are at once enmeshed in each other and deserving of respectful recognition in their difference, and that important case is persuasively argued through the memories and observations of a resolutely ordinary "father, husband, citizen" in a novel that recognises no such thing as a full stop.

Notes

1 In 1963 physicist Murray Gell-Man reciprocated this interest by using the word "quark" from *Finnegans Wake* to name a then-theoretical subatomic particle of matter smaller than the smallest elements known at the time.
2 As James Gallagher recently revealed, "Human cells make up only 43% of the body's total cell count. The rest are microscopic colonists." My thanks to Jim Ryder for bringing this article to my attention.

References

Barad, K. (2014). Diffracting Diffraction: Cutting Together-Apart. *Parallax* 20 (3), 168–187.
Barad, K. (2012a). Interview. In R. Dolphijn and I. van der Tuin (ed.) *New Materialism: Interviews and Cartographies* (pp. 48–70). London: Open Humanities.
Barad, K. (2007). *Meeting the Universe Halfway: Quantum Physics and the Entanglement of Matter and Meaning*. Durham, NC: Duke University Press.
Barad, K. (2012b). Nature's Queer Performativity. *Kvinder Køn og Forskning*, 1 (2), 25–53.
Barad, K. (2010). Quantum Entanglements and Hauntological Relations of Inheritance: Dis/continuities, SpaceTime Enfoldings, and Justice-to-Come. *Derrida Today*, 3 (2), 240–268.
Deckard, S. (2016, 21 October). S*olar Bones Is That Extraordinary Thing, an Accessible Experiment, Virtuosic yet Humane*. *The Irish Times*. Retrieved from https://www.irishtimes.com/culture/books/solar-bones-is-that-extraordinary-thing-an-accessible-experiment-virtuosic-yet-humane-1.2838095
Duszenko, A. (1994). The Joyce of Science: Quantum Physics in *Finnegans Wake*. *Irish University Review*, 24 (2), 272–282.
Gallagher, J. (2018, 10 April). More Than Half Your Body Is Not Human. *BBC News*. Retrieved from http://www.bbc.com/news/health-43674270
Geerts, E. and I. van der Tuin. (2016). Diffractions & Reading Diffractively. [Electronic version] *New Materialism: How Matter Comes to Matter, Almanac*. http://newmaterialism.eu/almanac/d/diffraction
Haraway, D. J. (1997). *Modest_Witness@Second_Millenium.FemaleMan©_Meets OncoMouse™: Feminism and Technoscience*. London: Routledge.
Haraway, D. J. (1992). *Promises of Monsters: A Regenerative Politics for Inappropriate/d Others*. London: Routledge.
Haraway, D. J. (2016). Staying with the Trouble: Anthropocene, Capitalocene, Cthulucene. In J.W. Moore (Ed.) *Anthropocene or Capitalocene?: Nature, History, and the Crisis of Capitalism* (pp. 34–77). Oakland, CA: PM Press.
Haraway, D. J. (2007). *When Species Meet*. Minneapolis: University of Minnesota Press.

Kaiser, B. M. and K. Thiele. (2014). Diffraction: Onto-Epistemology, Quantum Physics, and the Critical Humanities. *Parallax*, 20 (3), 165–167.

McCormack, M. (2017). *Solar Bones*. Edinburgh: Canongate.

Moore, J. W. (2016). The Rise of Cheap Nature. In J.W. Moore (Ed.) *Anthropocene or Capitalocene? Nature, History, and the Crisis of Capitalism* (pp. 78–115). Oakland, CA: PM Press.

Ryker, M. (2018, 5 January) A Stylistically Daring Novel Considers Fundamental Questions. *New York Times*. Retrieved from https://www.nytimes.com/2018/01/05/books/review/solar-bones-mike-mccormack.html

Thiele, K. (2014) Ethos of Diffraction: New Paradigms for a (Post)humanist Ethics. *Parallax*, 20 (3), 202–216.

Van der Tuin, I. (2014). Diffraction as a Methodology for Feminist Onto-Epistemology: On Encountering Chantal Chawaf and Posthuman Interpellation. *Parallax*, 20 (3), 231–244.

14 *The Rain Box*

Raining on the radio and other stories

Jools Gilson

Well now, let me whisper in those ears of yours, even as you read this. Let me wonder in print about processes of listening and embodiment. Best turn up the volume; am I a little indistinct? Tricky to catch through all that crackling? (or is it cackling?) Listening to the radio anchors us in space and place, it changes how we feel – the quotidian everyday of Morning Ireland, or the World Service, or the bop and pleasure of contemporary music, or the lull and lurch of classical radio, or a podcast global world of story and news and song and longing. Radio from the rectangle in your kitchen to the phone in your pocket, makes us feel at home, or comforts us on the long drive, it grips and chills and bores us, but it always surrounds us. It wrestles the dominance of the visual to the ground. Ireland is a nation of radio-listeners, and we are curious and lonely creatures, hungry for connection. Listen.

Introduction

This chapter proposes that creative practice disciplines have a powerful and critical role to play in collaborative research models for sustainable development. The creative exploration of meaning through metaphor and creative writing, dance, theatre, music, film, visual art (as well as multiple combinations of these) practices contribute blue sky/outside of the box thinking as a matter of course. If we want to retain our blue sky and not end up in boxes, we need the world-making intelligences of creative practices. These are knowledges which challenge the hegemony of critical analytical modes of meaning – production. They engage with the world differently than many academic disciplines, they are fluent at embodiment, affect, visual and sonic literacy, imagination, and engaging with communities. Using the case studies of creative works which use rain or water as a primary metaphor, the chapter asks how such work might change thinking about climate change.

But enough of that general stuff. Enough of that introductory paragraph lark, let's get on with the particular listen we're listening for here. Let's behave ourselves and write something appropriate for an academic audience. Or perhaps not. Perhaps that is the critical thing, to notice the assumption of form, and play at its edges, to remember longing and wonder as well as analysis, and know that change comes from being moved. Being moved: that strange expression

DOI: 10.4324/9781003143567-18

which describes a movement so profound, it's usually felt as unearthly stillness. Profound and lasting change happens on the level of embodiment, through an individual's felt sense of themselves. So if we want to change how people behave, we have to move them. Touch them. And the realms of movement and touch are mostly anathema to scientific disciplines. Or am I wrong about that?

The Rain Box was a collaboration between myself, writer and artist, Jools Gilson, and RTÉ Composer in Residence, Sebastian Adams. Commissioned by Lyric FM and broadcast in July 2017, *The Rain Box* is a rainy fairy tale for your ears. This drizzly, torrential, playful piece of radio explores the science and poetry of rain through the tale of a child who finds a hidden box with rain falling inside, who discovers underworlds of rainy possibility. *The Rain Box* also takes its cue from old Irish words for rain and rainy days, which often tangle weather and emotion, so that *gruamán* is a gloomy spell of weather, as well as a fit of despondency. *The Rain Box* listens to rainy science from Head of Geography at UCC, Dr. Kieran Hickey who tells us how raindrops form. We hear glimpses of rain wonder from artist Angela Ginn and the pleasures of running in the rain, from climate change researcher, Clare Watson. Voices and stories float in the haunting sound worlds written/performed by Gilson and composed by Adams. Moving from drizzle to downpour, and from curiosity to enchantment, *The Rain Box* is an innovative form of radio somewhere in the space between documentary, drama, and sound art. How might processes of imagining be at the heart of how to transform thinking about a sustainable future? Why do stories matter? Why is it so critical that artists collaborate with scientists in tangling tales of possibility?

In contemporary Ireland, where there has been so much recent and historical trauma related to unethical touch and abuse at an institutional level[1], we are powerfully in need of a philosophy of touch and movement that meets the Irish context. Imagine "a touch that doesn't want anything"[2]. I'm interested in the connection between the everyday and poetic performance of metaphor and the specificity of our Irish context. If we need to be moved, to be touched by metaphors and stories which assault us with their world-changing perspectives, then we need also to attend to the materiality of a context of fraught cultural touch and movement. It is not a simple accident of history that dance practice and practices of touch and embodiment are so often disavowed in Ireland. In a hierarchy of knowledges, dance as academic and practical discipline, as well as the first-person experience of embodiment (somatic practice) as sites of research and teaching are resisted as serious forms of academic/embodied learning or intelligence. So, let us turn to an exploration of blind spots.

Notes on Blindness (2016)

In 1983 after years of deteriorating sight, writer and theologian, John Hull, went blind. For the next three years, he kept a diary on audio cassettes, and used these as the basis for his book *Touching the Rock: An Experience of Blindness*, published in 1990[3].

In these audio tapes, Hull describes a charged moment where he opens a door to his garden during heavy rain, and the world is given to him again in a layered sonic gift: "If only there could be something equivalent to rain falling inside, then the whole of a room would take on shape and dimension" (Middleton & Spinney 2016).

For John Hull, falling rain brings the world into being. Like a bat's sonar, the sound of rain and its complexity of different tones, cadences, and rivulets makes present space, scale, trees, houses, surfaces, and textures. This is a sonic and affective geography. For Hull, this moment of listening to rain in his garden is an epiphany, a movement from an internal world of visual loss, to the physical world around him. *Notes on Blindness*, the film made in 2016 by Peter Middleton and James Spinney, uses these audio diaries as well as actors, to dramatise Hull's arresting enquiry into what it feels like to go blind. This particular depiction of the moment when his world comes into presence through the detail of rain's sonic cartography is a haunting watch[4].

Importantly, this is also a dramatisation of touch, as both literal sensory sensation, and the "touch" of experience; how moved Hull is by listening to the rain fall, and how, intonated in the visual realm of film, the film viewer is also moved. The scene opens with the blind man walking towards the door. His hand trails along the smooth edge of a table, until he opens the door, and the sound of rainfall swallows him. This is the full text from that moment in the film:

A note on the experience of hearing rain falling:

> This evening I came outside the front door of the house and it was raining. I stood for a few minutes, lost in the beauty of it. The rain brings out the contours of what's around you, in that it introduces a blanket of differentiated but specialized sound, which fills the whole of the audible environment. If only there could be something equivalent to rain falling inside, then the whole of a room would take on shape and dimension, instead of being isolated, cut off, preoccupied internally, you are presented with a world, you're related to a world, you are addressed by a world. Why should this experience strike one as being beautiful? Cognition is beautiful. It's beautiful to know.
>
> (Transcript from *Notes on Blindness*, Middleton & Spinney 2016)

The nuanced sound environment of this film layers the specific tonalities of rain falling through trees, on concrete, on a dustbin lid. These visual/sonic scenes are followed by slow panning shots of a domestic interior, where it's raining, just as John Hull imagines might be possible. Here the film makers collaborate with John Hull and engage in what Maria Pramaggiore calls "scandalising metaphoricity" (Pramaggiore 1992: 279); they confuse the distinction between literality and metaphor as a poetic strategy. Playfulness, or "mischief" as the poet Doreinn calls it, is a leitmotif of much artistic practice. Such "bouldness" as she also says, is something critical about her (and many other) artist's sensibility. Film makers, Peter Middleton and James Spinney, make John Hull's dream of an internal raining, literal. It is a stunning image, as Hull sits quietly at his kitchen table before a teacup, and the panning camera and sound, pick out distinctly the sound of rain on its delicate china surfaces. This scene brings magical realism to the docu-drama in a way which speaks directly to Hull's text about the beauty of cognition. We see, he hears, we understand, and in this visual, sonic, and affective composition, the longing to be connected to our environment is disarmingly and powerfully performed.

For me, John Hull's audio diary and Middleton and Spinney's film is a part of this discussion because it uses metaphor playfully and hauntingly to navigate the territory between internal felt worlds and the environment. For John Hull, rain is an alchemy of belonging to a world[5].

Metaphor is a linguistic, poetic, and embodied trope. Metaphor is playful meaning-making; playful in the sense of being light about the form of things – that this thing could be seen through the shape of this other thing, and at the heart of this gesture is an enquiry into form, into the ways in which the shape of a poetic idea, a discipline, an industry, a global crisis holds in place a range of possible meanings, of assumptions. We need to attend to the things we cannot see (like John Hull), we need to shift in our seats to see the blind spots (as does *Notes on Blindness*), we need to understand the profound ways in which we assume unconsciously.

Rain Room (2012)

rAndom International's *Rain Room*, an installation at the Barbican Centre's Curve Gallery in 2012, also made it rain indoors, but this was an altogether different kind of weather. Made in the tradition of installation art and presented in a gallery setting, *Rain Room* is a rectangle of continually falling rain. Its alchemy lies initially with the actuality of it raining indoors but then shifts to the discovery that it isn't possible to get wet under this rain. As visitors test the touch of falling water, they soon discover that it ceases to fall where a hand is held. And in this genre of participatory installation, the art-work itself is as much an invitation to play as it is a visual image. It isn't possible to do this once and not try again, and so visitors risk their feet under the torrents of water, and no, no water falls where they move, until they move inside the falling rectangle and wonder at the experience of being in the rain, and yet untouched by water.

> "Although you feel you have the measure of the work before you take the leap, nothing prepares you for the physical experience. It goes against our instincts to walk into water pouring so insistently, but somehow every movement is anticipated — quickly shoot out your arm and still it stays dry; change the speed or angle of your walking and still the rain stops."
>
> (*London Standard*, Oct 5th, 2012)

As a feat of engineering, *Rain Room* is driven by 3D depth cameras, which track bodies underneath the falling water and sensitively turn off valves above any interrupting object. While the engineering here is a spectacle, it isn't what's important. What drives visitors into the rain that doesn't touch them, is enchantment. What does it feel like to be in the rain, to have rain all around you, and not to be touched by its fall, however much you jive and jump and try to catch the system out? And so, this installation becomes a site for playfulness and wonder. We don't sit in a darkened room, or at our screens to watch this, as we do *Notes on Blindness*.[6]

We can imagine what it might feel like, but profoundly, it isn't the thing it was and is in the gallery space, it isn't the affective cartography of embodiment and space that it is if you were there. And so, this is a performative work, a work that *does* in the present tense; is not itself without an audience there engaging with its rainy technology, that is, of course, a metaphor of some slippery kind. *Rain Room* is an abstract canvas for the resonance and memory of raining. This rain doesn't drench, and in its failure to touch, we are touched differently, and might re-imagine space itself. Like Hull's presencing of the world through rain, *Rain Room* is an affective experience of space, but here rain resists our presence, and will not drench us. This environment can "see" moving bodies, and responds with ceasing the fall of rain. John Hull wants the opposite; a rain for interior rooms, that doesn't locate the perceiving individual, but allows the individual to be located – returns him to a geography of space and texture.

At the seminar which germinated this book, we were asked to present for ten minutes only. I put on my timer and performed my presentation, and when the alarm went off, I stopped. But I was the only one to do this. Many colleagues spoke for 15 or 20 minutes, and sometimes more. Artists wouldn't do this en masse in this way. This performance of presentations pointed to the assumptions of academia, to the bread and butter of what it means to academically perform, and this is to do with speaking our arguments, and our explorations, to do with PowerPoints and photographs and sometimes video. It is to do with disregarding the ten-minute rule. And in this disregard is the key to what it would take to shift profoundly our thinking and our doing to effect profound and global change. What comfort do we take in performing again the things we implicitly assume? What blind spots allowed a rule to be proposed that was then consistently ignored by (almost) all of us? This might be to do with a collaborative gesture of academic performance. Surprisingly, for someone profoundly interested in game changing, I stuck to my ten minutes. This is not an essay about how well behaved I am, but about noticing the things we do not discuss. Why did I do as I was told for ten minutes, and then within that ten minutes quietly transgress the form of the academic presentation?

Difficult Joys (1995)

Making an experimental radio feature about rain for Lyric FM in 2017 wasn't the first time I'd made work out of the materiality and metaphors of water and weather. As a young artist in 1995, I made a work called *Difficult Joys* [7](http://halfangel.info/?page_id=256) which was a choreographic installation performed underwater. I was exploring ideas about femininity, writing, and fluid mechanics. I wrote underwater, using special paper and pens I'd found in a fishing shop. At other places in the installation, my poetic text was present:

> Under here, there is yelloworange glasslight.
> Suddenly quiet, and lightful. I swim like a
> trooper a foot from the bottom, gathering
> eyefuls of browngreen shafts of light. Day

falls in and tries to get me. I gasp with
pleasure and plunge again. Swimming
upstream I watch the depth go greenerdeep.
It is like the only aloneness I desire.
Something in my bellypit turns faster than
I can arch in the water. Down
(Jools Gilson, *Difficult Joys,* 1995)

And in this way, *Difficult Joys* explored the space between performance and memory, and the ways in which both these things are refracted through femininity. This was a difficult piece to perform, and my first foray into the gallery space. We warmed the water as best we could, but I couldn't do more than an hour, and it was a strange plunging process of choreographic repetitions and coming up for air. Powerfully and profoundly framed in the glass tank, as performance rather than film, the water here acts as metaphor and materiality, as a site of critique, but also holds me in the chilly rectilinear cube of its liquid glass and watery frame.

The Lios (2004)

Nine years later, in collaboration with Richard Povall, our company *half/angel* made the installation work *The Lios* for galleries in Ireland and the UK. A *lios* (phonetically pronounced "liss") is the Irish word for a ring fort – probably built from the late Iron Age to the early Christian period. These circular constructions later became imbued with the folklore of the fairy fort. At the time of making, I lived in East Cork, and a *boreen* led from our cottage down to the sea. At the bottom of this *boreen*, there were two paths; one led down to the beach, the other up to the *lios*, a circular space which was the remains of an ancient ring fort. The first time I heard the word "*lios*" it was from a 95-year-old neighbour, who had been born in our house. She told us about dancing on the *lios* when she was young, and I became enchanted by this place, and the stories it held. I collected two sets of audio stories about the *lios*, the first were a series of walks I made down the *boreen*, by the *lios* and along the beach, the second were stories my neighbours told about this place.[8]

We made an installation which involved two parallel spaces. Visitors were asked to remove their shoes and socks, and in the first installation, they walked on real grass, we'd laid in the gallery, and a series of small speakers hung in rows, muttering the low audio of the walks I'd made down the *boreen* past the *lios*, but when they came close, so the audio rose to meet them (remember the rain stopping in *Rain Room*). This work is 14 years old now, but I've just listened to the extract of my walk from 4th February 2004, in which it's lashing, and the beach flooded, but I make my way across the rocks nonetheless, to get the fish box I've spied, and in which I'll later grow salads. In this installation work, I bring watery presence into the gallery, link this with the sensation of walking barefoot on grass, and the fragrance of damp sod. This is also a work about being held by place, about the connection of landscape to present tense and memory. In the second space, visitors walk on sand, and there is a soundscape of the beach. In the centre are two glass cubes of water, visitors are invited to move their hands in the water, and as they do so, their movement calls up the stories of my neighbours about the *lios*. One in particular, Madge, died not long after these recordings, and we loved her.[9] This work is a meditation on connection to place, through the charged location of the *lios*, through walking and through stories. We worked hard to make sure that these same neighbours were able to come to the opening of the exhibition at the Crawford Art Gallery in Cork in 2004. John Hull's rainfall connects him to a lost geography. *The Lios* tangled my presence in the Irish countryside, with the stories of my neighbours to conjure another kind of fairy fort, one with permeable walls, evoked by listening and the affect of grass, sand, and water.

> Out in the hinterlands things are different. Out beyond the rocks, beyond the birds, beyond the waves – even beyond the blackness of the lighthouse on its island, and its red wink. Beyond all that. Beyond the patina of living things. Beyond terrified bunnies sprinting, arse bouncing skywards. Beyond sheep and soot and ordinary things. Beyond rain in the afternoon. Beyond wet blackberries and aching sloes. Beyond rocks, purple and otherwise. Beyond sea-ridden plastic and other ocean ephemera. Beyond moving clouds. Beyond wind and hawks and all the seabirds she doesn't know the name of. Beyond lichen and beach herbs, she also doesn't know the name of. Beyond fingers picking up driftwood kindling. Beyond the hundreds of strange blue jellyfish that landed here in August. Beyond coldness. Beyond tripping on the rocks and looking into the hole in her leg. Beyond swimming in the cold sea to stop it from bleeding. Beyond lobster pots and carrageen and falling sheep. Beyond twilight. Beyond all tides. Beyond weeping and singing out to sea. Beyond fishing boats and passenger ferries and cargo ships. Beyond caravans. Beyond erosion. Beyond the smooth grey armchair rock at the head of the swimming beach. Beyond dancing. Beyond any photograph. Beyond unsteadiness. Beyond all this. Beyond. Beyond.
>
> (Jools Gilson, text from The Lios installation, Triskel Arts Centre, Cork, 2004)

The Knitting Map (2005)

In 2005, also in collaboration with Richard Povall and our company *half/angel*, we made a vast collaborative textile installation, that took a year to make, involved several thousand people (mostly older, working class women from Cork) and had as a major strand of its poetics – the metaphor of the women of a city rising up and knitting the weather. We connected a weather station with code written by Richard Povall, to generate the colour of yarn. *The Knitting Map* also involved five motion-sensing cameras around Cork City, translating the degree of movement into complexity of stitch – if the city was busy, the knitting stitch was complex, if it was quiet it was simple. The vision of *The Knitting Map* – the women knitting the weather for a year, has a revolutionary gesture at its core. Its poetic motion sought to find a quiet but profound way to give space to the astonishing in the everyday of so much feminine activity. It sought to give space to a profound politics of care, to ask if skills normally used for gift giving and solace, could be used for something of vast collaborative gorgeousness, something whose use-value (a thing which would so often trouble our critics and collaborators alike) was both poetic and political.

TKM, then, involved the culturally disenfranchised in the making of a vast artwork that was commissioned (and certainly perceived) as a flagship project for Cork's year as European Capital of Culture in 2005. Poetically and politically it was a work that sought to rework the urban territory of matter and meaning: knitting was used as something monumental – an abstract cartography of Cork generated by the city itself and its weather, and knitted every day for a year. To make such a gesture using feminine and female labour aspired to re-work the relationship between femininity and power in an Irish context: it gave cartographic authority to working-class older women from Cork, for a year.

The process of conjuring the energies of a city's climate into an abstract cartography meant that in an important sense the women involved in making TKM were knitting the weather. Such a communal gesture brought frosts and rain and heat into the domestic and ordinary act of knitting. It opened its close, domestic, and feminine associations to the literal and metaphorical sky. TKM also allowed the mathematical complexity of knitting difficult stitches to be brought into proximity to a frantic city, clogged with traffic and queues and crowded streets. In keeping track of shifting numerical combinations to produce, for example, an open honeycomb cable, these women re-worked the actual digital information about busyness being sent up to them from the city, and they did so by integrating this data with their hands (their digits) in processes of communal hand-knitting. TKM allowed the prevailing cultural peripherality of middle-aged women to make a collectively original and beautiful thing, and in doing so re-mapped their own apparently tangential geography.[10]

While it isn't only a watery thing, *The Knitting Map* marked the space and place of (primarily) older, working class women in Cork, through their knitting of the rain. And it caused so much trouble in the press, that it is the entire subject of a book to be published by Bloomsbury in 2019, called *Textiles, Community and Controversy*. At this stage, I've definitely got out of my glass tank, and am using the rain for revolution. Vast swathes of purple and deep blues are knitted in woolly collaboration to document the storms and floods of November 2005. What does it mean to knit the weather? And how might this mischievous metaphor make a connection with the urgencies of climate change?

Irish words for rain: *The Rain Box* (2017)

The Rain Box braids several threads; documentary footage, fairy tales, and Irish words for rain or rainy days, many of them no longer in use.[11] I owe knowledge of these words to the visual artist Carol Anne Connolly, who met with me and who made the beautiful artist's book, *The Water Glossary*. Connolly's book explores the ways in which words for water in old and contemporary Irish locate bodies and communities in a landscape of rain, light, and wind. I'm interested in the connection between linguistic and environmental gestures as a way to connect the individual to the vastness of global climate change, by focussing on local embodied knowledge. These words perform the relationship between internal emotional worlds and environment. Unless we can make this emotional connection in our art performance and broadcast practices, and by extension, in collaboration with climate science itself, then we cannot engage fruitfully with climate change issues.

The Water Glossary (like Robert Macfarlane's extraordinary *Landmarks*) documents examples of how words commemorate and perform again an embodied connection to place. What's important in these books is that these words mark the relationship between embodiment and place through *utterance*. Writing, ironically, is less important when the word is in use, but it becomes the only thing that will mark this relationship of embodied performance to landscape, once it no longer happens habitually; once the word, and the rural relationship has fallen out of use. *The written word* is the thing that commemorates absence in these books, and that's because the physical gesture of utterance is marked and recalled by the written word, but remains absent, even when resonant of its loss. And so voiced performance, here on the radio, places such language in a bricolage of rainy documentary and fiction. Utterance is itself a movement, a shape in the mouth, a flow of breath, of tongue to palate, of vibrating chords. And utterance enchants these words, and both Macfarlane and Connolly point to this, but don't say it out loud – that these are mouth spells, ways to world meaning, ways to have the land inside you, ways to spit it out, use it as metaphor, metonym, ways to wind yourself in the land. To ache for it as well as piss on it and everything in between. And so, to weave a thread in *The Rain Box* of these spoken words in Irish for rain, is to recall landscape as well as embodiment, is to drench listeners through that surrounding cartography that is radio.

Conclusion

And so here at the rainy end of this chapter, is another box of rain. In a drenching lineage from *Notes on Blindness* to *Rain Room*, and in my own work through *Difficult Joys*, through *The Lios*, and the knitted rain of *The Knitting Map*, comes a girl who finds an enchanted box. Artists, like academics, sometimes spend a lifetime worrying the same subject, or poetic idea. When we train young artists, we often speak about being willing not to know, about allowing the processes of curiosity and longing to work to meander and lurch and howl. This might seem like an odd thing, but it is at the core of this chapter's point, that there are other ways of knowing than the analytical. Why is it that I have been compelled by water and watery things? Why might it be important to imagine that there could be a box of rain that held stories? And what of this might contribute to a debate about climate change and artists working with scientists? Key to this is the shifting shadows of moving between (here artistic) disciplines. So that the glass cubes of ocean in *The Lios*, invited visitors to move their hands in the water as a way to stir up the audio layers of my neighbours' telling stories about the *lios* in Ballymacoda; as if memory itself could be dissolved in water. This was a performative installation in a gallery. While you could feel the sand between your toes, and listen to the layered sound-world of ocean if you just stood there, you would not hear the stories unless you got your hands wet. Such a gesture brings together the touch of water, and the listening of stories told; of place recollected, of other moments which touched, which curled in memories, which also made one thing of touch and experience. In our radio programme, the box of rain that Missy finds, also needs a touch to conjure stories. And the funny thing is, until I started writing this chapter, I'd forgotten I'd made a similar thing before in the gallery space of *The Lios*. So the rain box on the radio is a fairy tale of sorts. It makes of raining a thing we could hold and keep, and dissolves stories and memories in there. In here is a whole folklore of rainy tales, of a people who sing the rain out of the sky, who weep when it rains, who mark in its spaces all species of yearning. And we need stories, and disciplinary fluency. We need water to live.

If you were there on that day, there was a strange quiet that came about when I danced rather than spoke my presentation. My voice was present, my stories spoken; as ten minutes of *The Rain Box* played, I riffed. What might happen if we told a different story in a different way? If we remembered embodiment, not as something way out there belonging to me or to someone else given to be seen, but to ourselves. The astonishing and everyday blind spot of our own embodiment. And from such connectedness to ourselves, how might we map our care for our connected environment, and from what connection might we meet each other and begin to make material a different future?

Notes

1 The most recent example being the discovery of the bodies of 800 babies and toddlers in a sewage outlet in the grounds of a Mother and Baby home in Tuam, Co. Galway in 2017.

2 My teacher Brian Siddhartha Ingle used this phrase in one of my classes. I'm in the first year of a training in Clinical Somatic Education.

3 I'm grateful to Melissa Harrison's *Rain: Four Walks in English Weather* (2016), for introducing me to John Hull's comments about rain.

4 It is not lost on the film makers that they're using film (a visual medium) to explore the loss of the discipline's core defining principal, but this is also why this film is important, both generally, and for my argument here.

5 Notes on Blindness directed by Pete Middleton & James Spinney (2016)
Short film, part of official Sundance selection. Watch the two minute rain section here:http://www.nytimes.com/interactive/2014/01/16/opinion/16OpDoc-NotesOn-Blindness.html?_r=2 (Scroll forward to 9.20–11.16)

6 Watch the video at :https://www.youtube.com/watch?v=EkvazIZx-F0

7 http://halfangel.info/?page_id=256

8 http://halfangel.info/?page_id=380

9 Listen to Madge talking about dancing on the *lios* (http://halfangel.info/?page_id=380).

10 This section on *The Knitting Map* is adapted from a longer article on this work. See Gilson 2012.

11 Listen here: https://soundcloud.com/the-lyric-feature/the-rain-box

References

Anonymous (2012) 'The Rain Room in London's Barbican'. *London Standard*, Oct 5th.

Connolly, C. A. (2015) *The Water Glossary*. Book, Lithograph, 60 pages, ISBN 9780995636101. http://www.carolanneconnolly.com/index.php?/work/the-water-glossaryan-sanasan-uisce/

Gilson, J. (2012) 'Navigation, Nuance and half/angel's Knitting Map: A Series of Navigational Directions'. *Performance Research*, 17 (1): 9–20.

Harrison, M. (2016) *Rain: Four Walks in English Weather*. London: Faber & Faber.

Hull, J. M. (1990) *Touching the Rock: An Experience of Blindness*. London: Penguin.

MacFarlane, R. (2016) *Landmarks*. London: Penguin.

Middleton, P. & Spinney, J. (2016) *Notes on Blindness*. London: Artificial Eye/Arte, [film].

Pramaggiore, M.T. (1992) 'Resisting/Performing/Femininity: Words, Flesh, and Feminism in Karen Finley's *The Constant State of Desire*.' *Theatre Journal* 44: 269–290

Index

Page numbers in *Italics* refer to figures; Page numbers in **bold** refer to table; page numbers followed by 'n' refer to notes number